DYNAMICS AND BIOAVAILABILITY OF HEAVY METALS IN THE ROOTZONE

DYNAMICS AND BIOAVAILABILITY OF HEAVY METALS IN THE ROOTZONE

H. Magdi Selim

CRC Press
Taylor & Francis Group
Boca Raton London New York

CRC Press is an imprint of the
Taylor & Francis Group, an **informa** business

CRC Press
Taylor & Francis Group
6000 Broken Sound Parkway NW, Suite 300
Boca Raton, FL 33487-2742

First issued in paperback 2017

ISBN 13: 978-1-138-07391-3 (pbk)
ISBN 13: 978-1-4398-2622-5 (hbk)

Visit the Taylor & Francis Web site at
http://www.taylorandfrancis.com

and the CRC Press Web site at
http://www.crcpress.com

Contents

Preface

The nature of heavy metal dynamics in the rootzone depends largely on the interactions of heavy metals in the soil environment over time. It is generally recognized that heavy metals are strongly retained by the soil matrix and thus influence their transport and bioavailability in soils. Moreover, the mobility of heavy metals has been frequently observed as nonlinear or concentration dependent and is not clearly understood. Because soils are heterogeneous, heavy metals in soils can be involved in a series of complex chemical and biological interactions, including oxidation–reduction, precipitation and dissolution, volatilization, and surface and solution-phase complexation. The heterogeneous nature of the different soil constituents adds to the complexity of understanding the dynamics of heavy metal species with the soil environment. This book's contribution is interdisciplinary in nature with emphasis on heavy metal reactivity, bioavailability, transport, and environmental chemistry in soils, and is intended to attract a mix of both senior and junior scientists from a wide range of disciplines.

In the first three chapters, the primary emphasis is on different approaches that describe the dynamics and bioavailability of heavy metals in the soil system. Such knowledge is necessary because such approaches provide direct information on the concentration and thus on the dynamics of heavy metal bioavailability in soils. Chapter 1 discusses heavy metal interactions based on equilibrium type and kinetic type models. This is followed by retention models of the multiple reaction type, including the two-site equilibrium-kinetic models, the concurrent and consecutive multireaction models, and the second-order approaches. Processes that govern the transport processes of heavy metals in stratified layered systems are also presented, along with experimental results. Chapter 2 emphasizes nonequilibrium transport models of heavy metal ions in the critical zone that combine hydrological, geochemical, and biological processes. A general description and mathematical treatment of the hydrological processes that impact the fate and transport of heavy metals in the subsurface environment is presented in detail. This is followed by an overview of applications of coupled physical and chemical nonequilibrium models for the simulation of heavy metals in soils. In Chapter 3 the focus is on chemical processes that control soil solution concentrations and thus the bioavailability of heavy metals ions in soils. The topics discussed include soil solution chemistry, methylation and volatilization reactions, precipitation–dissolution reactions, oxidation–reduction reactions, and adsorption–desorption reactions. Modeling of heavy metals adsorption by soils based on empirical approaches of the Langmuir and kinetic type are first discussed, followed by modeling of adsorption by soils based on the constant capacitance model. Examples of model predictions based on the constant capacitance model of experimental results for arsenic(V) and selenium(IV) adsorption by several soils are discussed.

Rhizosphere microorganisms can strongly affect availability, plant uptake, translocation, and the volatilization of pollutants, as well as chelation and the immobilization of contaminants in root-explored soil. In Chapter 4 the focus is on reducing the impact of naturally occurring or anthropogenic heavy metals such selenium (Se) in soils by phytoremediation. Factors that affect Se bioavailability in soils are discussed—for example, pH, organic matter content, texture, and microbiological activity. Colloid generation and transport in soils is of significance because of potential colloid-facilitated transport of heavy metals to the groundwater. In Chapter 5, colloid-associated transport and metal speciation at reclaimed mine sites are discussed. A case study of enhanced mobilization due to colloid-facilitated transport for cadmium, chromium, cooper, nickel, lead, and zinc is presented for reclaimed mine sites. In Chapter 6 an overview of recent advancements in the biogeochemistry of trace elements in plant rhizospheres and rootzones and their environmental implications is presented.

Chapter 7 discusses the fate of heavy metals and their concentration in runoff waters at the watershed scale. Methods to manage nonpoint sources of heavy metals from agricultural watersheds are presented. The Natural Resources Conservation Service (NRCS) developed a technique to estimate. This technique assumes that dissolved inorganic chemicals are lost from a specific depth of surface soil and is applicable to essential plant nutrients (i.e., nitrogen, phosphorus, copper, zinc, etc.) as well as environmentally toxic elements such as lead, cadmium, nickel, and arsenic. Geographical Information Systems (GIS) for spatial presentations of data in watershed maps are an essential part of this estimation method. Results from a case study of the Wagon Train (WT) Watershed in Lancaster County, Nebraska, are discussed.

Wetland plants influence heavy metal uptake transport and mobility and serve as the overall theme of Chapters 8 and 9. Specifically, metal and metal speciations that determine their behaviors in the context of wetland environments are discussed. In Chapter 8, heavy metal transformation in wetlands with a focus on processes governing the availability or toxicity of heavy metals in wetlands is discussed. The processes presented include valence, sorption to sediments, complexation with organic matter, precipitation, and interaction with microorganisms. Wetland plants alter the redox conditions, pH, and organic matter content of sediments and thus affect the chemical speciation and mobility of metals. Chapter 9 discusses factors controlling the dynamics of trace metals in frequently flooded soils. Factors discussed include redox processes and the valence state of metals, the redox behavior of iron and manganese, sulfur cycling, carbonates and pH, adsorption and desorption, salinity, and organic matter. Metals may be mobilized or immobilized, depending on a combination of several factors, and it is difficult to predict what effects plants will actually have on metal mobility under a given set of conditions. Chapter 10 focuses on the assessment of the effect of heavy metals in biosolid-amended soils on crops and their relationship with various plant species. A summary of the types of heavy metals found in biosolids, in soils, and in biosolid-amended soils is given. In fact, knowledge of the forms of heavy metals in biosolids and in biosolid-amended soils is necessary in the risk assessment of their transfer to the subsoil and potential contamination of groundwater supplies.

The editor wishes to sincerely thank the contributors to this book for their diligence and cooperation in achieving our goal and making this volume a reality. We are most grateful for their time and effort in critiquing the various chapters, and in keeping with the dynamics of heavy metals in soils. The editor also expresses thanks to Irma Shagla, Jessica Vakili, and the Taylor & Francis staff for their help and cooperation in the publication of this book.

H. Magdi Selim
Baton Rouge, Louisiana

Editor

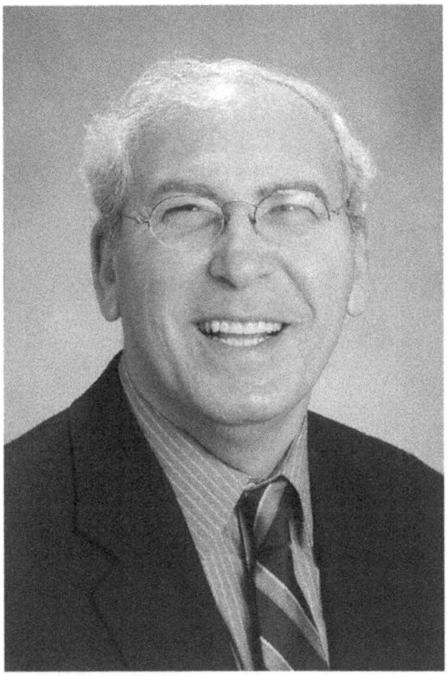

H. Magdi Selim is a professor of soil physics and the A. George and Mildred L. Caldwell Endowed Professor in the School of Plant, Environmental and Soil Sciences at Louisiana State University at Baton Rouge. He received his B.S. degree in soil science from Alexandria University, Alexandria, Egypt, and his M.S. and Ph.D. degrees in soil physics from Iowa State University, Ames, Iowa. Professor Selim is internationally recognized for his research in the areas of kinetics of reactive chemicals in heterogeneous porous media and transport modeling of dissolved chemicals in water-saturated and unsaturated soils. He is the original developer of the two-site and second-order concepts models for describing the retention processes of dissolved chemicals in soils and natural materials in porous media. Pioneering work also includes multistep and multireaction and nonlinear kinetic models for heavy metals, radionuclides, explosive contaminants, and phosphorus and pesticides in soils and subsurface media. His research interests also include saturated and unsaturated water flow in multilayered one- and two-dimensional systems.

Professor Selim is the author or co-author of numerous scientific papers, book chapters, reports, and research bulletins. He is also co-author of several books and monographs. He is the recipient of several professional awards and honors; he was named a Fellow of both the American Society of Agronomy and the Soil Science Society of America. Awards received include the Phi Kappa Phi Research Award, the Gamma Sigma Delta Award for Research, the Joe Sedberry Graduate Teaching Award, the First Mississippi Research Award for Outstanding Achievements in Louisiana Agriculture, the Doyle Chambers Career Achievements Award, and the EPA Regional Administrator's Environmental Excellence Award. He is a member of the American Society of Agronomy, Soil Science Society of America, International Society of Soil Science, International Society of Trace Element Biogeochemistry, Louisiana Association of Agronomy, American Society of Sugarcane Technology, Honor Society of Gamma Sigma Delta, and Sigma Xi. Professor Selim was elected chair of the Soil Physics Division (S-1) of the Soil Science Society of America. He has served on many committees of the Soil Science Society of America, the American Society of Agronomy, and the International Society of Trace Element Biogeochemistry. He also served as associate editor of the *Water Resources Research* and *Soil Science Society of America Journal* and as technical editor of the *Journal of Environmental Quality.*

Contributors

R.D. DeLaune
Department of Oceanography and Coastal Sciences
School of the Coast and Environment
Louisiana State University
Baton Rouge, Louisiana
rdelaune@aol.com

Gijs Du Laing
Laboratory of Analytical Chemistry and Applied Ecochemistry
Ghent University
Faculty of Bioscience Engineering
Ghent, Belgium
Gijs.DuLaing@UGent.be

Moustafa A. Elrashidi
U.S. Department of Agriculture - Natural Resources Conservation Service
National Soil Survey Center
Lincoln, Nebraska
moustafa.elrashidi@lin.usda.gov

Sabine Goldberg
U.S. Salinity Laboratory
Riverside, California
Sabine.Goldberg@ars.usda.gov

F. Gorini
CNR, Institute for Ecosystem Studies
Pisa, Italy
francesca.gorini@ise.cnr.it

A.D. Karathanasis
Department of Plant & Soil Sciences
University of Kentucky
Lexington, Kentucky
akaratha@uky.edu

J.O. Miller
Department of Plant & Soil Sciences
University of Kentucky
Lexington, Kentucky
Jarrod.Miller@ars.usda.gov

Eva Oburger
Rhizosphere Ecology and Phytotechnologies
University of Natural Resources and Life Sciences
Vienna, Austria
eva.oburger@boku.ac.at

G. Petruzzelli
CNR, Institute for Ecosystem Studies
Pisa, Italy
g.petruzzelli@ise.cnr.it

Beatrice Pezzarossa
CNR, Institute for Ecosystem Studies
Pisa, Italy
beatrice.pezzarossa@ise.cnr.it

Markus Puschenreiter
Rhizosphere Ecology and Phytotechnologies
University of Natural Resources and Life Sciences
Vienna, Austria
markus.puschenreiter@boku.ac.at

Jörg Rinklebe
Soil and Groundwater Management
University of Wuppertal
Wuppertal, Germany
rinklebe@uni-wuppertal.de

Jakob Santner
Rhizosphere Ecology and Phytotechnologies
University of Natural Resources and Life Sciences
Vienna, Austria
jakob.santner@boku.ac.at

H. Magdi Selim
School of Plant, Environmental, and Soil Sciences
Louisiana State University Agriculture Center
Baton Rouge, Louisiana
mselim@agctr.lsu.edu

Dong-Choel Seo
Department of Oceangraphy and Coastal Sciences
School of the Coast and Environment
Louisiana State University
Baton Rouge, Louisiana
dseo@lsu.edu

Christos D. Tsadilas
National Agricultural Research Foundation
Institute of Soil Mapping and Classification
Larissa, Greece
tsadilas@ismc.gr

Walter W. Wenzel
Rhizosphere Ecology and Phytotechnologies
University of Natural Resources and Life Sciences
Vienna, Austria
walter.wenzel@boku.ac.at

Hua Zhang
Department of Water Resources
North China University of Water Resources and Electric Power
36 Beihuan Road
Zhengzhou, China
hua.zhang@tetratech.com

1

Nonlinear Behavior of Heavy Metals in Soils: Mobility and Bioavailability

H. Magdi Selim

CONTENTS

The dynamics of heavy metals and their transport in the soil profile play a significant role in their bioavailability and leaching losses beyond the root-zone. The primary emphasis in this chapter is on different approaches that describe the dynamics of heavy metals in the soil system. Such knowledge is necessary because these approaches provide direct information on the concentration of heavy metals in the soil solution and thus on their dynamics and bioavailability in soils. Moreover, such predictive capability requires knowledge of the physical, chemical, as well as biological processes influencing heavy metal behavior in the soil environment.

A number of theoretical models describing the transport of dissolved chemicals in soils have been proposed over the past three decades. One class of models deals with solute transport in well-defined geometrical systems where one assumes that the bulk of the solute moves in pores and/or cracks of regular shapes or through intra-aggregate voids of known geometries. Examples of such models include those dealing with the soil matrix of uniform spheres, rectangular or cylindrical voids, and discrete aggregate or spherical size geometries. Solutions of these models are analytic, often complicated, and involve several numerical approximation steps. In contrast, the second group of transport models consists of empirical models that do not consider well-defined geometries of the pore space or soil aggregates. Rather, solute transport is treated on a macroscopic basis with the water flow velocity, hydrodynamic dispersion, soil moisture content, and bulk density as the associated parameters that describe the soil system. Refinements of this macroscopic approach are the "mobile and immobile" transport models where local nonequilibrium conditions are due to diffusion or mass transfer of solutes between the mobile and immobile regions. Another refinement includes the "dual and multiporosity" type models, which are often referred to as two-flow or multiflow domain models. Others refer to such models as bimodal types, wherein the soil system can best be characterized by a bimodal pore size distribution. Mobile and immobile models as well as dual- and multi-flow have been used with various degrees of success to describe the transport of several solutes in soils. These empirical or macroscopic models are widely used and far less complicated than the more exact approach for systems of well-defined porous media geometries mentioned above.

To quantify the transport of heavy metals in the soil, models that include reactivity or retention and release reactions of the various heavy metal species with the soil matrix are needed. Retention and release reactions in soils include ion exchange, adsorption–desorption, precipitation–dissolution, and other mechanisms such as chemical or biological transformations. Retention and release reactions are influenced by several soil properties, including bulk density, soil texture, water flux, pH, organic matter, redox reactions, and type and amount of dominant clay minerals. Adsorption is the process wherein solutes bind or adhere to soil matrix surfaces to form outer- or inner-sphere solute surface-site complexes. In contrast, ion exchange reactions represent processes where charged solutes replace ions on soil particle surfaces. Adsorption and ion exchange reactions are related in that an ionic solute species may form a surface complex and may replace another ionic solute species already on surface sites.

Surface-complexation models have been used to describe an array of equilibrium-type chemical reactions, including proton dissociation, metal cation and anion adsorption reactions on oxides and clays, organic ligand adsorption, and competitive adsorption reactions on oxide and oxide-like surfaces. The application and theoretical aspects of surface complexation models are extensively reviewed by Goldberg (1992) and Sparks (2003). Surface-complexation models are chemical models based on a molecular description of the electric double layer using equilibrium-derived adsorption data. They include the constant capacitance model, triple-layer model, and Stern variable surface charge models, among others. Surface-complexation models have been incorporated into various chemical speciation models. The MINEQL model was perhaps the first where the chemical speciation was added to the triple-layer surface-complexation model. Others include MINTEQ, SOILCHEM, HYDRAQL, MICROQL, and FITEQL (see Goldberg, 1992).

All of the above-mentioned chemical equilibrium models require knowledge of the reactions involved and associated thermodynamic equilibrium constants. Due to the heterogeneous nature of soils, extensive laboratory studies may be needed to determine these reactions. Thus, predictions from transport models based on a surface-complexation approach may not describe heavy metal sorption by a complex soil system. As a result, the need for direct measurements of the sorption and desorption and release behavior of heavy metals in soils is necessary. Consequently, retention or the commonly used term "sorption" should be used when the mechanism of heavy metals removal from soil solution is not known, and the term "adsorption" should be reserved for describing the formation of solute-surface site complexes.

In this chapter, several models that govern heavy metals transport and retention reactions in the soil are derived. Models of the equilibrium type are discussed first, followed by models of the kinetic type. Retention models of the multiple reaction type, including the two-site equilibrium-kinetic models, the concurrent and consecutive multireaction models, and the second-order approach are also discussed. This is followed by multicomponent or competitive type models where ion exchange is considered the dominant retention mechanism. Retention reactions of the reversible and irreversible types are incorporated into the transport formulation. Selected experimental data sets will be described for the purpose of model evaluation and validation, and the necessary (input) parameters are discussed.

Transport Equations

Dissolved chemicals present in the soil solution are susceptible to transport through the soil subject to the water flow constraints in the soil system. At any given point within the soil, the total amount of solute χ (µg.cm^{-3}) for a species *i* may be represented by

$$\chi_i = \theta C_i + \rho S_i \tag{1.1}$$

where S is the amount of solute retained by the soil (µg.g^{-1} soil), C is the solute concentration in solution (µg.mL^{-1}), θ is the soil moisture content (cm^3.cm^{-3}), and ρ is the soil bulk density (g.cm^{-3}). The rate of change of χ for the *i*-th species with time is subject to the law of mass conservation such that (omitting the subscript *i*)

$$\frac{\partial(\theta C + \rho S)}{\partial t} = - \; div \; J - Q \tag{1.2}$$

or

$$\frac{\partial(\theta C + \rho S)}{\partial t} = - \left(\frac{\partial J_x}{\partial x} + \frac{\partial J_y}{\partial y} + \frac{\partial J_z}{\partial z}\right) - Q \tag{1.3}$$

where t is time (h) and J_x, J_y, and J_z represent the flux or rate of movement of solute species i in the x-, y-, and z-directions (µg.cm^{-2}.h^{-1}), respectively. The term Q represents a sink (Q positive) or source that accounts for the rate of solute removal (or addition) irreversibly from the bulk solution (µg.cm^{-3}.h^{-1}). If we restrict our analysis to one-dimensional flow in the z-direction, the flux J_z, or simply J, in the soil may be given by

$$J = - \theta \left(D_m + D_L\right) \frac{\partial C}{\partial z} + qC \tag{1.4}$$

where D_m is the molecular diffusion coefficient (cm^2.h^{-1}), D_L is the longitudinal dispersion coefficient (cm^2.h^{-1}), and q is Darcy's flux (cm.h^{-1}). Therefore, the primary mechanisms for solute movement are due to diffusion plus dispersion and by mass flow or convection with water as the water moves through the soil. The molecular diffusion mechanism is due to the random thermal motion of molecules in solution and is an active process regardless of whether or not there is net water flow in the soil. The result of the diffusion process is the well-known Fick's law of diffusion where solute flux is proportional to the concentration gradient.

The longitudinal dispersion term of Equation (1.4) is due to the mechanical or hydrodynamic dispersion phenomena, which are due to the nonuniform flow velocity distribution during fluid flow in porous media. According to Fried and Combarnous (1971), nonuniform velocity distribution through the soil pores is a result of variations in pore diameters along the flow path, fluctuation of the flow path due to the tortuosity effect, and the variation in velocity from the center of a pore (maximum value) to zero at the solid surface interface (Poiseuille's law). The effect of dispersion is that of solute spreading, which is a tendency opposite to that of piston flow. Dispersion is effective only during fluid flow, that is, for a static water condition or when water flow is near zero; molecular diffusion is the dominant process for solute transport in soils. For multidimensional flow, longitudinal dispersion coefficients (D_L) and transverse dispersion coefficients (D_T) are needed to describe the dispersion mechanism. Longitudinal dispersion refers to that in the direction of water flow and that for the transverse directions for dispersion perpendicular to the direction of flow.

Apparent dispersion D is often introduced to simplify the flux Equation (1.4) such that

$$J = -\theta D \frac{\partial C}{\partial z} + qC \tag{1.5}$$

where D now refers to the combined influence of diffusion and hydrodynamic dispersion for dissolved chemicals in porous media. Incorporation of flux Equation (1.5) into the conservation of mass Equation (1.3) yields the following generalized form for solute transport in soils in one dimension:

$$\frac{\partial \theta C}{\partial t} + \rho \frac{\partial S}{\partial t} = \frac{\partial}{\partial z} \theta D \frac{\partial C}{\partial z} - \frac{\partial qC}{\partial z} - Q \tag{1.6}$$

The above equation is commonly known as the convection–dispersion equation (CDE) for solute transport and is valid for soils under transient and unsaturated soil water flow conditions. For conditions where steady water flow is dominant, D and θ are constants; that is, for uniform θ in the soil, we have the simplified form of the CDE as follows:

$$\frac{\partial C}{\partial t} + \frac{\rho}{\theta} \frac{\partial S}{\partial t} = D \frac{\partial^2 C}{\partial z^2} - v \frac{\partial C}{\partial z} - \frac{Q}{\theta} \tag{1.7}$$

where v (cm.h^{-1}) is commonly referred to as the pore-water velocity and is given by (q/θ).

Solutions of the above CDE Equations (1.6) or (1.7) yield the concentration distribution of the amount of solute in soil solution C and that retained by the soil matrix S with time and depth in the soil profile. To arrive at such a solution, the appropriate initial and boundary conditions must be specified. Several boundary conditions are identified with the problem of solute transport in porous media. The simplest is that of a first-order type boundary condition such that a solute pulse input is described by

$$C = C_s, \qquad z = 0, \qquad t < T \tag{1.8}$$

$$C = 0, \qquad z = 0, \qquad t \geq T \tag{1.9}$$

where C_s (μg.cm^{-3}) is the concentration of the solute species in the input pulse. The input pulse application is for a duration T, which was then followed by a pulse input that is free of such a solute. Such a boundary condition was used by Lapidus and Amundson (1952). The more precise third-type boundary condition at the soil surface was considered by Brenner (1962) in his classical work, where advection plus dispersion across the interface was considered. A continuous solute flux at the surface can be expressed as

$$vC_s = -D\frac{\partial C}{\partial z} + vC, \qquad z = 0, \quad t > 0 \tag{1.10}$$

and a flux-type pulse input as

$$vC_s = -D\frac{\partial C}{\partial z} + vC, \qquad z = 0, \quad t < T \tag{1.11}$$

$$0 = -D\frac{\partial C}{\partial z} + vC, \qquad z = 0, \quad t \geq T \tag{1.12}$$

The advantages of using third-type boundary conditions were discussed by Selim and Mansell (1976) and Kreft and Zuber (1978). The boundary condition at some depth L in the soil profile is often expressed as (Danckwerts, 1953; Brenner, 1962; Lindstrom et al., 1967; Kreft and Zuber, 1978)

$$\frac{\partial C}{\partial z} = 0, \qquad z = L, \quad t \geq 0 \tag{1.13}$$

which is used to deal with solute effluent from soils having finite lengths. However, it is often convenient to solve the CDE where a semi-infinite rather than a finite length (L) of the soil is assumed. Under such circumstances, the appropriate condition for a semi-infinite medium is

$$\frac{\partial C}{\partial z} = 0, \qquad z \to \infty, \quad t \geq 0 \tag{1.14}$$

Analytical solutions to the CDE subject to the appropriate boundary and initial conditions are available for a number of situations, whereas the majority of solute transport problems must be solved using numerical approximation methods. In general, whenever the form of the retention reaction is a linear one, a closed-form solution is obtainable. A number of closed-form solutions are available from Kreft and Zuber (1978), and Van Genuchten and Alves (1982). However, most retention mechanisms are nonlinear and time dependent in nature, and analytical solutions are not available. As a result, a number of numerical models using finite-difference or finite element have been utilized to solve nonlinear retention problems of multireaction and multicomponent solute transport in soils (see Selim, Selim, Amacher, Iskandar, 1990).

Retention Models

The form of heavy metal reactions in the soil system must be identified if prediction of their fate in the soil is sought. In general, heavy metals retention processes in soils have been quantified by scientists using several approaches. One approach represents equilibrium reactions and the second represents kinetic or time-dependent type reactions. Equilibrium models are those where heavy metal reaction is assumed to be fast or instantaneous in nature. Under such conditions, "apparent equilibrium" may be observed in a relatively short reaction time. Langmuir and Freundlich models are perhaps the most commonly used equilibrium models for the description of fertilizer chemicals such as phosphorus and for several heavy metals. These equilibrium models include the linear and Freundlich (nonlinear) and the one- and two-site Langmuir type. Kinetic models represent slow reactions where the amount of solute sorption or transformation is a function of contact time. Most common is the first-order kinetic reversible reaction for describing time-dependent adsorption-desorption in soils. Others include linear irreversible and nonlinear reversible kinetic models. Recently, a combination of equilibrium and kinetic type (two-site) models, and consecutive and concurrent multireaction type models has been proposed.

First-Order and Freundlich Models

The first-order kinetic approach is perhaps one of the earliest single forms of reactions used to describe sorption versus time for several dissolved chemicals in soils. This may be written as

$$\frac{\partial S}{\partial t} = k_f \left(\frac{\theta}{\rho}\right) C - k_b S \tag{1.15}$$

where the parameters k_f and k_b represent the forward and backward rates of reactions (h^{-1}) for the retention mechanism, respectively. The first-order reaction was originally incorporated into the classical CDE by Lapidus and Amundson (1952) to describe solute retention during transport under steady-state water flow conditions. Integration of Equation (1.15) subject to initial conditions of $C = C_i$ and $S = 0$ at $t = 0$, for several C_i values, yields a system of linear sorption isotherms. That is, for any reaction time t, a linear relation between S and C is obtained.

Linear isotherms are not often encountered except for selected cations and heavy metals at low concentrations. Isotherms that exhibit nonlinear or curve linear retention behavior are commonly observed for several reactive chemicals as depicted by the nonlinear isotherms for nickel and arsenic shown in Figures 1.1, 1.2, 1.3 respectively (Liao and Selim, 2010; Zhang and Selim, 2005; Selim and Ma, 2001). To describe such nonlinear behavior,

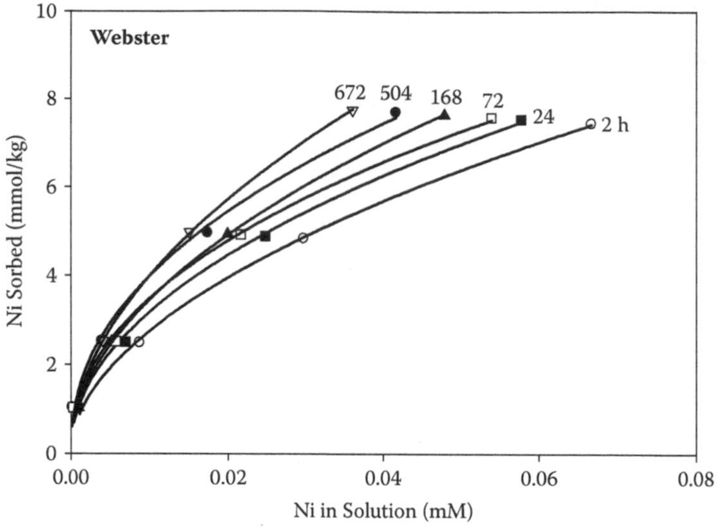

FIGURE 1.1
Adsorption isotherms for Ni on Webster soil at different retention times. The solid curves are based on the Freundlich equation.

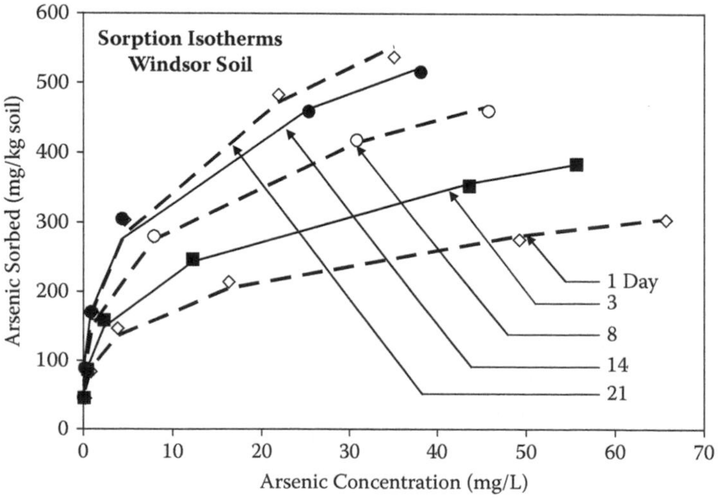

FIGURE 1.2
Adsorption isotherms for arsenic on Windsor soil at different retention times.

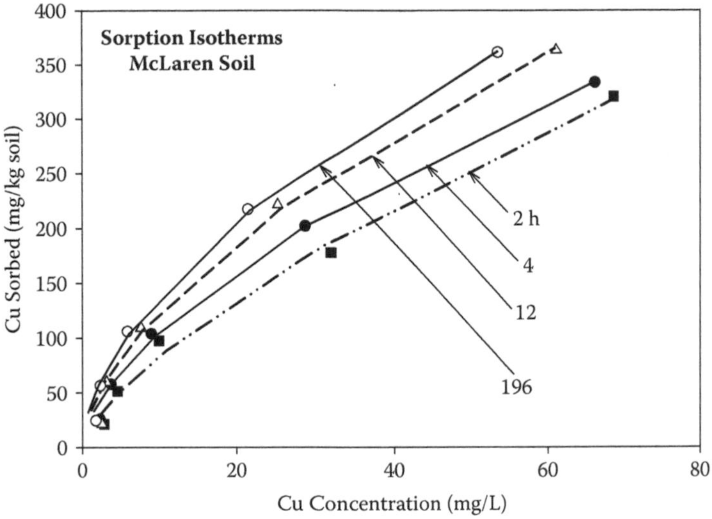

FIGURE 1.3
Adsorption isotherms for copper on McLaren soil at different retention times.

the single reaction given in Equation (1.15) is commonly extended to include nonlinear kinetics such that (Selim and Amacher, 1997)

$$\frac{\partial S}{\partial t} = k_f \left(\frac{\theta}{\rho} \right) C^b - k_b S \tag{1.16}$$

where b is a dimensionless parameter commonly less than unity, represents the order of the nonlinear or concentration-dependent reaction, and illustrates the extent of heterogeneity of the retention processes. This nonlinear reaction (Equation (1.16)) is fully reversible and the magnitudes of the rate coefficients dictate the extent of kinetic behavior of retention of the solute from the soil solution. For small values of k_f and k_b, the rate of retention is slow and strong kinetic dependence is anticipated. In contrast, for large values of k_f and k_b, the retention reaction is a rapid one and should approach quasi-equilibrium in a relatively short time. In fact, at large times ($t \rightarrow 4$), when the rate of retention approaches zero, Equation (1.16) yields

$$S = K_f \, C^b \qquad \textit{where} \qquad K_f = \left(\frac{\theta k_f}{\rho k_b} \right) \tag{1.17}$$

Equation (1.17) is analogous to the Freundlich equilibrium equation where K_f is the solute partitioning coefficient ($cm^3.g^{-1}$). Therefore, one may regard the parameter K_f as the ratio of the rate coefficients for sorption (forward reaction) to that for desorption or release (backward reaction).

The parameter b is a measure of the extent of the heterogeneity of sorption sites of the soil matrix. In other words, sorption sites have different affinities for heavy metal retention by matrix surfaces, where sorption by the highest energy sites takes place preferentially at the lowest solution concentrations.

For the simple case where $b = 1$, we have the linear form:

$$S = K_d\, C \qquad \textit{where} \qquad K_d = \left(\frac{\theta k_f}{\rho k_b}\right) \tag{1.18}$$

where the parameter K_d is the solute distribution coefficient ($cm^3.g^{-1}$) and of similar form as the Freundlich parameter K_f. There are numerous examples of cations and heavy metals retention, which were described successfully using the linear or the Freundlich equation (Sparks, 1989; Buchter et al., 1988). The lack of nonlinear or concentration-dependent behavior of sorption patterns as indicated by the linear case of Equation (1.18) is indicative of the lack of heterogeneity of sorption-site energies. For this special case, sorption-site energies for linear sorption processes of heavy metals are regarded as relatively homogeneous.

Second-Order and Langmuir

An alternative to the above first- and n-th order models is that of the second-order kinetic approach. Such an approach is commonly referred to as the Langmuir kinetic and has been used for the prediction of phosphorus retention (Van der Zee and Van Riemsdijk, 1986) and heavy metals (Selim and Amacher, 1997). Based on second-order formulation, it is assumed that the retention mechanisms are site specific, where the rate of reaction is a function of the solute concentration present in the soil solution phase (C) and the amount of available or unoccupied sites φ ($\mu g.g^{-1}$ soil), by the reversible process,

$$C + \phi \underset{k_b}{\overset{k_f}{\rightleftarrows}} S \tag{1.19}$$

where k_f and k_b are the associated rate coefficients (h) and S is the total amount of solute retained by the soil matrix. As a result, the rate of solute retention may be expressed as

$$\rho \frac{\partial S}{\partial t} = k_f\, \theta\, \phi\, C - k_b\, \rho\, S$$

or

$$\rho \frac{\partial S}{\partial t} = k_f\, \theta\, (S_T - S) C - k_b\, \rho\, S \tag{1.20}$$

where S_T (µg.g^{-1} soil) represents the total number of total sorption sites. As the sites become occupied by the retained solute, the number of vacant sites approaches zero ($\phi \rightarrow 0$) and the amount of solute retained by the soil approaches that of the total capacity of sites, that is, $S \rightarrow S_T$. Vacant specific sites are not strictly vacant. They are assumed occupied by hydrogen, hydroxyl, or by other specifically sorbed species. As $t \rightarrow 4$, that is, when the reaction achieves local equilibrium, the rate of retention becomes

$$k_f \theta \phi C - k_b \rho S = 0, \qquad or \qquad \frac{S}{\phi C} = \left(\frac{\theta}{\rho}\right)\frac{k_f}{k_b} = \omega \tag{1.21}$$

Upon further rearrangement, the second-order formulation, at equilibrium, obeys the widely recognized Langmuir isotherm equation:

$$\frac{S}{S_T} = \frac{KC}{1+KC} \tag{1.22}$$

where the parameter $K\,[= \theta k_f/k_b\rho]$ *is* now equivalent to ω in Equation (1.21) and represents the Langmiur equilibrium constant. Sorption–desorption studies showed that highly specific sorption mechanisms are responsible for solute retention at low concentrations. The general view was that metal ions have a high affinity for sorption sites of oxide minerals surfaces in soils. In addition, these specific sites react slowly with reactive chemicals such as heavy metals and are weakly reversible.

Hysteresis

Adsorption–desorption results are presented as isotherms in the traditional manner in Figures 1.4 and 1.5 and clearly indicate considerable hysteresis for nickel and arsenic retention in two different soils, respectively (Liao and Selim, 2010; and Zhang and Selim, 2005). Other examples for copper and zinc are shown in Figures 1.6 and 1.7 (Selim and Ma, 2001; Zhao and Selim). As seen from the family of curves, desorption did not follow the same path (i.e., nonsingularity) as the respective adsorption isotherm. This nonsingularity or hysteresis may result from the failure to achieve equilibrium adsorption prior to desorption. If adsorption as well as desorption were carried out for times sufficient for equilibrium to be attained, or the kinetic rate coefficients were sufficiently large, such hysteretic behavior would perhaps be minimized (Selim et al., 1976). Such hysteretic behavior resulting from a discrepancy between adsorption and desorption isotherms was not surprising in view of the strong kinetic retention behavior of these heavy metals in soils. Several studies indicated that observed hysteresis in batch experiments may be due to kinetic retention behavior and slow release and/or irreversible adsorption

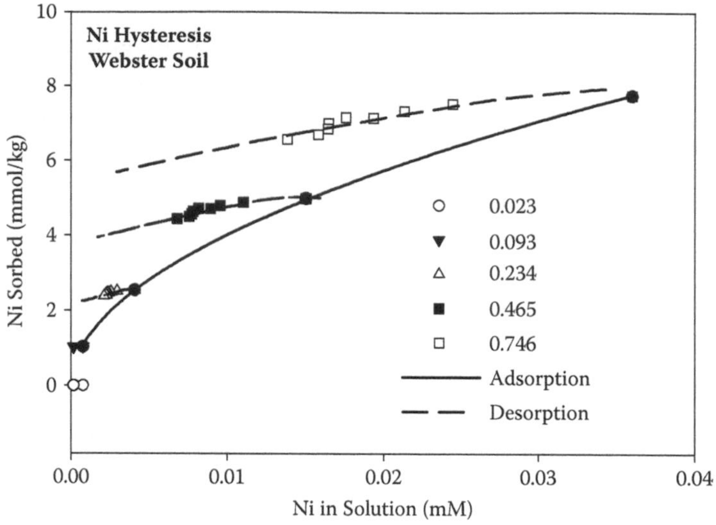

FIGURE 1.4
Adsorption and desorption isotherms illustrating hysteresis behavior of Ni retention on Webster soil.

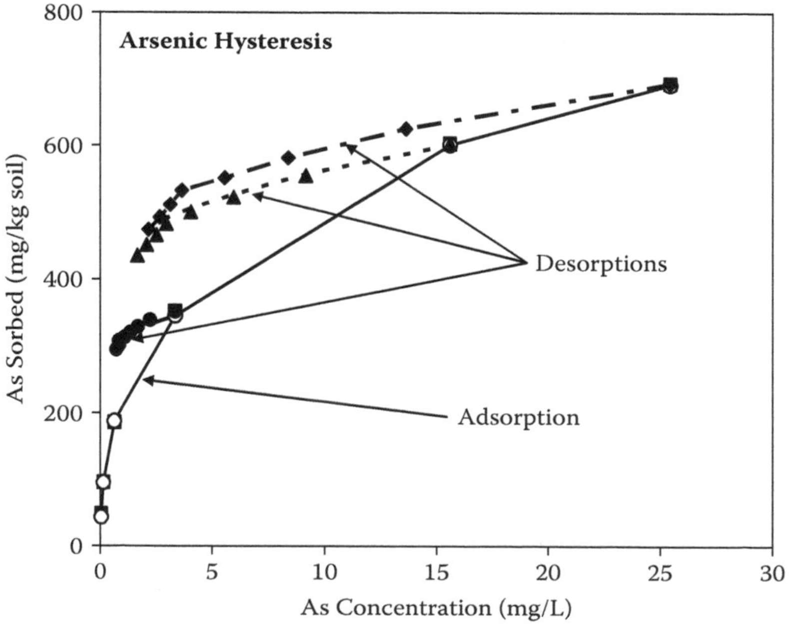

FIGURE 1.5
Adsorption and desorption isotherms illustrating hysteresis behavior of As retention in Sharkey clay soil.

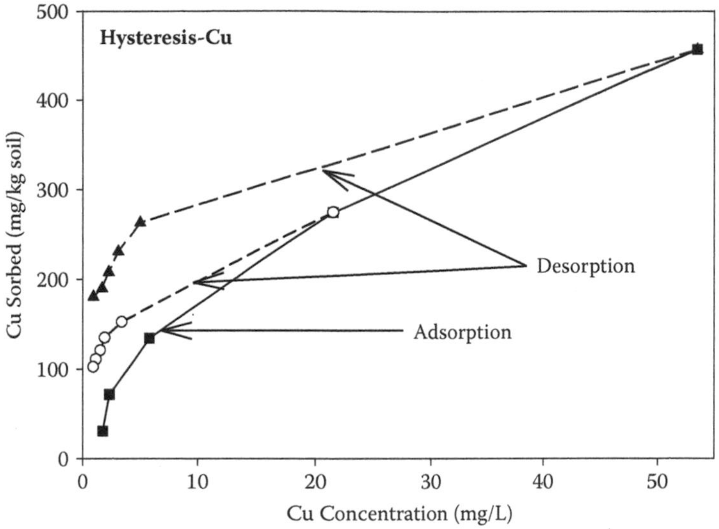

FIGURE 1.6
Adsorption and desorption isotherms illustrating hysteresis behavior of Cu retention in McLaren soil.

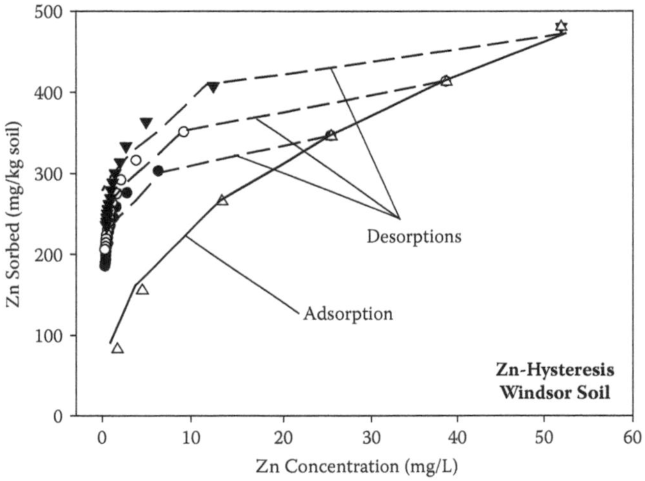

FIGURE 1.7
Adsorption and desorption isotherms illustrating hysteresis behavior of Zn retention in Windsor soil.

conditions. Adsorption–desorption isotherms indicate that the number of irreversible or nondesorbable phases increased with time of reaction. Heavy metals may be retained by heterogeneous type sites having a wide range of binding energies. At low concentrations, binding may be irreversible. The irreversible amount almost always increased with time. It is suggested that hysteresis for heavy metals is probably due to extremely high energy bonding with organic matter and layer silicate surfaces. The fraction of nondesorbable solutes is often referred to as specifically sorbed. Others suggested that several solutes were fixed in a nonexchangeable form, which resulted in a lack of reversibility as well as hysteretic behavior. Moreover, several researchers reported that the magnitude of hysteresis increases with longer sorption incubation periods. Increasing hysteretic behavior upon aging has been observed consistently for several pesticides and heavy metals.

Hysteresis has also been observed in ion exchange reactions for several cations, where the exchange of one sorbed cation for another is not completely reversible, that is, the forward and reverse exchange reactions do not result in the same isotherms. The hysteretic behavior of cation exchange is abundantly reported in the literature; a critical review of this literature was published by Verburg and Baveye (1994). From a survey of the literature, they were able to categorize several elements into three categories. The elements in each category were found to show hysteretic exchange *between* groups, but not *within* groups. They proposed that exchange reactions are most likely multistage kinetic processes in which the later rate-limiting processes are a result of physical transformation in the system (e.g., surface heterogeneity, swelling hysteresis, and formation of quasi-crystals) rather than simply a slow kinetic exchange process where there exists a unique thermodynamic relationship for forward and reverse reactions. While this may be true in some circumstances, an apparent (pseudo) hysteresis also can result from slow sorption and desorption reactions, that is, lack of equilibrium (Selim et al. 1976). Regardless of the different reasons for hysteresis, it is evident that kinetic models such as those proposed in this study need to be complemented by detailed information on the mechanism(s) responsible for the slow kinetic reaction(s).

Multiple Reaction Models

Several studies showed that the use of single reaction models, such as those described above, is not adequate because such models are of the equilibrium or kinetic type. The failure of single reaction models is not surprising as they only describe the behavior of one species with no consideration for the simultaneous reactions of others in the soil system. Multicomponent models consider a number of processes governing several species, including ion exchange, complexation, precipitation and dissolution, and competitive adsorption, among others. Multicomponent models rely on the basic assumption of local equilibrium of the governing reactions where possible kinetic reactions are ignored.

Multisite or multireaction models deal with the multiple interactions of one species in the soil environment. Such models are empirical in nature and based on the assumption that a fraction of the total sites are highly kinetic, whereas the remaining fraction of sites interact slowly or instantaneously with that in the soil solution (Selim et al., 1976; Selim and Amacher, 1997). Nonlinear equilibrium (Freundlich) and first- or *n*-order kinetic reactions were the associated processes. Such a two-site approach proved successful in describing observed extensive tailing of breakthrough results. Amacher, Selim, and Iskandar (1988) developed a multireaction model that includes concurrent and concurrent–consecutive processes of the nonlinear kinetic type. The model was capable of describing the retention behavior of Cd and Cr(VI) with time for several soils. In addition, the model predicted that a fraction of these heavy metals was irreversibly retained by the soil. A schematic representation of the multireaction model is shown in Figure 1.8. In this model we consider the solute to be present in the soil solution phase (C) and in four phases representing solute retained by the soil matrix as S_e, S_1, S_2, S_3, and S_{irr}. We further assume that S_e, S_1, and S_2 are in direct contact with the solution phase and are governed by concurrent type reactions. Here we assume that S_e is the amount of solute sorbed reversibly and is in equilibrium with C at all times. The governing equilibrium retention and release mechanism was of the nonlinear Freundlich type, as discussed previously.

The retention and release reactions associated with S_1 and S_2 were considered in direct contact with C and reversible processes of the (nonlinear) kinetic type govern their reactions:

$$S = S_e + S_1 + S_2 + S_3 + S_{irr} \tag{1.23}$$

$$S_e = K_e\left(\frac{\theta}{\rho}\right)C^b \tag{1.24}$$

$$\frac{\partial S_1}{\partial t} = k_1\left(\frac{\theta}{\rho}\right)C^n - k_2 S_1 \tag{1.25}$$

$$\frac{\partial S_2}{\partial t} = k_3\left(\frac{\theta}{\rho}\right)C^m - k_4 S_2 \tag{1.26}$$

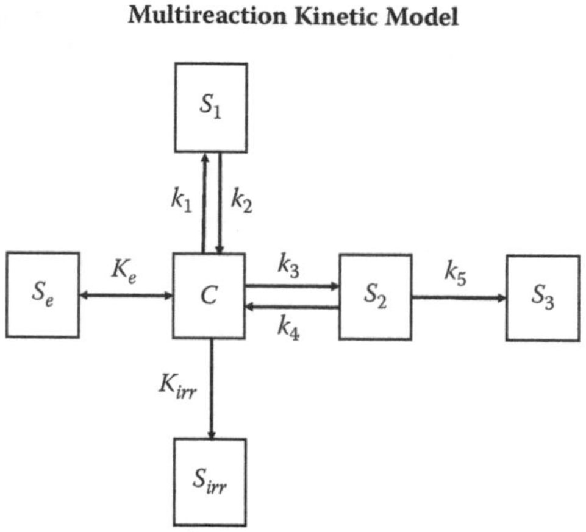

FIGURE 1.8
A schematic representation of the multireaction kinetic model.

where k_1 to k_4 are the associated rates coefficients (h^{-1}). These two phases (S_1 and S_2) may be regarded as the amounts sorbed on surfaces of soil particles and chemically bound to Al and Fe oxide surfaces or other types of surfaces, although it is not necessary to have a priori knowledge of the exact retention mechanisms for these reactions to be applicable. Moreover, these phases may be characterized by their kinetic sorption and release behavior to the soil solution and thus are susceptible to leaching in the soil. In addition, the primary difference between these two phases not only lies in the difference in their kinetic behavior, but also in the degree of nonlinearity as indicated by the parameters n and m. The multireaction model also considers irreversible solute removal via a retention sink term Q in order to account for irreversible reactions such as precipitation and dissolution, mineralization, and immobilization, among others. We expressed the sink term as a first-order kinetic process:

$$Q = \rho \frac{\partial S_{irr}}{\partial t} = k_{irr}\, \theta\, C \tag{1.27}$$

where k_{irr} is the associated rate coefficient (h^{-1}).

The multireaction model also includes an additional retention phase (S_3) that is governed by a consecutive reaction with S_2. This phase represents the amount of solute strongly retained by the soil that reacts slowly and reversibly with S_2 and may be a result of further rearrangements of the solute retained on matrix surfaces. Thus, inclusion of S_3 in the model allows the description of the frequently observed very slow release of solute from the soil. The reaction between S_2 and S_3 was considered to be of the kinetic first-order type, that is,

$$\frac{\partial S_3}{\partial t} = k_5 \ S_2 \ - k_6 \ S_3 \tag{1.28}$$

where k_5 and k_6 (h^{-1}) are the reaction rate coefficients. If a consecutive reaction is included in the model, then Equation (1.26) must be modified to incorporate the reversible reaction between S_2 and S_3. As a result, the following equation

$$\rho\frac{\partial S_2}{\partial t} = k_3\theta C^n + \rho(k_4 + k_5)S_2 \ - \rho k_6 S_3 \tag{1.29}$$

must be used in place of Equation (1.26). The above reactions are nonlinear in nature and represent initial-value problems that are typically solved based on numerical approximations. In addition, the above retention mechanisms were incorporated, in a separate model, into the classical convection–dispersion equation in order to predict solute retention as governed by the multireaction model during transport in soils (Zhang and Selim, 2005).

The capability of the multireaction approach discussed above in describing experimental batch data for arsenic retention is shown by the solid curves of Figure 1.9. The results and model predictions are given for the various initial concentrations (C_i). Overall, good model predictions were observed for the wide range of input concentrations values considered. The multireaction model used here accounts for several interactions of the reactive solute species (As) within the soil system. Specifically, the model assumes that a fraction of the total sites is highly kinetic whereas the remaining fraction interacts slowly or instantaneously with solute in the soil solution. As illustrated in Figure 1.8, the model also accounts for irreversible reactions of the concurrent (S_{irr}) and consecutive type (S_3). As a result, different versions of the multireaction model shown in Figure 1.8 represent different reactions from which one can deduce possible retention mechanisms.

In our simulations, the multireaction model was fitted to arsenic versus time for all input concentrations (C_i) simultaneously. As a result, an *overall* set of model parameters for the appropriate rate coefficients, applicable for the entire data set, was achieved. The examples shown in Figure 1.9 are for two different model versions. In the first version, the simulations in Figure 1.9 (top), a kinetic phase (S_1), and an irreversible phase (S_{irr}) were considered, where the necessary model parameters were n, k_1, k_2, and k_{irr}. In the second version, the simulations in Figure 1.9 (bottom), a kinetic phase (S_2), as well as a consecutive irreversible reaction represented by (S_3) were considered. The presence of a consecutive S_3 phase may occur as a result of further surface rearrangement of the adsorbed phase (see Figure 1.8). For this version, model parameters considered were m, k_3, k_4, and k_5, where all other model parameters were set equal to zero. It is obvious from the simulations shown

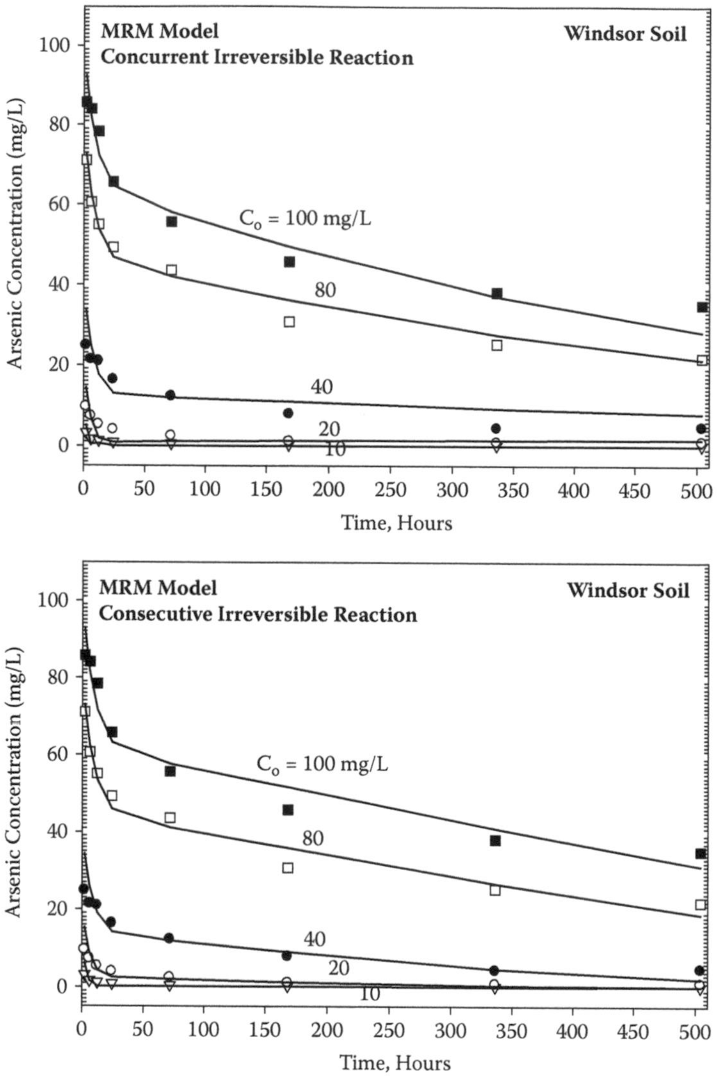

FIGURE 1.9
Experimental results of As(V) concentration in soil solution for Windsor soil versus time for all C_o's. Dashed curves were obtained using the MRM with concurrent (top figure) and consecutive (bottom figure) irreversible reactions.

in Figure 1.9 that a number of model versions were capable of producing indistinguishable simulations of the data. Similar conclusions were made by Amacher, Selim, and Iskandar (1988) for Cd and Cr(VI) for several soils. They also stated that it was not possible to determine whether the irreversible reaction is concurrent or consecutive, as both model versions provided similar fit of their batch data. For the example shown in Figure 1.9, the use of a consecutive irreversible reaction provided an improved fit of Cu retention over other model versions. This finding is based on goodness-of-fit (r^2 and root mean square errors) as well as visual observation of measured data and model simulations.

Second-Order Models (SOMs)

The basic assumption of the second-order modeling approach is that there exist at least two types of retention sites for heavy metals on soil matrix surfaces. Moreover, the primary difference between these two types of sites is based on the rate of the proposed kinetic retention reactions. Furthermore, the retention mechanisms are site specific, wherein the rate of reaction is a function of not only the solute concentration present in the soil solution phase, but also the number of available retention sites on matrix surfaces.

The original second-order model (SOM) was first proposed by Selim and Amacher (1988) to describe Cr retention and transport in several soils. Here, two types of sites were considered: the first was of the equilibrium type and the second was kinetically controlled type sites. Moreover, S_{max} (µg.g^{-1} soil) was considered to represent the total retention capacity or total number of sites on matrix surfaces. It is also assumed that S_{max} is an intrinsic soil property that is time invariant. Therefore, based on the two-site approach, the total adsorption sites are given by

$$S_{max} = (S_e)_{max} + (S_k)_{max} \tag{1.30}$$

where S_{max} is the adsorption maximum, and $(S_e)_{max}$ and $(S_k)_{max}$ are the total amounts or adsorption maxima for equilibrium and kinetic type sites, respectively (µg.g^{-1} soil). If f represents the fraction of equilibrium type sites $(S_e)_{max}$ to the total sites, we thus have

$$(S_e)_{max} = f\, S_{max} \qquad and \qquad (S_k)_{max} = (1 - f)\, S_{max} \tag{1.31}$$

Assuming φ_e and φ_k as the vacant or available sites (µg.g^{-1} soil) for adsorption on equilibrium and kinetic type sites (S_e and S_k), respectively, we have

$$\phi_e = (S_e)_{max} - S_e = f\, S_{max} - S_e$$

$$\phi_k = (S_k)_{max} - S_k = (1 - f)\, S_{max} - S_k \tag{1.32}$$

with the total available sites equal to $\varphi = \varphi_e + \varphi_k$. As the sites become filled or occupied by the retained solute, the number of vacant sites approaches zero, $(\varphi_e + \varphi_k)$ ® S. In the meantime, the amount of solute retained by the soil matrix approaches the total capacity or sorption maxima $(S_e + S_k)$ ® S_{max}.

The second-order approach was successfully used for Cr retention and transport predictions by Selim and Amacher (1988) and for Zn retention by Hinz, Buchter, and Selim (1992). This model was recently modified such that the total adsorption sites S_{max} were not partitioned between S_e and S_k phases based on a fraction of sites (Selim and Amacher, 1997; Ma and Selim, 1998). Instead, it was assumed that the vacant sites were available to both types of S_e and S_k. Therefore, f is no longer required and the amount of solute adsorbed on each type of sites is only determined by the rate coefficients associated with each type of site. As a result, sites associated with equilibrium or instantaneous type reactions will compete for available sites prior to slow or kinetic type sites being filled. Perhaps such a mechanism is in line with observations where rapid (equilibrium type) sorption is encountered first, followed by slow types of retention reactions. We are not aware of the use of this second-order approach to describe heavy metal retention kinetics and transport in soils.

In the following analysis we followed an overall structure for the second-order formulation similar to that described for the multireaction approach of Figure 1.8 where three types of retention sites are considered with one equilibrium-type site (S_e) and two kinetic-type sites, namely S_1 and S_2. Therefore, we have φ now related to the sorption capacity (S_{max}) by

$$S_{max} = \phi + S_e + S_1 + S_2 \tag{1.33}$$

The governing retention reactions can be expressed as follows (Ma and Selim, 1998):

$$S_e = K_e \theta C \phi \tag{1.34}$$

$$\frac{\partial S_1}{\partial t} = k_1 \theta C \phi - k_2 S_1 \tag{1.35}$$

$$\frac{\partial S_2}{\partial t} = \left[k_3 \theta C \phi - k_4 S_2\right] - K_5 S_2 \tag{1.36}$$

$$\frac{\partial S_3}{\partial t} = k_5 S_2 \tag{1.37}$$

$$\frac{\partial S_{irr}}{\partial t} = k_{irr}\,\theta C \qquad (1.38)$$

The units for K_e are $cm^3.\mu g^{-1}$; k_1 and k_3 have a derived unit of $cm^3.\mu g^{-1}.h^{-1}$; and k_2, k_4, k_5, and k_{irr} are assigned units of h^{-1}.

The input parameter S_{max} of the second-order model is a major parameter and represents the total sorption of sites. S_{max}, which is often used to characterize heavy metal sorption, can be quite misleading if the experimental data do not cover a sufficient range of solution concentration and if other conditions such as the amounts initially sorbed prevail (Houng and Lee, 1998). In an arsenic adsorption study, Selim and Zhang (2007) used S_{max} in the SOM model based on average values as determined from the Langmuir isotherm equation. This is a simple approach to obtain S_{max} estimated when a direct measurement of the sorption capacity is not available. Figures 1.10, 1.11, and 1.12 show simulated adsorption results using the SOM model based on two model versions for three different soils. Based on visual observations of the overall of fit of the model to the experimental data, SOM provided good overall predictions of the kinetic adsorption data for arsenic.

The question arises whether SOM model improvements can be realized when one relaxes the assumption of the use of Langmuir S_{max} and utilizes parameter optimization to arrive at a best estimate of the rate coefficients (e.g., k_1, k_2, and k_{irr} or k_3, k_4, and k_5) as well as S_{max}. Based on these results, Selim and Zhang (2007) concluded that the use of Langmuir S_{max} as an input parameter provided good predictions of the adsorption results. Moreover, the retention kinetics predictions for As(V) shown are in agreement with the biphasic arsenic adsorption behavior observed on several soil minerals (Fuller et al., 1993; Raven et al., 1998; Arai and Sparks, 2002) as well as whole soils (Elkhatib et al., 1984; Carbonell-Barrachina et al., 1996) over different time scales (minutes to months).

A comparison of the multireaction (MRM) and second-order two-site (SOTS) models for their capability to predict arsenic concentration with time is given in Figure 1.13. Selim and Zhang (2007) found that several model versions fit the data equally well, but the sorption kinetics prediction capability varied among the soils investigated. MRM was superior to SOM, and the use of irreversible reaction into the model formulations was essential. They also found that incorporation of an equilibrium sorbed phase into the various model versions for As(V) predictions should be avoided.

The success of the second-order approach in describing As(V) retention results is significant because, to our knowledge, the SOM formulation described in this chapter has not been applied to metalloid elements like As. Previous use of the second-order formulation, which included a partitioning of the sites, indicated that for Cr and Zn the rate coefficients were highly concentration dependent (Selim and Amacher, 1988; Hinz, Buchter, and Selim, 1992). Selim and Ma (2001) successfully utilized the SOM model

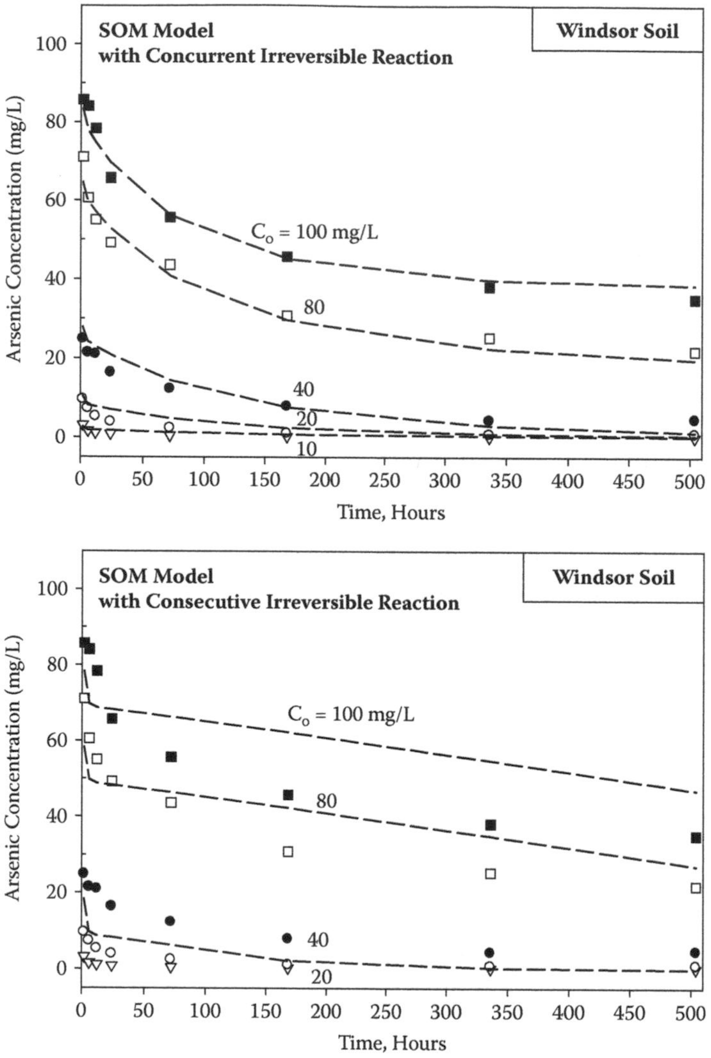

FIGURE 1.10
Experimental results of As(V) concentration in soil solution for Windsor soil versus time for all C_o's. Dashed curves were obtained using SOM with concurrent (top figure) and consecutive (bottom figure) irreversible reactions.

to describe Cu adsorption as well as desorption or release following sorption. They concluded that the use of consecutive irreversible reaction (k_5 in Figure 1.1) provided improvements in the description of the kinetic sorption and desorption of Cu compared to the concurrent irreversible reaction (k_{irr}). This finding is contrary to that from this study for As adsorption for all three soils. Such contradictions are not easily explained and are thus subjects for future research.

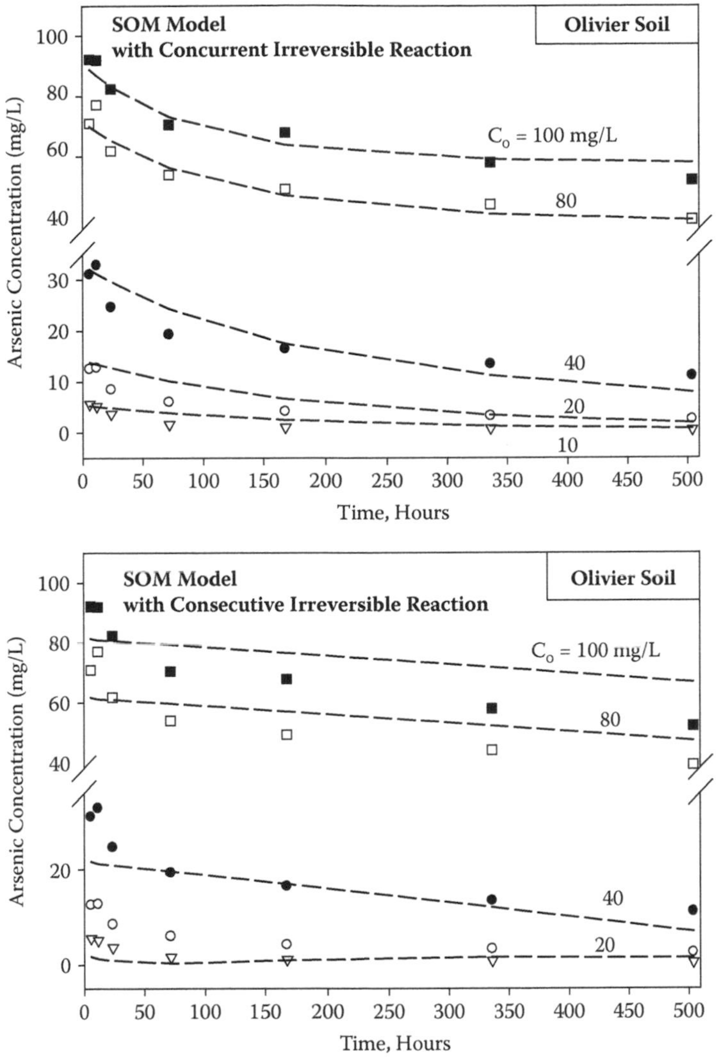

FIGURE 1.11
Experimental results of As(V) concentration in soil solution for Olivier soil versus time for all C_0's. Dashed curves were obtained using second-order model (SOM) with concurrent (top figure) and consecutive (bottom figure) irreversible reactions.

Transport in Layered Soils

A major source of soil heterogeneity is soil stratifications or layering. The phenomenon of stratifications or layering of the soil profile is an intrinsic part of soil formation processes and has been documented for several decades by soil survey work. These properties reflect various processes and

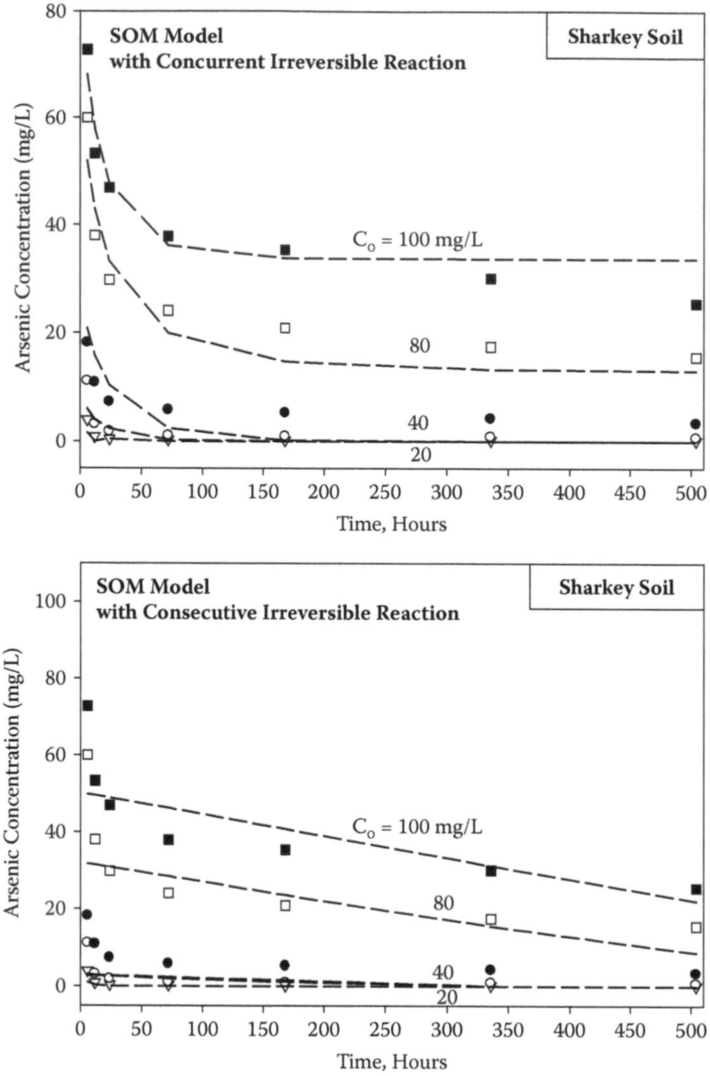

FIGURE 1.12
Experimental results of As(V) concentration in soil solution for Sharkey soil versus time for all C_0's. Dashed curves were obtained using SOM with concurrent (top figure) and consecutive (bottom figure) irreversible reactions.

their soil genesis. Layers are commonly of uniform thickness and stratified horizontally. The transport processes of dissolved chemicals in stratified or layered soils have been studied for several decades by Shamir and Harleman (1967); Selim, Davidson, and Rao (1977); Bosma and Van der Zee (1992); and Wu, Kool, and Huyakorn (1997), among others. Solute transport in layered soils can be investigated through numerical methods as well as approximate

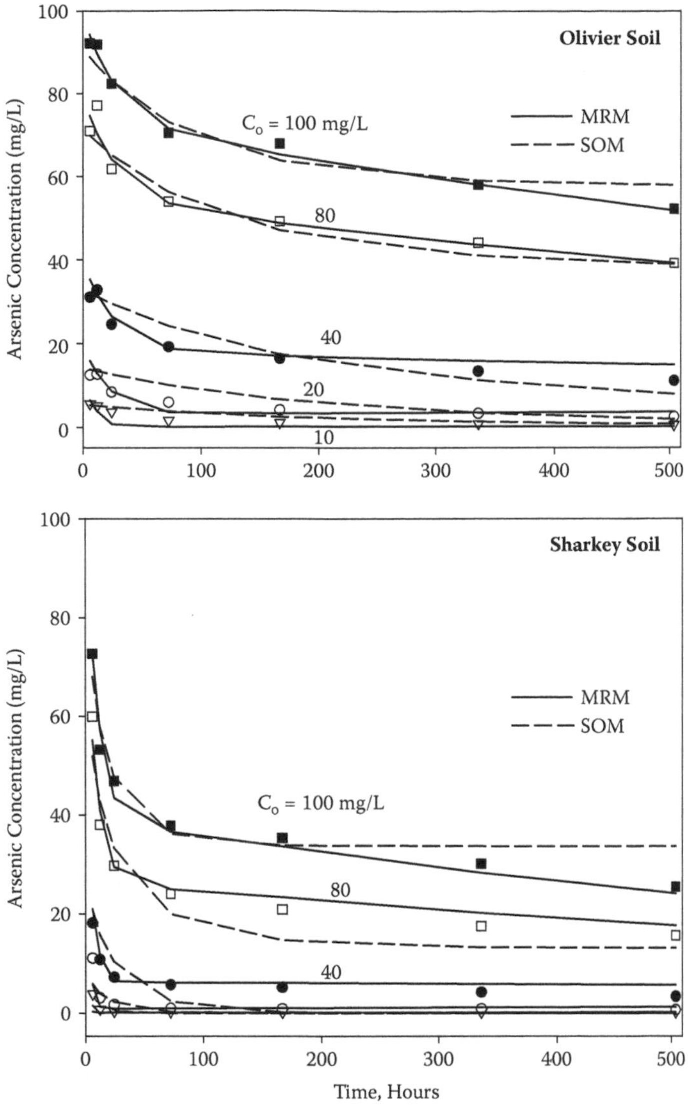

FIGURE 1.13
Experimental and predicted As(V) concentration versus time for Olivier and Sharkey soil and several input (initial) concentrations (C_0's). Predictions were obtained using MRM and SOM models.

analytical solutions. An early analytical method was proposed by Shamir and Harleman (1967) who used a systems analysis approach. They assumed that different layers were independent with regard to solute travel time. Each layer's response served as the boundary condition for the downstream layer and so on. Later, Selim, Davidson, and Rao (1977) discussed the movement of reactive solutes through layered soils using finite difference numerical methods. They considered both equilibrium and kinetic sorption models of the linear and nonlinear types. In the late 1980s, Leij and Dane (1989) developed analytical solutions for the linear sorption type models using Laplace transforms. Their solutions were based on the assumption that each layer was semi-infinite. Bosma and Van der Zee (1992) also proposed an approximate analytical solution for reactive solute transport in layered soils using an adaptation of the traveling wave solution. More recently, Wu, Kool, and Huyakorn (1997) developed another analytical model for nonlinear adsorptive transport through layered soils ignoring the effects of dispersion. In addition, Guo et al. (1997) showed that the transfer function approach was a very powerful tool to describe the nonequilibrium transport of reactive solutes through layered soil profiles with depth-dependent adsorption.

When we study the transport process of dissolved chemicals in layered soils, it is of interest to investigate whether soil layering affects breakthrough of reactive chemicals. When flow remains one-dimensionally vertical, which is the case when horizontal stratification is dominant, it is of interest whether the layering order affects breakthrough results at the groundwater level (Van der Zee, 1994). The early results from Shamir and Harleman (1967) showed that the order of layering did not affect breakthrough significantly. This interesting result was further elaborated upon by Barry and Parker (1987) based on various analytical approaches. Results from various linear and nonlinear numerical simulations for several sorption model types also supported this conclusion (Selim, Davidson, and Rao, 1977). Furthermore, Selim, Davidson, and Rao (1977) concluded that layering order was also unimportant for Freundlich adsorption. Their experimental results also supported this conclusion. However, Van der Zee (1994) attributed Selim, Davidson, and Rao's (1977) results to the small Peclet number assumed for the nonlinear layer, which prevents clear manifestation of any nonlinearity effects. Van der Zee (1994) used a hypothetical result to illustrate that layering sequence should have an effect. However, what Van der Zee (1994) used to support his conclusion was the traveling wave, which was the curve of concentration versus depth at different times, that is, a concentration profile. Recently, Zhou and Selim (2001) accounted for several nonlinear and kinetic retention mechanisms for multilayered soils. For individual soil layers, Zhou and Selim (2001) considered solute retention mechanisms of the nonlinear (Freundlich), Langmuir, first- and n-order kinetic, second-order kinetic, and irreversible reactions. For all retention mechanisms used, their simulation results indicated that solute breakthrough curves (BTCs) were similar, regardless of the layering sequence in a soil profile. This finding is consistent with earlier findings of Selim et al. (1977) and contrary to those of Bosma and Van der Zee (1992) for nonlinear adsorption. The convection–dispersion equation (CDE) governing solute transport in the i-th layer is given by (Selim, Davidson, and Rao, 1977)

$$\rho_i \frac{\partial S_i}{\partial t} + \theta_i \frac{\partial C_i}{\partial t} = \frac{\partial}{\partial x} \theta_i D_i \frac{\partial C_i}{\partial x} - q \frac{\partial C_i}{\partial x} - Q_i$$
$$(0 \le x \le L_i, i = 1, 2) \tag{1.39}$$

where the subscript i denotes the i-th layer.

An important boundary condition needed in the analysis of multilayered soils is at the interface between layers. It should be noted that both first-type and third-type boundary conditions are applicable at the interface. Leij, Dane, and Van Genuchten (1991) showed that although the principle of solute mass conservation is satisfied, a discontinuity in concentration develops when a third-type interface condition is used. On the other hand, a first-type interface condition will result in a continuous concentration profile across the boundary interface at the expense of solute mass balance. To overcome the limitations of both first- and third-type conditions, a combination of first- and third-type conditions was implemented. The first-type condition can be written as

$$C_I \Big|_{x \to L_1^-} = C_{II} \Big|_{x \to L_1^+}, \qquad t > 0 \tag{1.40}$$

where $x \to L_1^-$ and $x \to L_1^+$ denote that $x = L_1$ is approached from upper and lower layer, respectively. Similarly, the third-type condition can be written as

$$\left(qC_I - \theta_I D_I \frac{\partial C_I}{\partial x} \right)\Bigg|_{x \to L_1^-} = \left(qC_{II} - \theta_{II} D_{II} \frac{\partial C_{II}}{\partial x} \right)\Bigg|_{x \to L_1^+}, \quad t > 0 \tag{1.41}$$

Incorporation of Equation (1.6) into Equation (1.7) yields

$$\theta_I D_I \frac{\partial C_I}{\partial x}\Bigg|_{x \to L_1^-} = \theta_{II} D_{II} \frac{\partial C_{II}}{\partial x}\Bigg|_{x \to L_1^+}, \qquad t > 0 \tag{1.42}$$

The boundary condition (BC) of Equation (1.4) was first proposed by Zhou and Selim (2001) and resembles that for a second-type BC as indicated earlier by Leij, Dane, and Van Genuchten (1991).

The form of solute retention reactions in the soil system must be identified if prediction of the fate of reactive solutes in the soil using the CDE is sought. The reversible term $(\partial s / \partial t)$ is often used to describe the rate of sorption or exchange reactions with the solid matrix. Sorption or exchange has been described by either instantaneous equilibrium or a kinetic reaction where concentrations in solution and sorbed phases vary with time. Linear, Freundlich, and one- and two-site Langmuir equations are perhaps most commonly used to describe equilibrium reactions. In the subsequent sections we discuss Freundlich and Langmuir reactions and their use in describing equilibrium retention.

Figure 1.14 shows a comparison of BTCs for a two-layered soil column with reverse layering orders. Here we report results for a layered soil column where one layer is nonreactive ($R = 1$) and the other is linearly adsorptive. For the case of linear adsorption, a dimensionless retardation factor can be obtained from Equation (1.38) and is given by

$$R = 1 + \frac{\rho K_d}{\theta} \tag{1.43}$$

Here, for the case where $K_d = 0$, the retardation factor R equals 1 and the solute is considered nonreactive. The BTC for the case R1→R2 where the nonreactive layer was first encountered (top layer) was similar to that when the layering sequence was reversed (R2→R1) and the reactive layer (R2) was the top layer. Therefore, for the linear adsorption case, one concludes that the order of soil stratification or layering sequence fails to influence solute BTCs and is consistent with those reported earlier by Shamir and Harleman (1967) and Selim, Davidson, and Rao (1977) for systems with two or more layers. Based on these results, a layered soil profile could be regarded as homogeneous with an average retardation factor used to calculate effluent concentration distributions. An average retardation factor $\bar{R}$ for N-layered soil can simply be obtained from

$$\bar{R} = \frac{1}{L} \sum_{i=1}^{N} R_i L_i \tag{1.44}$$

BTCs identical to those in Figure 1.14 were obtained using the solution to the CDE Equation (1.7) presented by Van Genuchten and Alves (1982) and an average retardation factor. This averaging procedure (Equation (1.43)) can also be used to describe the BTCs from a soil profile composed of three or more layers. However, if solute distribution within the profile is desired, the use of an average retardation factor is no longer valid and the problem must be treated as a multilayered case.

Simulated BTCs of solutes from a two-layered soil system with one as a nonlinear (Freundlich) adsorptive layer are given in Figure 1.15. Here, R1 represents a nonreactive layer whereas R2 stands for a nonlinear (Freundlich) adsorptive layer. Our simulations were carried out for a wide range of the Peclet or Brenner number ($\mathbf{B} = qL/\theta D$). We also examined the influence of the nonlinear Freundlich parameter b (see Equation (1.17)) on the shape of the BTCs. For most reactive chemicals, including pesticides and trace elements, b is always less than unity (Selim and Amacher, 1997). Based on our

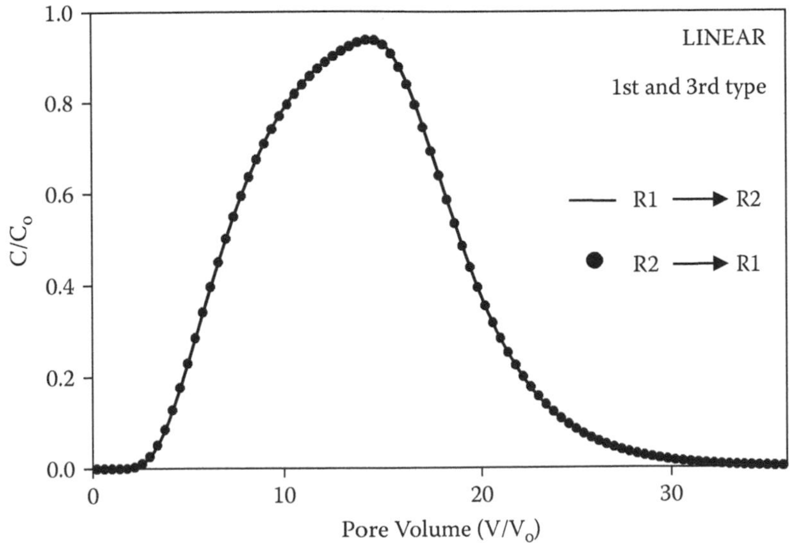

FIGURE 1.14
Simulated breakthrough results for a two-layered soil column under different layering orders (R1→R2 and R2→R1). Here R1 is for a nonreactive layer and R2 is for a reactive layer with kinetic adsorption.

simulations, BTCs were not influenced by the layering sequences regardless of the Brenner number **B** when nonlinear Freundlich adsorption was considered. This result is similar to that of Selim, Davidson, and Rao (1977). Dispersion is dominant for the case where the Brenner number is small, whereas convection becomes the dominant process for large **B** values. The BTCs exhibit increasing retardation or delayed arrival, and excessive tailing of the right-hand side of the BTCs for increasing values of nonlinear adsorption parameter *b*. In addition, the BTCs become less spread (i.e., a sharp front) with increasing Brenner numbers. All such cases provide similar observations, that is, the effects of nonlinearity of adsorption are clearly manifested. Nevertheless, for all combinations of *b* and the Brenner number **B** used in our simulations, the BTCs under reverse layering orders showed no significant differences. In other words, layering order is not important for solute breakthrough in layered soils with nonlinear adsorptive as the dominant mechanism in one of the layers. Zhou and Selim (2001) arrived at similar conclusions for layered soil profiles when several retention mechanisms of the kinetic reversible and irreversible type were considered.

To illustrate that the above finding is universally valid, other solute adsorption processes of the nonlinear type were investigated. The Langmuir adsorption model is perhaps one of the most commonly used equilibrium formulations for describing various reactive solutes in porous media. We only considered simulated columns consisting of one nonreactive layer and

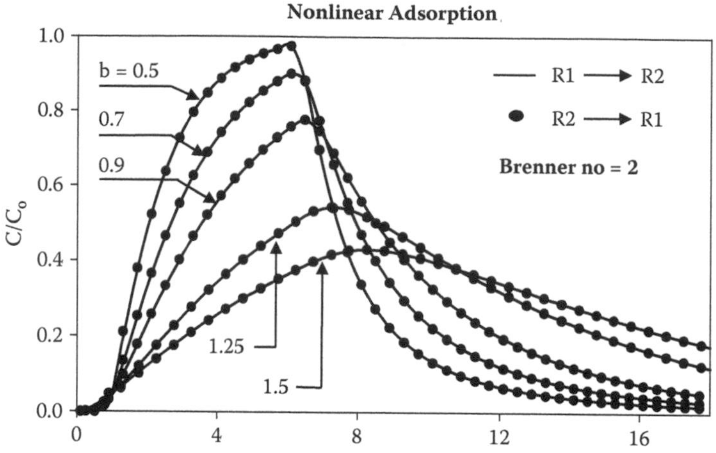

FIGURE 1.15
Simulated breakthrough results for a two-layered soil column under different layering orders (R1→R2 and R2→R1). Here, R1 is for a non-reactive layer and R2 is a reactive layer with nonlinear adsorption with b values (eq.[11]) of 0.5, 0.7, 0.9, 1.25, and1.5. The Brenner numbers **B** used was 2.

one reactive layer with Langmuir-type adsorption mechanism. The simulation results are shown in Figure 1.16. The combined first- and third-type BC was used at the interface between the layers. Consistent with the above finding, we found that for all parameters used in this study, the layering sequence had no effect on the BTCs when Langmuir-type adsorption was the dominant mechanism.

In this section, first- and *n*-order reversible kinetics were considered the dominant retention mechanisms as illustrated in Equation (1.16), where *n* is now substituted for *b*. Values of the reaction order *n* used were 0.3, 0.7, and 1.0. The BTCs under reverse layering orders showed a very good match regardless of the value of the nonlinear parameter *n*. For all cases, a good match was also realized (see Figure 1.17). We also carried out simulations where both layers were assumed reactive. Other retention mechanisms considered included irreversible reaction as well as second-order mechanism. Regardless of the retention mechanism, simulation results indicated that layered soils with reverse layering orders showed no significant differences. For example, as illustrated in Figure 1.18, the order of soil layers did not influence the shape or the position of the BTCs when a sink term was present. This finding is consistent with those of Selim, Davidson, and Rao (1977) where a similar sink term was implemented.

Examples of experimental and predicted BTC results based on miscible displacements from packed soil columns having two layers are shown in Figure 1.19 and Figure 1.20. Predictions and experimental BTC results for tritium, shown in Figure 1.19 (solid and dashed lines) for a sand–clay and

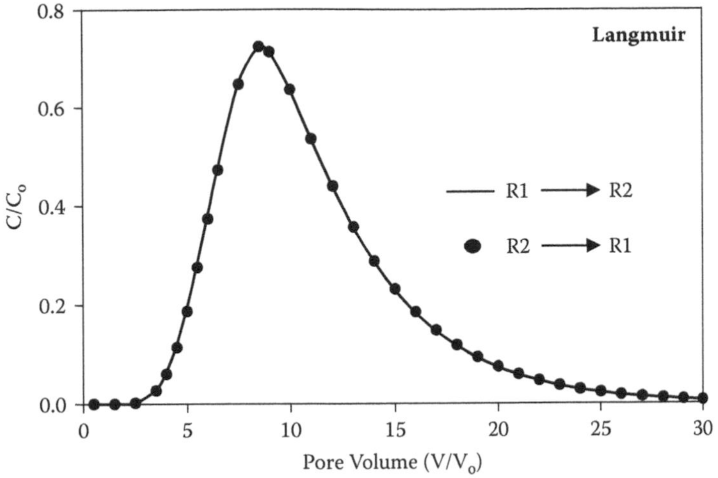

FIGURE 1.16
Simulated breakthrough results for a two-layered soil column under different layering orders (R1→R2 and R2→R1). Here, R1 is for a non-reactive layer and R2 is a reactive layer with Langmuir adsorption.

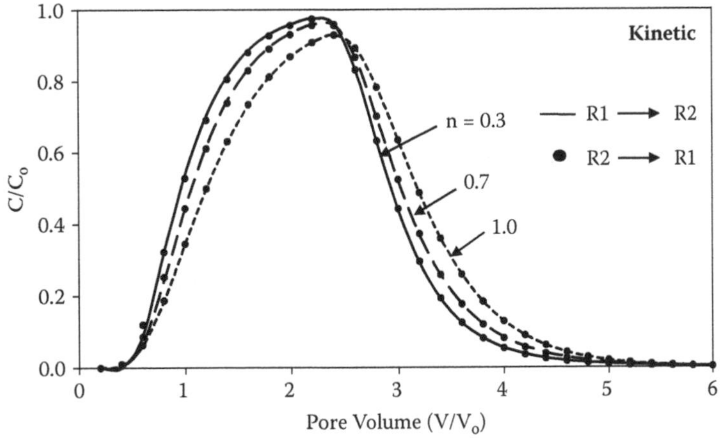

FIGURE 1.17
Simulated breakthrough results for a two-layered soil column under different layering orders (R1→R2 and R2→R1). Here, R1 is for a non-reactive layer and R2 is for a reactive layer with kinetic adsorption with n = 0.3, 0.7, and 1.0, respectively.

clay–sand sequence, indicate a good match of the experimental data. All input parameters were directly based on our experimental measurements and support our earlier findings. Zhou and Selim (2001) simulated the breakthroughs of Ca and Mg using a two-layered model for reactive solutes. The simulation results are shown in Figure 1.16 (solid and dashed lines

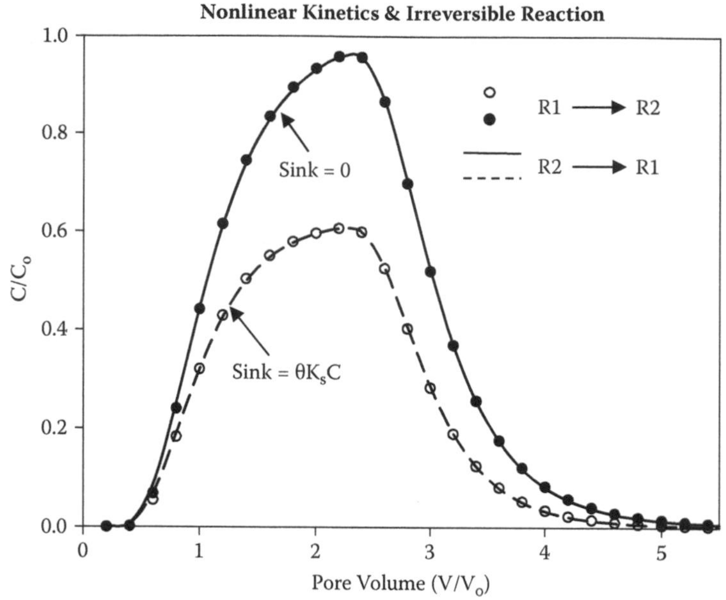

FIGURE 1.18
Simulated breakthrough results for a two-layered soil column under different layering orders (R1→R2 and R2→R1). Here, R1 is for a non-reactive layer adsorption and R2 is a reactive layer with n-th order kinetic adsorption and irreversible sink.

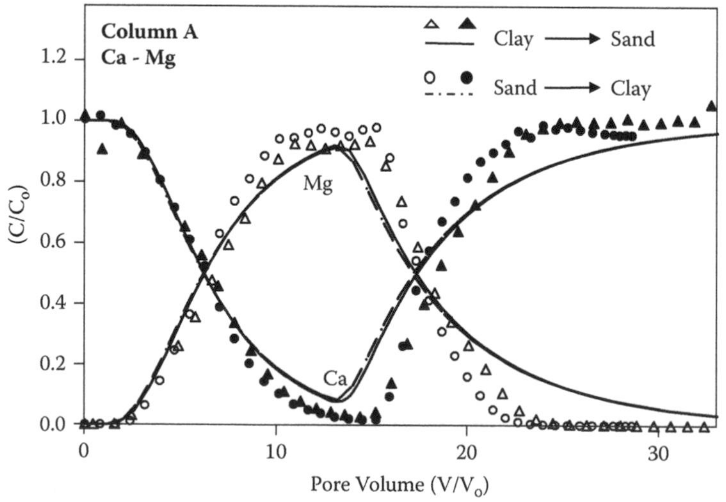

FIGURE 1.19
Experimental (symbols) and simulated (dashed and solid lines) breakthrough results for tritium in a Sharkey clay→sand column (column C) under different layering orders.

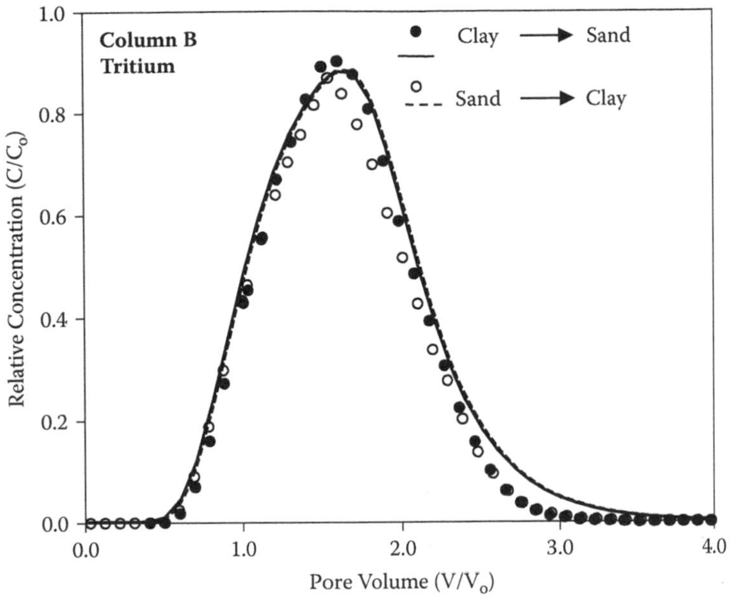

FIGURE 1.20
Experimental (symbols) and simulated (dashed and solid lines) breakthrough results for tritium in a two-layered soil column (Sharkey clay→sand, column B) under different layering orders.

for different layering arrangements) where the reaction mechanism for the Ca-Mg system was assumed to be governed by simple ion exchange for a binary system:

$$S = \frac{K_{12} c S_T}{1 + (K_{12} - 1)c} \tag{1.45}$$

where c is the relative concentration (C/C$_o$), K_{12} is the selectivity coefficient (dimensionless), and S_T is the cation exchange capacity (CEC). The simulation curves agree with the experimental data, especially for the adsorption front, and also exhibit tailing of the release curve for both Ca and Mg. Such tailing was not observed based on the experimental data, however. Results for nonreactive tracer solute are shown in Figure 1.20. Here, experimental (symbols) and simulated (dashed and solid lines) breakthrough results for tritium in a two-layered soil column (Sharkey clay–sand, column B) under different layering orders are presented and clearly illustrate that the order of layering does not influence solute transport under steady flow conditions.

Concluding Remarks

We have presented the framework covering the basics of the transport processes of heavy metals in soils. Relevant boundary conditions associated with the general convection–dispersion equation were presented. The processes that govern retention reactions of heavy metals in the soils were presented in the order of their complexity. Commonly used single retention models of the equilibrium type were discussed, along with kinetic type approaches. Retention models of the multiple reaction type, including the two-site equilibrium-kinetic models, the concurrent and consecutive multireaction models, and the second-order approach were subsequently presented. This was followed by multicomponent or competitive type models where ion exchange is considered the dominant retention mechanism. Selected experimental data sets were described for the purpose of model evaluation and validation, and necessary (input) parameters were discussed. The governing equations for the transport processes of reactive solutes in layered soil were presented. Emphasis was placed on nonlinear reactivity with the soil matrix where the physical and chemical properties of each soil layer were assumed to differ significantly from one another. Results from miscible displacement experiments for competitive adsorption in two-layered columns of sand over clay and clay over sand were well described based on multilayered modeling approaches.

References

Amacher, M.C., H.M. Selim, and I.K. Iskandar. 1988. Kinetics of chromium(VI) and cadmium retention in soils: A nonlinear multireaction model. *Soil Sci. Soc. Am. J.* 52: 398–408.

Arai Y., and D.L. Sparks. 2002. Residence time effect on arsenate surface precipitation at the aluminum oxid-water interface. *Soil Sci. 167*:303–314.

Barry, D.A., and J.C. Parker. 1987. Approximations for solute transport through porous media with flow transverse to layering. *Transp. Porous Media* 2: 65–82.

Bosma, W.J.P., and S.E.A.T.M. Van der Zee. 1992. Analytical approximations for nonlinear adsorbing solute transport in layered soils. *J. Contam. Hydrol.* 10: 99–118.

Brenner, H. 1962. The diffusion model of longitudinal mixing in beds of finite length: Numerical values. *Chem. Eng. Sci.* 17: 220–243.

Buchter, B., B. Davidoff, M.C. Amacher, C. Hinz, I.K. Iskandar, and H.M. Selim. 1988. Correlation of Freundlich *Kd* and *n* retention parameters with soils and elements. *Soil Sci.* 148: 370–379.

Carbonell-Barrachina, A., F.B. Carbonell, and J.M. Beneyto. 1996. Kinetics of arsenite sorption and desorption in Spanish soils. *Commun. Soil. Sci. Plant Anal.* 27:3101–3117.

Chen, J.S., R.S. Mansell, P. Nkedi-Kizza, and B.A. Burgoa. 1996. Phosphorus transport during transient, unsaturated water flow in an acid sandy soil. *Soil Sci. Soc. Am. J.* 60: 42–48.

Danckwerts, P.V. 1953. Continuous flow systems: Distribution of residence times. *Chem. Eng. Sci.* 2: 1–13.

Elkhatib, E.A., O.L. Bennett, and R.J. Wright. 1984a. Kinetics of arsenite adsorption in soils. *Soil Sci. Soc. Am. J. 48*:758–762.

Elkhatib, E.A., O.L. Bennett, and R.J. Wright. 1984b. Arsenite sorption and desorption in soils. *Soil Sci. Soc. Am. J. 48*:1025–1029.

Fried, J.J., and M.A. Combarnous. 1971. Dispersion in porous media. *Adv. Hydrosci.* 7: 169–282.

Fuller, C.C., J.A. Davis, and G.A. Waychunas. 1993. Surface chemistry of ferrihydrite: Part 2. Kinetics of arsenate adsorption and coprecipitation. *Geochim. Cosmochim. Acta. 57*:2271–2282.

Goldeberg, S. 1992. Use of surface complexation models in soil chemical systems. *Adv. Agron.* 47: 233–329.

Guo, L., R.J. Wagenet, J.L. Hutson, and C.W. Boast. 1997. Nonequilibrium transport of reactive solutes through layered soil profiles with depth-dependent adsorption. *Environ. Sci. Technol.* 31: 2331–2338.

Hinz, C., B. Buchter, and H.M. Selim. 1992. Heavy metal retention in soils: application of multisite models to zinc sorption. pp. 141–170. In I.K. Iskandar and H.M. Selim, Eds., *Engineering aspects of metal-waste management.* Lewis Publishers. Boca Raton, FL.

Houng, K-H., and D-Y. Lee. 1998. Comparison of linear and nonlinear Langmuir and Freundlich corve-fit in the study of CU, Cd, and Pb adsorption on Taiwan soils. *Soil Sci. 163*:115–121.

Kreft, A., and A. Zuber. 1978. On the physical meaning of the dispersion equation and its solutions for different initial and boundary conditions. *Chem. Eng. Sci.* 33: 1471–1480.

Lapidus, L., and N.R. Amundson. 1952. Mathematics of adsorption in beds. VI. The effect of longitudinal diffusion in ion exchange and chromatographic column. *J. Phys. Chem.* 56: 984–988.

Lee, K. 1983. Vanadium in the aquatic ecosystem. *Adv. Environ. Sci. Technol.* 13: 155–187.

Leif, F.J., and J.H. Dane. 1989. Analytical and numerical solutions of the transport equation for an exchangecable solute in a layered soil. *Agronomy and Soils Departmental Series No. 139.* Alabama Agricultural Experimental Station, Auburn University, Alabama.

Leij, F.J., J.H. Dane, and M.Th. van Genuchten. 1991. Mathematical analysis of one-dimensional solute transport in a layered soil profile. *Soil Sci. Soc. Am. J.* 55: 944–953.

Liao, Lixia, and H.M. Selim. 2010. Reactivity of nickel in soils: Evidence of retention kinetics. *J. Environ. Qual.* 39: 1290–1297.

Lindstrom, F.T., R. Hague, V.H. Freed, and L. Boersma. 1967. Theory on movement of some herbicides in soils: Linear diffusion and convection of chemicals in soils. *Environ. Sci. Technol.* 1: 561–565.

Ma, L., and H.M. Selim. 1998. Coupling of retention approaches to physical nonequilibrium models. pp. 83–115. In Selim, H.M. and L. Ma, Eds., *Physical nonequilibrium in soils: modeling and application.* Ann Arbor Press, Chelsea, MI.

Selim, H.M., and L. Ma. 2001. Modeling nonlinear kinetic behavior of copper adsorption-desorption in soil. In Selim, H.M. and D.L. Sparks, Eds., *Physical and chemical processes of water and solute transport/retention in soil.* Spec. Publ. 56, pp. 189–212. Soil Science Society of America, Madison, WI.

Selim, H.M., and M.C. Amacher. 1997. *Reactivity and transport of heavy metals in soils.* CRC/Lewis, Boca Raton, FL (240 pp.).

Selim, H.M., and M.C. Amacher. 1988. A second-order kinetic approach for modeling solute retention and transport in soils. *Water Resources Res.* 24: 2061–2075.

Selim, H.M., M.C. Amacher, and I.K. Iskandar. 1990. Modeling the transport of heavy metals in soils. U. S. Army Corps of Engineers, Cold Regions Research & Engineering Laboratory, Monograph 90-2.

Selim, H.M., B. Buchter, C. Hinz, and L. Ma. 1992. Modeling the transport and retention of cadmium in soils: Multireaction and multicomponent approaches. *Soil Sci. Soc. Am. J.* 56: 1004–1015.

Selim, H.M., J.M. Davidson, and R.S. Mansell. 1976. Evaluation of a two-site adsorption–desorption model for describing solute transport in soils. *Proc. Summer Computer Simulation Conf.*, Washington, D.C. 12–14 July, 1976 (La Jolla, CA, Simulation Councils Inc., La Jolla, CA.). pp. 444–448.

Selim, H.M., J.M. Davidson, and P.S.C. Rao. 1977. Transport of reactive solutes through multilayered soils. *Soil Sci. Soc. Am. J.* 41: 3–10.

Selim, H.M., and R.S. Mansell. 1976. Analytical solution of the equation of reactive solutes through soils, Reply. *Water Resourc. Res.* 13: 703–704.

Selim, H.M., and H. Zhang. 2007. Arsenic adsorption in soils: Second-order and multireaction models. *Soil Sci.* 72: 144–458.

Shamir, U.Y., and D.R.F. Harleman. 1967. Dispersion in layered porous media. *Proc. Am. Soc. Civil. Eng. Hydr. Div.* 93: 237–260.

Sparks, D.L. 1989. *Kinetics of soil chemical processes.* Academic Press, San Diego, CA.

Sparks, D.L. 2003. *Environmental soil chemistry.* Academic Press, San Diego, CA (352 pp.).

Van der Zee, S.E.A.T.M. 1994. Transport of reactive solute in soil and groundwater. pp. 27–87. In D.C. Adriano, I.K. Iskandar, and I.P. Muraka, Eds., *Contamination of groundwaters.* St. Lucie Press, Boca Raton, FL.

Van der Zee, S.E.A.T.M., and W.H. Van Riemsdijk. 1986. Sorption kinetics and transport of phosphate in soils. *Geoderma* 38: 293–309.

Van Genuchten, M.Th., and W.J. Alves. 1982. Analytical solutions of the one-dimensional convective-dispersive solute transport equation. U.S. Dept. Agric. Bull. No. 1661.

Verburg K., and P. Baveye. 1994. Hysteresis in the binary exchange of cations on 2:1 clay minerals: A critical review. *Clays and Clay Minerals* 42: 207–220.

Wu, Y.S., Kool, J.B., and P.S. Huyakorn. 1997. An analytical model for nonlinear adsorptive transport through layered soils. *Water Resources Res.* 33: 21–29.

Zhang, H., and H.M. Selim. 2005. Kinetics of arsenate adsorption-desorption in soils. *Environ. Sci. Technol.* 39: 6101–6108.

Zhao, K., and H.M. Selim. 2010. Adsorption-desorption kinetics of Zn in soils: Influence of phosphate. *Soil Sci.* 175: 145–153.

Zhou, L., and H.M. Selim. 2001. Solute transport in layered soils: Nonlinear kinetic reactivity. *Soil Sci. Soc. Am. J.* 65: 1056–1064.

2

Nonequilibrium Transport of Heavy Metals in Soils: Physical and Chemical Processes

Hua Zhang

CONTENTS

Introduction

Soil contamination of heavy metals from mining, industrial, agricultural, and geological sources poses serious environmental risk around the world because of its high toxicity to human health as well as the ecosystem. The transport of toxic metals in the vadose zone and aquifers may lead to the further contamination of surface and groundwaters (NRC, 2003). The widespread contamination caused by Cd, Cu, Hg, Pb, Ni, Zn, As, and Cr has elevated these heavy metals to the top of the priority pollutants list (Cameron, 1992).

Focused research efforts in the past few decades have been largely successful in revealing the dominant physical and chemical mechanisms controlling the fate and behavior of toxic metals in the environment (Carrillo-Gonzalez et al., 2006). Geochemical reactions, including adsorption and desorption, precipitation and dissolution, and reduction and oxidation, are governing processes of the bioavailability and toxicity of reactive metal in soils. The hydrological conditions impact their movements by controlling the water flow that carries the mobile phase of the solutes in the heterogeneous subsurface environments. In addition, the interaction between hydrological and biogeochemical processes in the critical zone has also been revealed to influence the fate and transport of reactive contaminants (Jardine, 2008).

Because of the heterogeneity in the physical and chemical soil properties at scales ranging from molecular to watershed, nonequilibrium conditions are found to be prevalent in both laboratory and field studies of heavy metal transport. A wide range of research in the fields of hydrology, geochemistry, biogeochemistry, soil physics, and environmental engineering has demonstrated that rate-limited processes play a dominant role in the transport of heavy metals in soils. The traditional solute transport models based on local equilibrium assumptions are found to be generally inadequate for the realistic simulation of reactive metals under natural environmental conditions. As a result, there has been a gradual shift in the past few decades from empirical equilibrium models to nonequilibrium models based on the accurate description of the time-dependent physical and chemical processes (Amacher et al., 1986; Selim, 1992; Brusseau, 1994; Ma and Selim, 1997).

The rapidly developing reactive transport models have provided a general framework for studying the behavior of reactive metals (Steefel, DePaolo, and Lichtner, 2005). Advancements in reactive transport modeling have enabled researchers to investigate the coupled geochemical, microbiological, and hydrological processes at various spatial and temporal scales. The development and application of numerical models have substantially enhanced our capability to evaluate and predict the environmental risk of heavy metal contamination (Selim and Amacher, 1988; Selim, 1992; Steefel, DePaolo, and Lichtner, 2005). Increasingly popular in the processes of remedial investigation and feasibility study, the reactive transport model has become an essential tool for environmental scientists and engineers in developing technology solutions for contaminated sites.

In this chapter we emphasize the nonequilibrium transport modeling of toxic trace elements in the critical zone that combines hydrological, geochemical, and biological processes. The equilibrium solute transport models are also briefly overviewed in the "Equilibrium Solute Transport" section because they are the mathematical basis for developing more sophisticated nonequilibrium models. The "Physical Nonequilibrium" section provides a general description and mathematical treatment of the hydrological processes that impact the fate and transport of contaminants in the subsurface environment. In the "Chemical Nonequilibrium" section, the kinetic models of geochemical reactions are summarized in detail in the context of reactive metal transport. And the "Multiprocess Nonequilibrium Transport" section provides an overview of the application of coupled physical and chemical nonequilibrium model for the simulation of reactive trace elements in soils. At the end of the chapter, the "Summary" provides concluding remarks on the challenges and directions of nonequilibrium reactive transport modeling of heavy metals.

Equilibrium Solute Transport

The transport of solutes in homogeneous soils under steady-state flow conditions can be described using the one-dimensional convection–dispersion equation (CDE) in the form of (Selim, 1992)

$$\theta\frac{\partial C}{\partial t}+\rho\frac{\partial S}{\partial t}=\frac{\partial}{\partial x}\left(D\theta\frac{\partial C}{\partial x}\right)-v\theta\frac{\partial C}{\partial x} \tag{2.1}$$

where C is the solute concentration in aqueous phase ($M.L^{-3}$), S is the amount of solute retention on solid matrix ($M.M^{-1}$), t is time (T), x is distance (L), θ is the volumetric water content ($L^3.L^{-3}$), ρ is the bulk density of soil ($M.L^{-3}$), D is the apparent dispersion coefficient ($L^2.T^{-1}$), v (= q/θ) is the average pore water velocity ($L.T^{-1}$), and q is Darcy's water flux density ($L.T^{-1}$).

The convective movement of solute with Darcian water flow is represented by the second term on the right-hand side of Equation (2.1). The convective transport of solutes is controlled by average flow velocity v, which is the straight-line distance of the path traversed within soil pores in unit time. Because of the significant heterogeneity in soil pore geometry, the coefficient v is only a gross approximation of the actual travel velocity of the water flow occurring inside soils. As we discuss in the next section, the inhomogeneous flow field is the root of the physical nonequilibrium processes in the transport of both reactive and nonreactive solutes.

The dispersion coefficient D in Equation (2.1) is an *apparent* coefficient representing the combined effect of several different physical processes. The molecular diffusion of aqueous chemical species across the concentration gradient is described by Fick's laws of diffusion. The intrinsic diffusion rate (D_o) is a function of the temperature and viscosity of the fluid as well as the size of the solute particles, and can be calculated using the Stokes-Einstein equation. In porous media, the effective molecular diffusion coefficient (D_w) is a macroscopic parameter that depends on the pore space. The effective diffusion coefficient can be calculated as (Brusseau, 1993)

$$D_w=\frac{D_o\varepsilon}{\tau} \tag{2.2}$$

where ε is the porosity available for solute transport (dimensionless) and τ is the tortuosity (dimensionless).

Hydrodynamic dispersion is an important solute transport process that arises from the inhomogeneous flow field in geological porous media. Because of the velocity gradient from the center to the walls inside pores and, more importantly, between narrower and wider pores, there is a dynamic concentration gradient as a result of the uneven water flow that carries solutes in porous media. The hydrodynamic dispersion coefficient has been found to depend more or less linearly on the average flow velocity (v):

$$D_h = av \tag{2.3}$$

where a is the apparent longitudinal dispersivity (L). As a result of the heterogeneity of the physical property of soils, the dispersivity is a scale-dependent parameter that increases with the mean travel distance (Zhou and Selim, 2003).

The dispersion coefficient D in the CDE Equation (2.1) accounts for the combined processes of molecular diffusion and hydrodynamic dispersion processes. It can be calculated by

$$D_w = \frac{D_o \varepsilon}{\tau} + \alpha v \tag{2.4}$$

As shown in Equation (2.4), the dispersion coefficient increases with flow velocity and hydrodynamic dispersion becomes the dominant process at high flow velocity (Ma and Selim, 1994).

Under normal hydrology conditions encountered in the field, the effect of molecular diffusion is generally minimal when compared with hydrodynamic dispersion. Therefore, the same dispersion coefficient can be used for different solutes. For laboratory and field transport studies, the dispersion coefficients of soils are commonly characterized using nonreactive tracers such as tritium (^3H) and subsequently used for the simulation of the transport of chemically reactive solutes. As shown in Figure 2.1a, the observed breakthrough curve (BTC) of ^3H transport in Windsor soil was successfully simulated using the CDE with the dispersion coefficient estimated by curve fitting.

The interactions on the surfaces of solid minerals and organic matter are the dominant mechanisms of heavy metal attenuation in the vadose zone and aquifers. Assuming that the soils are homogeneous materials and the sorption is a uniform and equilibrium process, the simplest method for describing the amount of solute retention on solid phase is the use of the linear sorption isotherm:

$$S = K_d C \tag{2.5}$$

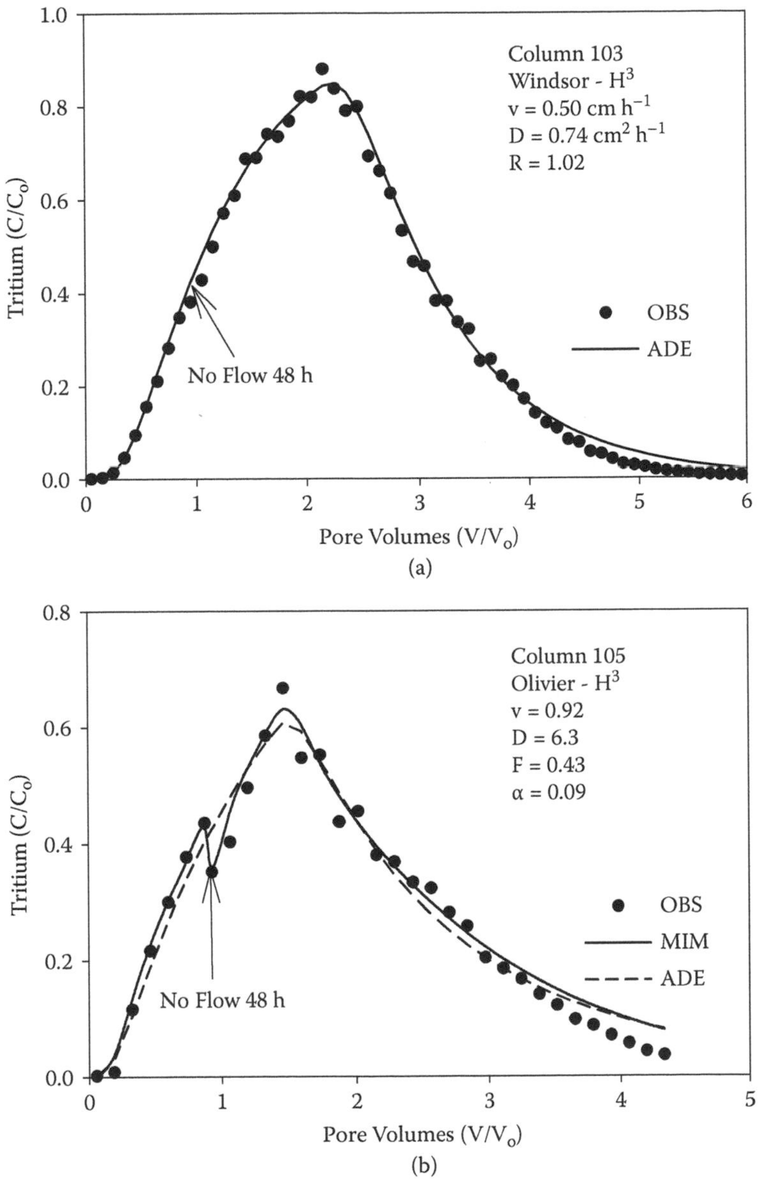

FIGURE 2.1
Experimental BTCs of tritium (^{3}H) in (a) Windsor soil and (b) Olivier soil. The observed BTCs are simulated using the CDE model and mobile–immobile model (MIM) with parameters obtained from least-squares optimization.

where K_d is the equilibrium distribution coefficient ($M^{-1}.L^3$). Using Equation (2.5) to represent the amount of solute retention, the CDE model of Equation (2.1) can be transformed into

$$R\frac{\partial C}{\partial t}=\frac{\partial}{\partial x}\left(D\frac{\partial C}{\partial x}\right)-v\frac{\partial C}{\partial x} \tag{2.6}$$

where the retardation factor is given by

$$R=1+\frac{\rho}{\theta}K_d$$

Equation (2.6) has been solved analytically in combination with several sets of initial and boundary conditions (Van Genuchten & Alves, 1982) and the analytical solution provided by computer software such as CXTFIT (Toride, Leij, and Van Genuchten, 1995). These equations can also be solved using numerical algorithms of finite difference and finite element methods (Selim, Amacher, and Iskandar, 1990). The analytical and numerical solutions of the CDE are commonly applied models for describing the transport of metals in porous media.

Because of the highly nonlinear sorption isotherms of many toxic metals, the simple linear sorption model in Equation (2.5) is inadequate for simulating heavy metal transport in most cases (Buchter et al., 1989). Therefore, the equilibrium model of the Freundlich or Langmuir type is often applied for describing the transport of reactive solutes in soils. The Freundlich equation is an empirical adsorption model that can be expressed as

$$S=K_F C^N \tag{2.7}$$

where K_F is the Freundlich distribution or partition coefficient, and N is the dimensionless reaction order, which is commonly less than one. The Langmuir equation has the advantage of providing a sorption maximum S_{max} ($M.M^{-1}$) that can be correlated to soil sorption properties. It has the form

$$S=S_{max}\frac{K_L C}{1+K_L C} \tag{2.8}$$

where K_L ($L^3.M^{-1}$) is a Langmuir coefficient related to the binding strength. Langmuir and Freundlich models can be incorporated into solute transport models to predict the transport of As in soils. However, because of the nonlinear nature of the equations, it is difficult to obtain the analytical solutions to the CDE models coupled with the Freundlich- or Langmuir-type sorption isotherms. Instead, the nonlinear solute transport models are frequently solved using numerical algorithms such as the Crank-Nicholson method (Hinz, Gaston, and Selim, 1994). Hinz and Selim (1994) evaluated eight different empirical equilibrium equations as to their capability of predicting the transport of Zn and Cd transport in soils. Their simulation results demonstrated that local equilibrium appears to be dominant for sandy Windsor soil while the isotherm equations failed to predict the solute transport in the loamy Olivier soil.

Physical Nonequilibrium

In general, the nonequilibrium transport of solute in soil medium is attributed to the following two processes: (1) the rate-limited mass transfer of aqueous species to and from the regions with limited or no advective flow, and (2) the time-dependent chemical reaction at the surfaces of solid materials. The first process is often referred as transport-related nonequilibrium or physical nonequilibrium, and the second process is referred as reaction-related nonequilibrium or chemical nonequilibrium (Brusseau, Jessup and Rao, 1992).

As a result of pedogenic processes, soils are porous media with extensive chemical and physical heterogeneities. Soil heterogeneity has a profound effect on the transport of contaminants. The physical heterogeneity is mainly attributed to the highly nonuniform pore space as a result of the soil aggregates, soil layers, as well as cavities developed from natural and anthropogenic activities. The highly irregular flow field results from the spatial variance of pore sizes, and connectivity is a major driving force of the nonequilibrium transport of solutes in soils (Biggar and Nielsen, 1976; Dagan, 1984). Conceptually, the nonequilibrium transport of nonreactive solutes in structured soils can be described as a rapid movement of water flow and solutes within preferential flow paths and a diffusive mass transfer of solutes to a stagnant region. This mobile–immobile dual region concept is the foundation of the mathematical models used for simulating the physical nonequilibrium transport (Van Genuchten, 1981; Reedy et al., 1996; Simunek et al., 2003).

Preferential Flow

Preferential flow occurring via distinct flow pathways in fractures and macropores can significantly influence the movement of heavy metals through soils. Preferential and nonequilibrium flow and transport are often difficult to characterize and quantify because of the highly irregular flow field as a result of the high degree of spatial variance in soil hydraulic conductivity and other physical characteristics (Biggar and Nielsen, 1976). A significant feature of preferential flow is the rapid movement of water and solute to significant depths while bypassing a large part of the matrix pore-space. As a result, water and solutes may move to far greater depths, and much faster, than would be predicted using averaged flow velocity (Jury and Fluhler, 1992).

The consequence of preferential flow is the nonequilibrium transport processes where solute in the fast flow region does not have sufficient time to equilibrate with slowly moving resident water in the bulk of the soil matrix. Therefore, the BTCs of the solute transport through preferential flow often exhibit a sharp front, indicating early breakthrough in the rapid flow region and long tailing as a result of the rate-limited transfer of solutes from the slow flow region to the fast flow region (Jury and Fluhler, 1992).

Conventional laboratory transport studies using column leaching techniques with homogenized soil samples are not capable of capturing the increased mobility of heavy metals through preferential flow. Studies using undisturbed soil samples where the pore structure remains intact are useful in evaluating the potential of transport through preferential flow. For example, the column leaching experiment of Cd, Zn, Cu, and Pb conducted by Camobreco et al. (1996) showed that the metals were completely retained in homogenized soil columns. In contrast, preferential flow paths in undisturbed soil columns allowed metals to pass through the soil profile. Distinctively different characteristics of metal BTCs were observed for the four undisturbed soil columns, reflecting different preferential flow paths. Similarly, Corwin, David, and Goldberg (1999) demonstrated with a lysimeter column study that without accounting for preferential flow, 100% of the applied As was isolated in the top 0.75 meters over a 2.5-year period. However, when preferential flow was considered and a representative bypass coefficient was used, about 0.59% of applied As moved beyond 1.5 meters.

Matrix Diffusion

In general, the nonequilibrium mass transfer of solutes in soils proceeds in two stages: (1) external or film diffusion from flow water to the solid interface, and (2) internal diffusion through the porous network of soil aggregates (Brusseau, 1993). Film diffusion is of importance only at low flow velocities. Under environmental conditions, diffusive mass transfer in the micropores inside soil aggregates is an important process influencing the mobility of reactive metals in soils. The intra-aggregate diffusion serves as a reservoir in the transport of contaminants by spreading them from flowing water to the immobile region of stagnant pore water. Furthermore, the diffusion of reactive solutes inside the soil matrix increases their contact with reaction sites on the surfaces of minerals and organic matter. The time-dependent process can significantly impact the extent and rate of geochemical and microbial reactions with the solid phase.

Matrix diffusion can be manifested in the laboratory by examining the asymmetric BTCs from miscible displacement experiments with nonreactive tracers. Experimental and modeling studies have demonstrated that the mass transfer process can be influenced by aggregate shape, particle size distribution, and pore geometry. Brusseau (1993) performed column experiments using four porous media with different physical properties to investigate the influence of solute size, pore water velocity, and intraparticle porosity on solute dispersion and transport in soil. He concluded that solute dispersion in aggregated soil was caused by hydrodynamic dispersion, film diffusion, intraparticle diffusion, as well as axial diffusion at low pore water velocities. While for sandy soils, the contribution of intraparticle diffusion and film diffusion is negligible, hydrodynamic dispersion is the predominant source of dispersion for most conditions.

The flow interruption technique has been employed to detect and quantify the rate and extent of physical nonequilibrium during solute transport in heterogeneous porous media. A flow interruption stops the injection of fluids and tracers, allowing more time for reactive tracers to interact with the solid phase and for all tracers to diffuse into or out of the matrix. Reedy et al. (1996) conducted nonreactive tracer (Br^-) miscible displacement experiments with a large, undisturbed soil column of weathered, fractured shale from a proposed waste site on the Oak Ridge Reservation. The tracer flow was interrupted for a designated time to quantify the diffusive mass transfer of a nonreactive solute between the matrix porosity and preferential flow paths in fractured subsurface media. Decreased and increased tracer concentrations were observed after flow interruption during tracer infusion and displacement, respectively, when flow was reinitiated. The concentration perturbations increased with increasing interrupt duration as well as water flux as a result of velocity and concentration gradient. As shown in Figure 2.1, a decrease in tracer (^3H) concentration after 48 hours of flow interruption was observed for the nonequilibrium transport in Oliver soil. In comparison, the 48-hour flow interruption did not affect tracer transport in sandy Windsor soil, indicating the lack of physical nonequilibrium.

Physical Nonequilibrium Model

Previous studies have often suggested that the simple CDE model was not adequate to describe solute transport in heterogeneous soils (Van Genuchten and Wierenga, 1976; Nkedi-Kizza et al., 1984; Brusseau, Jessup, and Rao, 1989; Reedy et al., 1996; Ma and Selim, 1997). The physical nonequilibrium model was proposed to simulate the transport nonreactive solutes across multiple regions within the soils. It is also referred as the two-region model, two-compartment model, dual porosity model, or dual permeability model. The model assumes that the porous medium consists of two interacting regions: (1) the mobile region associated with the inter-aggregate, macropore, or fracture system, and (2) the immobile region comprising micropores (or intra-aggregate pores) inside soil aggregates. In this model, the liquid phase of the porous media is partitioned into a mobile region where advective flow takes place and an immobile region that has no advective flow. The rate-limited mass transfer between the mobile and immobile regions is assumed to be a diffusion-controlled process where the transfer rate is proportional to the difference in concentration between the two liquid regions. The governing equations for the mobile-immobile model (MIM) are

$$\theta^m \frac{\partial C^m}{\partial t} + f\rho \frac{\partial S^m}{\partial t} + \theta^{im} \frac{\partial C^{im}}{\partial t} + (1-f)\rho \frac{\partial S^{im}}{\partial t} = \frac{\partial}{\partial x}\left(D^m \theta^m \frac{\partial C^m}{\partial x} \right) - v^m \theta^m \frac{\partial C^m}{\partial x} \quad (2.9)$$

$$\theta^{im} \frac{\partial C^{im}}{\partial t} + (1-f)\rho \frac{\partial S^{im}}{\partial t} = \alpha_c \left(C^m - C^{im} \right) \quad (2.10)$$

where superscripts *m* and *im* indicate the mobile and immobile liquid regions, respectively. The dimensionless parameter *f* represents the fraction of soil that is in direct contact with the mobile region, and α_c is a first-order mass transfer coefficient between the mobile and immobile regions ($M.L^{-3}.T^{-1}$).

Assuming that retention on solid surfaces can be described using linear equilibrium isotherm Equation 2.1, the commonly employed two-region flow model has the form of

$$(\theta^m + f\rho K_d)\frac{\partial C^m}{\partial t} + [\theta^{im} + (1-f)\rho K_d]\frac{\partial C^{im}}{\partial t} = \frac{\partial}{\partial x}\left(D^m \theta^m \frac{\partial C^m}{\partial x} \right) - v^m \theta^m \frac{\partial C^m}{\partial x} \quad (2.11)$$

$$[\theta^{im} + (1-f)\rho K_d]\frac{\partial C^{im}}{\partial t} = \alpha_c \left(C^m - C^{im} \right) w \quad (2.12)$$

A single distribution coefficient K_d is applied to depict retention on both mobile and immobile domains.

The BTC shown in Figure 2.1b for tracer transport in Oliver soil was successfully simulated using the MIM with model parameters obtained from nonlinear curve fitting. The MIM quantitatively predicted the concentration decrease after flow interruption, which proved that the diffusive mass transfer into immobile flow regions inside soil aggregates is the cause of the physical nonequilibrium.

Chemical Nonequilibrium

The governing chemical processes of heavy metal transport in soils may include reactions such as ion exchange, formation of inner-sphere surface complexes, precipitation into distinct solid phases, or surface precipitation on minerals. Moreover, the speciation and mobility of multivalent elements such as As, Hg, and Cr may also be influenced by the kinetics of oxidation–reduction reactions of various electron acceptor–donor pairs. As a result, the biogeochemistry of heavy metals in heterogeneous soil systems is rather complex, comprising a large array of chemical and microbiological reactions that are affected by a series of environmental conditions such as pH, soil redox potential (Eh), soil constituents, electrolytes, microbial activity, temperature, and residence time (Carrillo-Gonzalez et al., 2006).

The local equilibrium assumption commonly employed for describing heavy metal transport is only valid for solute transport in homogenous porous medium where the reaction is linear and reversible. In reality, soils are heterogeneous materials composed of various chemical components and reaction sites. Heavy metals may react with different mineral phases (clay minerals, oxides, hydroxides, sulfide minerals, carbonate minerals, etc.) and organic materials that exist in a wide range of particle sizes. Reactions may occur at different rates, depending on the reactant transport in the complex pore space and chemical reaction kinetics on the surface sites. Specifically, the transport processes include (1) diffusion in the aqueous solution, (2) film diffusion at the solid–liquid interface, (3) intraparticle diffusion in micropores and along pore-wall surfaces, and (4) interparticle diffusion inside solid particles. The kinetic processes, including cation exchange reactions, surface complexation, surface precipitation, reductive or oxidative dissolution, and reductive or oxidative precipitation, may occur on time scales varying from milliseconds to years. Chemical reactions such as reductive or oxidative precipitation as well as surface precipitation last longer than the residence times of solute transport in vadose zone and aquifer. In many cases, the slow reactions determine the spatial and temporal distribution of reactive trace elements (Sparks, 1989).

Kinetic Ion Exchange

Exchange between aqueous solution and sites with static electric charges on surfaces of minerals is an important reaction influencing the reaction of ionic species in soils. Due to the negative charge borne by soil colloids, cation exchange reactions can heavily influence the retention and transport of toxic metal elements. Ion exchange was considered an instantaneous process where the exchange between aqueous ions and ions on a charged surface was fully reversible. However, physically and chemically heterogeneous soil systems contain a wide range of exchange sites with varying degrees of reactivity and accessibility. The reactions on readily accessible external binding sites rapidly reach equilibrium, while an extended reaction time is needed for the metal cations to reach and react with the interlattice exchange sites in the interlayers of clay minerals. The kinetic rates of cation exchange generally depend on the diffusion to exchange sites (Jardine and Sparks, 1984;

Sparks, 1989). Selim et al. (1992) developed a multicomponent approach that accounts for both kinetic ion exchange and specific sorption. A mass transfer coefficient similar to the mobile–immobile approach (Equation (2.10)) was used to describe the diffusion of ions between solid and solution phase. They concluded that the kinetic ion exchange model successfully described the transport of Cd in Windsor soil and Eustis soil with kinetic parameters independently obtained from time-dependent sorption experiments.

Kinetic Adsorption and Desorption

The elementary chemisorption reaction of aqueous metal ions on mineral surfaces through outer-sphere or inner-sphere surface complexation is generally a rapid process with equilibration times on the order of seconds. Because of their rapid reaction rates, surface complexation is not considered a rate-limiting step of adsorption in soils. However, different types of surface complexes (e.g., outer sphere, inner sphere, monodentate, bidentate, mononuclear, binuclear) can be formed on oxide surfaces at high or low surface coverage. This heterogeneity of sorption sites may contribute to observed adsorption kinetics where sorption takes place preferentially on high-affinity sites followed subsequently by slow sorption on sites of low sorption affinity. Diffusion of aqueous metal ions to reaction sites within the soil matrix is possibly a rate-limiting step of adsorption kinetics. For heterogeneous soil systems, the complex network of macro- and micropores may further limit the access of solute to the adsorption sites and cause the time-dependent adsorption (Zhang and Selim, 2008). Many studies have demonstrated that adsorption on mineral surfaces has a rapid initial adsorption phase followed by a long plateau phase that can extend to years (Sparks, 1989). A two-phase process was generally assumed for diffusion-controlled adsorption, with the reaction occurring instantly on liquid–mineral interfaces during the first phase, whereas slow penetration or intraparticle diffusion is responsible for the second phase (Fuller, Davis, and Waychunas, 1993).

A wide variety of empirical kinetic rate expressions were developed over the past three decades to describe the results from sorption and desorption kinetic experiments (Selim and Amacher, 1997). Reversible nonequilibrium sorption models have been proposed by many researches to describe the sorption–desorption kinetics involving multiple chemical and physical reaction processes. The reversible first-order kinetic sorption equation has the form

$$\frac{\partial S}{\partial t} = \alpha_1 (K_d C - S) \tag{2.13}$$

where α_1 is the first-order kinetic rate coefficient (T^{-1}). Under equilibrium conditions (i.e., $\frac{\partial S}{\partial t} = 0$), Equation (2.13) yields a linear adsorption equation of

$S = K_d C$. The reversible *n*-order (Freundlich-type) kinetic sorption equation is in the form of

$$\frac{\partial S}{\partial t} = k_{fF} \frac{\theta}{\rho} C^b - k_{bF} S \tag{2.14}$$

where k_{fF} and k_{bF} are the forward and backward reaction rate coefficients (T^{-1}), respectively, and *b* is a dimensionless nonlinear parameter usually less than 1. Under equilibrium conditions

$$\text{(i.e., } \frac{\partial S}{\partial t} = 0w\text{)}$$

Equation (2.14) yields the Freundlich Equation (2.7) assuming that

$$K_F = \frac{k_{fT}}{k_{bF}} \frac{\theta}{\rho}$$

and $N = b$.

The second-order or Langmuir kinetic equation is another reversible type where a concentration maximum (S_{max}) is assumed:

$$\frac{\partial S}{\partial t} = k_{fL} \frac{\theta}{\rho} (S_{max} - S) C - k_{bL} S \tag{2.15}$$

where k_{fL} ($L^3.M^{-1}.T^{-1}$) and k_{bL} are the forward and backward reaction rate coefficients (T^{-1}), respectively. Under equilibrium conditions

$$\text{(i.e., } \frac{\partial S}{\partial t} = 0\text{)}$$

Equation (2.15) yields the Langmuir equilibrium Equation (2.8) where

$$K_L = \frac{k_{fL}}{k_{bL}} \frac{\theta}{\rho}$$

The Freundlich- and Langmuir-type reversible kinetic sorption equations are commonly solved using numerical algorithms such as the fourth-order Runge-Kutta methods. The numerical solutions give us the flexibility of simulating a wide range of initial conditions. The nonequilibrium kinetic equations are incorporated into the dispersion–advection transport model for simulating the dynamics of solute concentration across space and time. For example, the batch kinetic sorption data of Darland and Inskeep (1997) were described using first-order and *n*-order reversible adsorption. The first-order forward and backward rate constants (k_f and k_b) were 2.65×10^{-1} and 8.75×10^{-3} h^{-1} for arsenate sorption on acid-washed sand. Yin et al. (1997) successfully described the adsorption and desorption data of Hg(II) on different soils using a one-site, second-order model. The fitted adsorption and desorption rates were found to inversely correlate with the soil organic carbon content, indicating the strong affinity of Hg(II) for organic matter. Amacher et al. (1986) evaluated several kinetic reaction models for their capability to fit the time-dependent retention of Cr, Cd, and Hg by the Windsor soil. They found that the reversible first-order, Freundlich, and Langmuir models all failed to fit the observed retention data and concluded that the multireaction model is required to describe the metal retention reactions in chemically heterogeneous soils.

Irreversible Retention

Hysteretic behavior resulting from a discrepancy between adsorption and desorption has been frequently observed for heavy metals. Observed desorption hysteresis might be due to kinetic retention behavior, such as slow diffusion, as well as irreversible retention. Field studies have demonstrated that prolonged metal–soil contact times may lead to increased amounts of irreversible or nondesorbable phases of these compounds in the environment. This long-term aging phenomenon was attributed to the increased bonding energies on the surface sites as well as to the internal diffusion into mineral lattice. Ainsworth et al. (1994) found that despite increasing the desorption time from 16 hours to 9 weeks, hysteresis persisted for Co and Cd sorbed on hydrous ferric oxide (HFO). Increased desorption hysteresis was observed when the aging time was increased from 2 weeks to 16 weeks. Yin et al. (1997) attributed the irreversible adsorption of Hg(II) on soils to intraparticle micropores of soil organic carbon and/or the binding on high-affinity sites (such as S-containing groups) on soil organic matter.

The adsorption–desorption studies of heavy metals suggested that surface precipitation (i.e., three-dimensional growth of a particular surface phase) may occur for heavy metals at relatively low concentrations (Sposito, 1989). The development of surface precipitates is a slow process involving multiple reaction steps and may explain, in part, the slow heavy metal retention kinetics in soils. The formation and aging of metal oxides and hydroxides may have a substantial influence on the release of heavy metals (Sparks, 2003).

Several kinetic equations can be utilized to describe the irreversible retention of metals in soils (Sparks, 2003). The pseudo-first-order equation assumes a fixed amount of retention at equilibrium (S_{eq}). It has the following form:

$$\frac{dS}{dt} = k(S_{eq} - S) \tag{2.16}$$

where k is the retention rate (h^{-1}). For the initial condition $S = 0$ at $t = 0$, the integrated form of the pseudo-first-order equation for adsorption is

$$\frac{S}{S_{eq}} = 1 - e^{-kt}w \tag{2.17}$$

Another kinetic adsorption equation is that of the fraction power equation, where

$$\frac{dS}{dt} = kt^{n-1} \tag{2.18}$$

and n is a parameter between 0 and 1.

Retention of solute by soil is commonly described with the empirical Elovich equation of the form

$$\frac{dS}{dt} = \alpha e^{-\beta S} \tag{2.19}$$

where α is the initial adsorption rate and β is a constant. Assuming that $\alpha\beta t >> 1$ and initial condition $S = 0$ at $t = 0$, the above rate equation yields this model expressing a linear relationship between S and ln t:

$$S = \frac{1}{\beta}1\mathrm{n}(\alpha\beta) + \frac{1}{\beta}1\mathrm{n}(t) \tag{2.20}$$

Mulitreaction Model

The simple kinetic models discussed previously are often inadequate for describing chemical reactions in highly heterogeneous soils where a wide range of particle sizes and reaction sites exists (Sparks, 2003). Several models based on physical kinetic reactions have been proposed to simulate the nonequilibrium transport of solutes in soils. The two-site model describes surface reactions such as adsorption on heterogeneous surfaces, assuming that a fraction of sorption occurs rapidly and another fraction is rate limited (Selim, Davidson, and Mansell, 1976). The combined equilibrium and kinetic models assume two types of surface reaction sites: one achieves instantaneous equilibrium and a slow reaction site with time-dependent kinetic reactions. The reaction on equilibrium reaction sites ($S_1 = FK_dC$) is described using linear equilibrium isotherm Equation (2.5) while the slow reaction on kinetic sites

$$(\frac{\partial S_2}{\partial t} = \alpha_1\left[(1-F)K_dC - S_2\right])$$

is simulated with the reversible first-order kinetic Equation (2.13). The F term denotes the fraction of equilibrium reaction sites in the total reaction sites. The two-site model can be incorporate into the CDE model (Equation (2.1)) in the following form:

$$[1 + F\frac{\rho K_d}{\theta}]\frac{\partial C}{\partial t} + \frac{\rho}{\theta}\frac{\partial S_2}{\partial t} = \frac{\partial}{\partial x}\left(D\frac{\partial C}{\partial x}\right) - v\frac{\partial C}{\partial x} \quad (2.21)$$

By introducing dimensionless variables, Nkedi-Kizza et al. (1984) showed that the two-site model is mathematically equivalent to the mobile–immobile two-region model (Equations (2.11) and (2.12)) discussed in the "Physical Nonequilibrium Model" section. The parameters in the model can be estimated from column BTC data by nonlinear least-squares parameter optimization using computer programs such as CXTFIT (Toride et al., 1995).

Using the two-site chemical nonequilibrium model, Pang et al. (2002) simulated the sorption and transport of reactive metals (Cd, Zn, and Pb) in uniformly packed columns of coarse alluvial gravel aquifer material where physical nonequilibrium did not exist, as evidenced by the symmetrical tritium BTC. The observed metal BTCs exhibited significant tailing and decreasing concentrations during flow interruption, suggesting that chemical nonequilibrium processes affected transport. The estimated two-site model parameters correlated with changes in pore water velocity, indicating diffusion-related reaction kinetics.

The concept of multiple reaction sites was further developed into a multireaction model (MRM) based on soil heterogeneity and kinetics of adsorption–desorption for the purpose of describing the time-dependent sorption of heavy metals in the soil environment. The MRM kinetic approach considers several interactions of heavy metals with soil matrix surfaces (Amacher, Kotubyamacher, Selim, and Iskandar, 1986; Amacher, Selim, and Iskandar, 1988; Selim, 1992). Specifically, the model assumes that a fraction of the total sorption sites is kinetic in nature, whereas the remaining fractions interact rapidly or instantaneously with solute in the soil solution. The model accounts for reversible as well as irreversible sorption of the concurrent and consecutive type. The model can be presented in the following formulations:

$$S_e = K_e C^n \tag{2.22}$$

$$\frac{\partial S_k}{\partial t} = k_1 \frac{\theta}{\rho} C^m - (k_2 + k_3) S_k \tag{2.23}$$

$$\frac{\partial S_i}{\partial t} = k_3 S_i \tag{2.24}$$

$$\frac{\partial S_s}{\partial t} = k_s \frac{\theta}{\rho} C \tag{2.25}$$

where S_e is the amount retained on equilibrium sites ($M.M^{-1}$); S_k is the amount retained on kinetic type sites ($M.M^{-1}$); S_i is the amount retained irreversibly by consecutive reaction ($M.M^{-1}$); S_s is the amount retained irreversibly by concurrent type of reaction ($M.M^{-1}$); n and m are dimensionless reaction orders commonly less than 1; K_e is a dimensionless equilibrium constant; k_1 and k_2 (T^{-1}) are the forward and backward reaction rates associated with kinetic sites, respectively; k_3 (T^{-1}) is the irreversible rate coefficient associated with the kinetic sites; and k_s (T^{-1}) is the irreversible rate coefficient associated with solution. For the case $n = m = 1$, the reaction equations become linear. In the above equations, we assumed that $n = m$ because there is no known method for independently estimating n or my. According to model formulation, the total amount of solute retention (S) by the soil is

$$S = S_e + S_k + S_i + S_s \tag{2.26}$$

Retention mechanisms associated with equilibrium and kinetic adsorption sites (Se and S_k) may include the formation of outer-sphere complexes, inner-sphere complexes, and diffuse ions in electrical double layers (Sposito, 1989). The reactions on the consecutive and concurrent irreversible reaction sites (S_i and S_s) include different types of surface precipitation that account for the formation of metal polymers, solid solutions or co-precipitates, as well as homogeneous mineral precipitates. The above retention mechanisms were incorporated into the CDE (Equation (2.1)) to predict solute retention and transport in soils. The numerical scheme and computer code of the multireaction and transport model (MRTM) are documented in the report by Selim, Amacher, and Iskandar (1990).

Similar to the MRM, the kinetic second-order model (SOM) proposed by Selim and Amacher (1988) assumes that a fraction of the total sorption sites is rate limited, whereas the remaining fractions interact rapidly or instantaneously with solute. The second-order or Langmuir type of reaction was represented in the model to describe retention mechanisms. The equilibrium and kinetic retention in the model can be presented by the following formulations:

$$S_e = K_e\left(\Gamma - S_e - S_k\right)C \tag{2.27}$$

$$\frac{\partial S_k}{\partial t} = k_1 \frac{\theta}{\rho}(\Gamma - S_e - S_k)C - (k_2 + k_3)S_k \tag{2.28}$$

where Γ is the adsorption capacity of the soil surface. In addition, the irreversible retentions of the consecutive or concurrent type as described by Equations (2.24) and (2.25) were also incorporated in the model.

The MRM was first developed by Amacher, Selim, and Iskandar (1988) to describe the kinetics of Cr(VI) and Cd retention in soils. They evaluated a number of model variations and found that the model with both irreversible and kinetic reversible reactions adequately described the experimental data. Several reaction mechanisms were postulated to explain the kinetic retention mechanisms that are compatible with the MRM (Figure 2.2). Selim et al. (1989) further expanded the multireaction concept to describe the nonequilibrium transport of Cr(V) in several soils. The model with kinetic retention parameters independently obtained from batch experiments overestimated Cr(V) mobility in soils with high retention capacity but an excellent fit of the experimental BTCs was obtained using kinetic parameters from nonlinear optimization (Figure 2.3). Selim and Ma (2001) employed the MRM to describe the kinetic adsorption–desorption of Cu in McLaren soil. The goodness of fit of their experimental results indicated that irreversible retention of the consecutive type was the dominant mechanism of the observed desorption hysteresis, which was explained in terms of the transition of adsorbed Cu from low-energy sites to high-energy sites. Moreover, the MRM with nonlinear equilibrium and kinetic sorption successfully described the kinetic data of AsO_4 adsorption on Olivier loam and Windsor sand. The model was also capable of predicting AsO_4 desorption kinetics for both soils. However, for Sharkey clay, which exhibited the strongest affinity for arsenic, an additional irreversible reaction phase was required to predict AsO_4 desorption or release with time (Zhang and Selim, 2005).

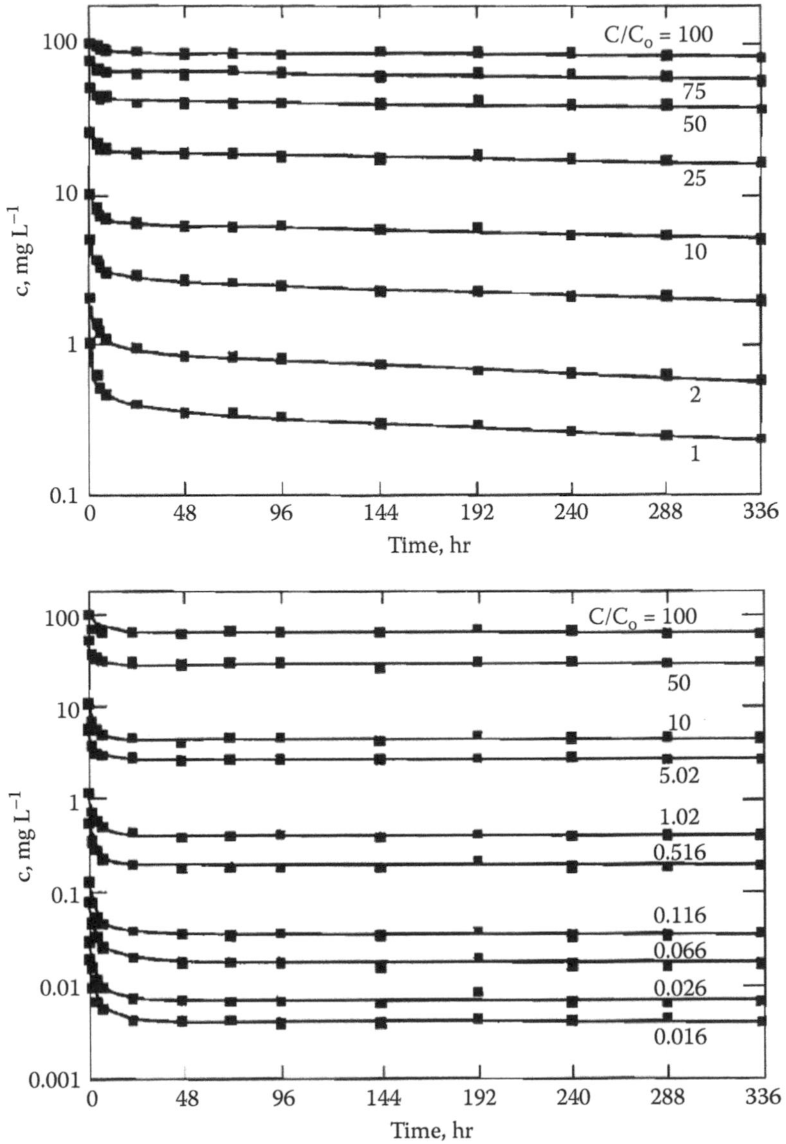

FIGURE 2.2
Experimental data on time-dependent retention of (a) Cr(VI) and (b) Cd by Windsor soil. The solid curves are model predictions using the multireaction model (MRM). (*Source:* Reprinted from Selim, H.M., M.C. Amacher, and I.K. Iskandar. 1990. Modeling the transport of heavy metals in soils. *CRREL Monogr. 2,* U.S. Government Printing Office.)

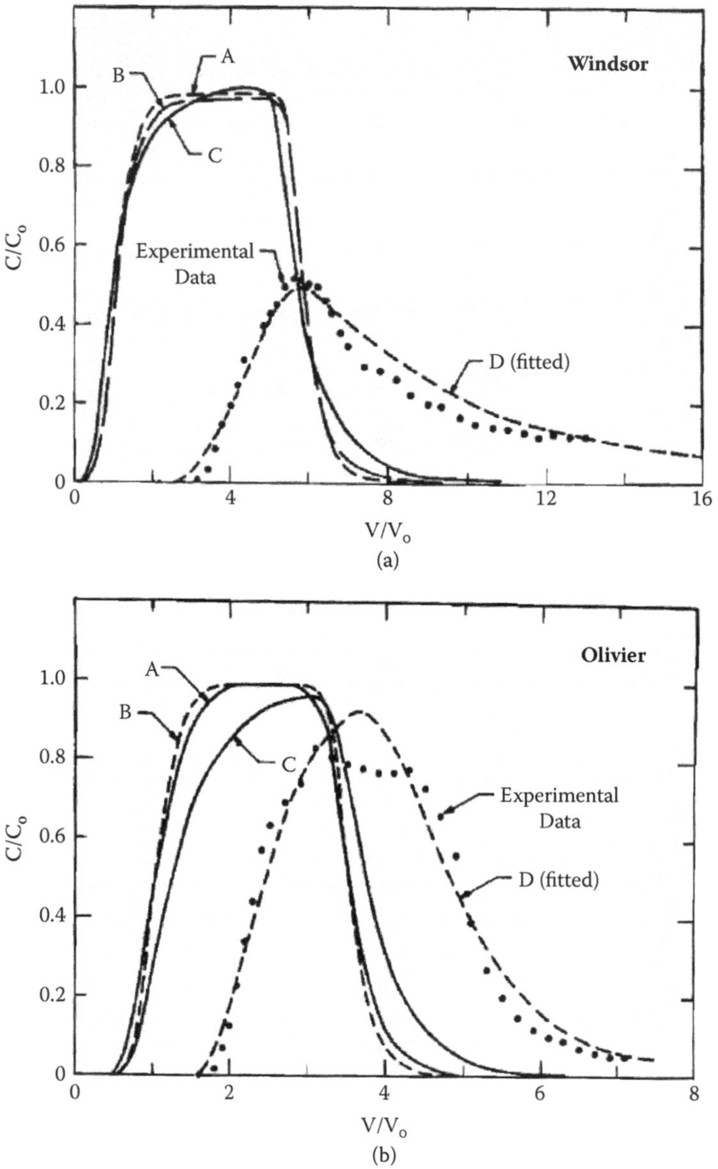

FIGURE 2.3
Measured and predicted BTCs for Cr(VI) in (a) Windsor soil and (b) Olivier soil. The curves A, B, and C are model predictions using batch rate coefficient for initial concentration C_o = 100, 10, and 1 mg.L^{-1}, respectively. Curve D is for fitted BTC using parameters obtained from least-squares optimization (*Source:* Reprinted from Selim, Amacher, and Iskandar, 1989. With permission.)

Multiprocess Nonequilibrium Transport

In modeling the transport of reactive trace elements, physical nonequilibrium models need to be coupled with equilibrium or kinetic chemical reactions. The MIM described by Equations (2.9) and (2.10) has been coupled to chemical nonequilibrium models to explicitly account for multiple sources of nonequilibrium (Brusseau, Jessup, and Rao, 1989; Ma and Selim, 1997). Brusseau, Jessup, and Rao (1989) developed a multiprocess nonequilibrium with transformation (MPNET) model that explicitly accounts for the rate-limited mass transfer between mobile and immobile regions as well as first-order abiotic and biotic kinetic transformations. Using a combined model where second-order two-site (SOTS) approaches were used to account for chemical nonequilibrium and the MIM approach was used to account for the physical nonequilibrium, Selim, Ma, and Zhu (1999) achieved considerably improved BTC predictions for pesticide transport in columns with different soil aggregate sizes. Leij and Bradford (2009) provided an analytical solution to the combined physical–chemical nonequilibrium (PCNE) model and successfully applied the model to describe the transport of colloidal particles.

To account for the rate and extent of physical nonequilibrium processes, the parameters associated with the mobile–immobile equations are often estimated from nonreactive tracer studies. In addition, the nonequilibrium chemical reactions are simulated using a series of equilibrium and/or kinetic expressions, as summarized in previous sections. The rate coefficients or equilibrium constants in the chemical model components can be obtained using two different methods: (1) independently estimated by conducting batch experiments; or (2) optimized by fitting the coupled model to the BTCs from transport experiments (Ma and Selim, 1997).

Using two-region flow and two-site sorption models, Gamerdinger et al. (2001) described the asymmetrical BTCs of U(VI) in an unsaturated silt loam. Ma and Selim (1997) compared the prediction capabilities of several formulations of coupled chemical–physical nonequilibrium models and found that an MIM-SOM best described atrazine BTCs in Sharkey soil. To our knowledge, the coupled physical–chemical nonequilibrium models have not been tested for the simulation of reactive transport of metal elements. Figure 2.4 presents a multiprocess nonequilibrium model based on the concepts of the MIM and MRM. Because there is no practical approach for separating the chemical reactions in the dynamic and stagnant flow regions, it is assumed that the same rate coefficients apply to both soil regions (Ma and Selim, 1997).

Figure 2.5 presents the BTC from column experiments of arsenite transport in Olivier loam, which displays diffusive fronts followed by extensive tailing or slow release during leaching. Sharp decreases or increases in arsenite concentration after flow interruption further verified the extensive nonequilibrium condition. The arsenite BTC was simulated using coupled physical and chemical nonequilibrium approaches. Here, physical transport

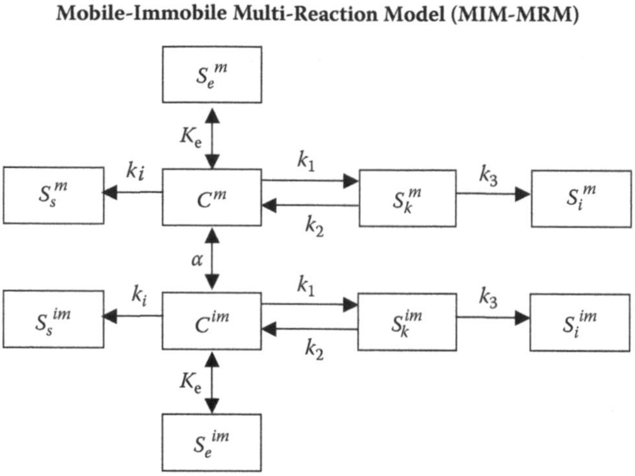

FIGURE 2.4
A schematic diagram of the MIM-MRM model with equilibrium, kinetic, and irreversible adsorption sites. Here, C is concentration in solution, S_e is the amount sorbed on equilibrium sites, S_k is the amount sorbed on kinetic sites, S_i is the amount retained on consecutive irreversible sites, and S_s is amount retained on concurrent irreversible sites. The parameters K_e, k_1, k_2, k_3, and k_s are the respective rates of reactions. Superscripts *m* and *im* indicate mobile and immobile liquid regions, respectively, and α is a first-order mass transfer coefficient between mobile and immobile regions.

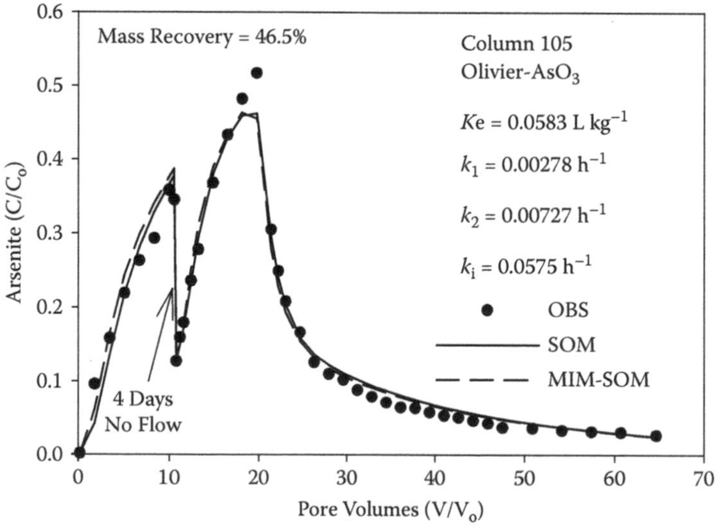

FIGURE 2.5
BTCs of arsenite [As(III)] from column of Olivier soil. Transport of arsenite was described using the multireaction model (MRM) as well as the coupled mobile–immobile and multireaction model (MIM-MRM) with equilibrium, kinetic, and irreversible retention.

parameters (D, f, and α) were based on the estimated values from tritium BTCs, whereas kinetic retention parameters (K_e, k_1, k_2, k_i) were obtained from inverse modeling. The incorporation of a physical nonequilibrium approach resulted in a slightly sharper front, reflecting the effect of preferential flow path. The coupled model successfully predicted the concentration drop during flow interruption on the adsorption side. Based on the results of model predictions, we concluded that the kinetics of arsenite reaction on soil surfaces is the dominant process of arsenite transport in soils. The utilization of the physical nonequilibrium parameters in the combined model is expected to increase even further for modeling contaminant transport in heterogeneous natural porous media than in columns repacked in the laboratory with homogenized soils.

Significance to Site Remediation

Because of the widespread soil pollution of heavy metals, there is an urgent need for effective engineering approaches to remediate contaminated sites. Both physical and chemical nonequilibrium transport processes are important for the design and implementation of ex situ or in situ cleanup technologies, including isolation, immobilization, toxicity reduction, physical separation, and extraction (Mulligan, Yong, and Gibbs, 2001).

In situ isolation or capping of contaminated soils through the construction of physical barriers is a common technique employed at landfills and superfund sites (Mulligan, Yong, and Gibbs, 2001). Such barriers made of steel, cement, bentonite clay, and grout walls reduce the permeability of the waste and limit the movement of groundwater through the contaminated area. Additional layers of sandy soils are employed to prevent upward movement of groundwater by capillary action. The most commonly encountered problem of capping is that the aging of the liner eventually leads to a preferential flow path in the fractures. The failure of physical barriers may result in the rapid release of contaminated water through connected cracks. In addition, the slow diffusion of heavy metal ions through lining material is a possible pathway of contamination over extended time periods. Preferential flow and matrix diffusion processes are critical factors in determining the successfulness of site remediation through isolation.

Pumped flushing followed by groundwater extraction is one of the most widely used in situ remediation techniques for contaminated sites with relatively high permeability. Nonideal behavior such as long tailing has been commonly observed and attributed to nonequilibrium. The nonequilibrium adsorption behavior of heavy metals limited the suitability of this technique for the treatment of contaminated soils. The rate-limited sorption and transport processes made it practically impossible to achieve the cleanup goal in a reasonable time frame with flushing and extraction processes. A high residual content of toxic metals may remain after treatment because of the highly hysteretic adsorption behavior of toxic metals.

A permeable reactive barrier (PRB) consists of installing a reactive material in the aquifer to induce sequestration and/or transformations of the contaminants and reduce contaminant concentrations in groundwater (Scherer et al., 2000). The effective design and operation of the PRB must take into account the preferential flow path and kinetic reactions of metals in the reactive barrier. A long hydraulic residence time is necessary for the heavy metals to absorb and react. Careful construction with consideration of the hydraulic field is required to avoid groundwater bypassing the PRB.

Monitored natural attenuation (MNA) is proposed as an alternative remediation strategy for soils contaminated with inorganic contaminants. Long-term retention and transformation is frequently the dominant process of natural attenuation of toxic elements in contaminated sites. A key issue of natural attenuation is the reversibility of sequestration (sorption and precipitation) into the solid phase because the intrinsic toxicity of toxic metals is not affected by the immobilization process (Reisinger, Burris, and Hering, 2005). Therefore, successful implementation of MNA for As-contaminated sites requires detailed, site-specific characterization, including source identification, plume boundary delineation, and time-series monitoring in soil and groundwater.

Summary

Numerous laboratory and field works have elucidated the significance of rate-limited processes in the transport of reactive trace elements such as As, Hg, Pb, Cu, Co, Cd, Cr, and Zn in soils. A thorough understanding of the physical and chemical nonequilibrium processes is prerequisite for accurately modeling heavy metal transport and scientific design of effective remediation technologies. Physical nonequilibrium (transport-related nonequilibrium behavior) due to slow exchange of solute between mobile and less mobile flow regions is a critical transport phenomenon in the vadose zone with heterogeneous pore space. The concept of mobile–immobile flow regions is the most widely used model for simulating nonequilibrium transport caused by preferential flow and matrix diffusion. A range of chemical reactions, including ion exchange, surface complexation, precipitation–dissolution, biotransformation, and oxidation–reduction, may have substantial influence on the transport of reactive metals. Local equilibrium assumptions are inadequate in simulating the transport of heavy metals when the reactions occur at time scales longer than the hydraulic residence time. The multireaction model that takes into account the heterogeneity of the chemical reaction sites and time-dependent reaction behavior has been developed for simulating the nonideal transport of heavy metals in soil. The kinetic multireaction model can be coupled with the mobile–immobile model for a multiprocess model that describes both physical nonequilibrium and chemical nonequilibrium.

References

Ainsworth, C.C., J.L. Pilon, P.L. Gassman, and W.G. Vandersluys. 1994. Cobalt, cadmium, and lead sorption to hydrous iron-oxide — Residence time effect. *Soil Sci. Soc. Am. J.* 58: 1615–1623.

Amacher, M.C., H.M. Selim, and I.K. Iskandar. 1988. Kinetics of chromium(VI) and cadmium retention in soils — A nonlinear multireaction model. *Soil Sci. Soc. Am. J.* 52: 398–408.

Amacher, M.C., J. Kotubyamacher, H.M. Selim, and I.K. Iskandar. 1986. Retention and release of metals by soils — Evaluation of several models. *Geoderma* 38: 131–154.

Biggar, J.W., and D.R. Nielsen. 1976. Spatial variability of leaching characteristics of a field soil. *Water Resour. Res.* 12: 78–84.

Brusseau, M.L. 1993. The influence of solute size, pore water velocity, and intraparticle porosity on solute-dispersion and transport in soil. *Water Resour. Res.* 29: 1071–1080.

Brusseau, M.L. 1994. Transport of reactive contaminants in heterogeneous porous-media. *Rev. Geophysics* 32: 285–313.

Brusseau, M.L., R.E. Jessup, and P.S.C. Rao. 1989. Modeling the transport of solutes influenced by multiprocess nonequilibrium. *Water Resour. Res.* 25: 1971–1988.

Brusseau, M.L., R.E. Jessup, and P.S.C. Rao. 1992. Modeling solute transport influenced by multiprocess nonequilibrium and transformation reactions. *Water Resour. Res.* 28: 175–182.

Buchter, B., B. Davidoff, M.C. Amacher, C. Hinz, I.K. Iskandar, and H.M. Selim. 1989. Correlation of Freundlich K_d and N retention parameters with soils and elements. *Soil Sci.* 148: 370–379.

Cameron, RE. 1992. Guide to Site and Soil Description for Hazardous Waste Site Characterization. Volume 1: Metals. Environmental Protection Agency EPA/600/4-91/029.

Camobreco, V.J., B.K. Richards, T.S. Steenhuis, J.H. Peverly, and M.B. McBride. 1996. Movement of heavy metals through undisturbed and homogenized soil columns. *Soil Sci.* 161: 740–750.

Carrillo-Gonzalez, R., J. Simunek, S. Sauve, and D. Adriano. 2006. Mechanisms and pathways of trace element mobility in soils. *Adv. Agron.* 91: 111–178.

Corwin, D.L., A. David, and S. Goldberg. 1999. Mobility of arsenic in soil from the Rocky Mountain Arsenal area. *J. Contam. Hydrol.* 39: 35–58.

Dagan, G. 1984. Solute transport in heterogeneous porous formations. *J. Fluid Mech.* 145: 151–177.

Darland, J.E., and W.P. Inskeep. 1997. Effects of pore water velocity on the transport of arsenate. *Environ. Sci. Technol.* 31: 704–709.

Fuller C.C., J.A. Davis, and G.A. Waychunas. 1993. Surface chemistry of ferrihydrite. 2. Kinetics of arsenate adsorption and coprecipitation. *Geochim. Cosmochim. Acta* 57: 2271–2282.

Gamerdinger, A.P., D.I. Kaplan, D.M. Wellman, and R.J. Serne. 2001. Two-region flow and decreased sorption of uranium (VI) during transport in Hanford groundwater and unsaturated sands. *Water Resour. Res.* 37: 3155–3162.

Hinz, C., and H.M. Selim. 1994. Transport of zinc and cadmium in soils — Experimental evidence and modeling approaches. *Soil Sci. Soc. Am. J.* 58: 1316–1327.

Hinz, C., L.A. Gaston, and H.M. Selim. 1994. Effect of sorption isotherm type on predictions of solute mobility in soil. *Water Resour. Res.* 30: 3013–3021.

Jardine, P.M. 2008. Influence of coupled processes on contaminant fate and transport in subsurface environments. *Adv. Agron.* 99: 1–99.

Jardine, P.M., and D.L. Sparks. 1984. Potassium-calcium exchange in a multireactive soil system. 1. Kinetics. *Soil Sci. Soc. Am. J.* 48: 39–45.

Jury, W.A., and H. Fluhler. 1992. Transport of chemicals through soil — Mechanisms, models, and field applications. *Adv. Agron.* 47: 141–201.

Leij, F.J., and S.A. Bradford. 2009. Combined physical and chemical nonequilibrium transport model: Analytical solution, moments, and application to colloids. *J. Contam. Hydrol.* 110: 87–99.

Ma, L., and H.M. Selim. 1994. Tortuosity, mean residence time, and deformation of tritium breakthroughs from soil columns. *Soil Sci. Soc. Am. J.* 58: 1076–1085.

Ma, L.W., and H.M. Selim. 1997. Physical nonequilibrium modeling approaches to solute transport in soils. *Adv. Agron.* 58: 95–150.

Mulligan, C.N., R.N. Yong, and B.F. Gibbs. 2001. Remediation technologies for metal-contaminated soils and groundwater: an evaluation. *Engin. Geol.* 60: 193–207.

Nkedi-Kizza, P., J.W. Biggar, H.M. Selim, M.T. van Genuchten, P.J. Wierenga, J.M. Davidson, and D.R. Nielsen. 1984. On the equivalence of two conceptual models for describing ion exchange during transport through an aggregated oxisol. *Water Resour. Res.* 20: 1123–1130.

NRC (National Research Council). 2003. Bioavailability of contanminants in soils and sediments: Processes, tools and applications. National Academies Press, Washington, D.C.: 240.

Pang, L.P., M. CLose, D. Schneider, and G. Stanton. 2002. Effect of pure-water velocity on chemical nonequillibrium transport of Cd, Zn, and Pb in alluvial gravel columns. *J. Contam. Hydrol.* 57:241–258.

Reedy, O.C., P.M. Jardine, G.V. Wilson, and H.M. Selim. 1996. Quantifying the diffusive mass transfer of nonreactive solutes in columns of fractured saprolite using flow interruption. *Soil Sci. Soc. Am. J.* 60: 1376–1384.

Reisinger, H.J., D.R. Burris, and J.G. Hering. 2005. Remediating subsurface arsenic contamination with monitored natural attenuation. *Environ. Sci. Technol.* 39: 458A–464A.

Scherer, M.M., S. Richter, R.L. Valentine, and P.J.J. Alvarez. 2000. Chemistry and microbiology of permeable reactive barriers for in situ groundwater cleanup. *Crit. Rev. Environ. Sci. Technol.* 30: 363–411.

Selim, H.M., and L. Ma. 2001. Modeling nonlinear kinetic behavior of copper adsorption-desorption in soil. In *Physical and chemical processes of water and solute transport/retention in soil.* Selim, H.M., Sparks, D.L., Eds.. SSSA Special Publication No. 56. SSSA, Madison, WI, p. 189–212.

Selim, H.M., and M.C. Amacher. 1997. *Reactivity and transport of heavy metals in soils.* CRC/Lewis, Boca Raton, FL.

Selim, H.M. 1992. Modeling the transport and retention of inorganics in soils. *Adv. Agron.* 47: 331–384.

Selim, H.M., M.C. Amacher, and I.K. Iskandar. 1989. Modeling the transport of chromium (VI) in soil columns. *Soil Sci. Soc. Am. J.* 53:966–1004.

Selim, H.M., and M.C. Amacher. 1988. A second-order kinetic approach for modeling solute retention and transport in soils. *Water Resour. Res.* 24: 2061–2075.

Selim, H.M., L.W. Ma, and H.X. Zhu. 1999. Predicting solute transport in soils: Second-order two-site models. *Soil Sci. Soc. Am. J.* 63: 768–777.

Selim, H.M., B. Buchter, C. Hinz, and L. Ma. 1992. Modeling the transport and retention of cadmium in soils — Multireaction and multicomponent approaches. *Soil Sci. Soc. Am. J.* 56: 1004–1015.

Selim, H.M., M.C. Amacher, and I.K. Iskandar. 1990. Modeling the transport of heavy metals in soils. *CRREL Monogr. 2,* U.S. Government Printing Office.

Selim, H.M., J.M. Davidson, and R.S. Mansell. 1976. Evaluation of a two-site adsorption-desorption model for describing solute transport in soils. In *Proc. Summer Computer Simulation Conf., Washington D.C.* La Jolla, CA: Simulation Councils, Inc. pp. 444–448.

Simunek, J., N.J. Jarvis, M.T. van Genuchten, and A. Gardenas. 2003. Review and comparison of models for describing non-equilibrium and preferential flow and transport in the vadose zone. *J. Hydrol.* 272: 14–35.

Sparks, D.L. 2003. *Environmental soil chemistry (2nd ed).* Academic Press, San Diego, CA.

Sparks, D.L. 1989. *Kinetics of soil chemical processes.* Academic Press, San Diego, CA.

Sposito, G. 1989. *The chemistry of soils.* Oxford University Press, New York.

Steefel, C.I., D.J. DePaolo, and P.C. Lichtner. 2005. Reactive transport modeling: An essential tool and a new research approach for the Earth sciences. *Earth Planet. Sci. Lett.* 240: 539–558.

Toride, N., F.J. Leij, and M.T. van Genuchten. 1995. The CXTFIT code for estimating transport parameters from laboratory or field tracer experiments version 2.0. U.S. Salinity Lab., USDA, ARS, Riverside, CA.

Van Genuchten, M. Th., and W. J. Alves. 1982. Analytical solutions of the one-dimensional convective-dispersive solute transport equation, Tech. Bull. USDA, 1661.

Van Genuchten, M. Th. 1981. Non-equilibrium transport parameters from miscible displacement experiments, Res. Rep. No. 119, U. S. Salinity Lab., USDA, ARS, Riverside, CA.

Van Genuchten, M.T., and P.J. Wierenga. 1976. Mass transfer studies in sorbing porous media. 1. Analytical solutions. *Soil Sci. Soc. Am. J.* 40: 473–480.

Yin, Y.J., H.E. Allen, C.P. Huang, D.L. Sparks, and P.F. Sanders. 1997. Kinetics of mercury(II) adsorption and desorption on soil. *Environ. Sci. Technol.* 31: 496–503.

Zhang, H., and H.M. Selim. 2005. Kinetics of arsenate adsorption-desorption in soils. *Environ. Sci. Technol.* 39: 6101–6108.

Zhang, H., and H.M. Selim. 2008. Reaction and transport of arsenic in soils: Equilibrium and kinetic modeling. *Adv. Agron.* 98: 45–115.

Zhou, L., and H.M. Selim. 2003. Scale-dependent dispersion in soils: An overview. *Adv. Agron.* 80: 223–263.

3

Chemical Equilibrium and Reaction Modeling of Arsenic and Selenium in Soils

Sabine Goldberg

CONTENTS

High concentrations of the trace elements arsenic (As) and selenium (Se) in soils pose a threat to agricultural production and the health of humans and animals. As is toxic to both plants and animals. Se, despite being an essential micronutrient for animal nutrition, is potentially toxic because the concentration range between deficiency and toxicity in animals is narrow. Seleniferous soils release enough Se to produce vegetation toxic to grazing animals. Such soils occur in the semiarid states of the western United States (Lakin, 1961). Concentrations of As and Se in soils and waters can become elevated as a result of discharge from petroleum refineries, disposal of fly ash, mining activities, geothermal discharge, and mineral oxidation and dissolution. Enrichment of As can also occur through the application of arsenical pesticides (Wauchope, 1983) and poultry manures from chickens fed the arsenical additive, roxarsone (Garbarino et al., 2003). Elevated concentrations of As and Se are found in agricultural drainage waters from some soils in arid regions. In recognition of the hazards that these trace elements pose to the welfare of humans and animals, the U.S. Environmental Protection Agency (EPA) has set the drinking water standard at 10 ppb for As and 50 ppb for Se.

The concentrations of As and Se in soil solution may be affected by various chemical processes and soil factors. These include soil solution chemistry, methylation and volatilization reactions, precipitation–dissolution reactions, oxidation–reduction reactions, and adsorption–desorption reactions. The objective of this chapter is to discuss the soil factors and chemical processes that may control soil solution concentrations of the trace elements As and Se.

Inorganic Chemistry of As and Se in Soil Solution

The dominant inorganic As solution species are As(III) (arsenite) and As(V) (arsenate), while Se(IV) (selenite) and Se(VI) (selenate) are the dominant inorganic Se species in soil solution (Adriano, 1986). For both elements, toxicity depends on the oxidation state, with the lower redox state considered more toxic: that is, arsenite is more toxic than arsenate (Penrose, 1974) and selenite is more toxic than selenate (Harr, 1978).

At most natural pHs, arsenite is present in solution predominantly as H_3AsO_3 because the pK_a values for arsenious acid are high: $pK_a^1 = 9.2$, $pK_a^2 = 12.7$. Solution arsenate occurs as $H_2AsO_4^-$ and $HAsO_4^{2-}$ because the pK_a values for arsenic acid are $pK_a^1 = 2.3$, $pK_a^2 = 6.8$, and $pK_a^3 = 11.6$. Solution selenite occurs as $HSeO_3^-$ and SeO_3^{2-} because the pK_a values for selenious acid are $pK_a^1 = 2.5$ and $pK_a^2 = 7.3$. Selenate is present in solution as SeO_4^{2-} because the pK_a values for selenic acid are low: $pK_a^1 \approx -3$ and $pK_a^2 = 1.9$. Because the kinetics of As and Se redox transformations are relatively slow, both oxidation states are often found in soil solution, regardless of redox conditions (Masscheleyn, Delaune, and Patrick, 1990, 1991a). Under highly reducing conditions, Se occurs in solution primarily as HSe^- because the pK_a values for hydrogen selenide are $pK_a^1 = -1.1$ and $pK_a^2 = 15.0$ (Elrashidi et al., 1987).

In typical waters, As forms few aqueous arsenite or arsenate complexes (Cullen and Reimer, 1989). The formation of metal arsenate ion-pairs has been suggested from ion chromatography but their role is expected to be minor (Lee and Nriagu, 2007). In anaerobic aquatic environments, arseno-carbonate solution complexes $As(CO_3)_2^-$, $As(CO_3)^+$, and $As(CO_3)(OH)_2^-$ were proposed to be the most stable inorganic As species (Kim et al., 2000). Subsequently, Neuberger and Helz (2005) showed that As(III) carbonate solution complexes would be negligible at carbonate concentrations found in most natural waters but could be significant, although minor, in extremely carbonate-rich, high ionic strength waters, such as those of evaporative basins. Solution complexes of selenite and selenate appear to make minor contributions to dissolved Se in agricultural soils (Elrashidi et al., 1987).

Methylation and Volatilization Reactions

Methylation reactions involve the addition or substitution of methyl groups, $-CH_3$, to substrates. Methylated compounds of As that can form in soils include monomethylarsenic acid ($CH_3AsO_3H_2$), dimethylarsenic acid ($(CH_3)_2AsO_2H$), trimethylarsine oxide, ($(CH_3)_3AsO$), and the methylarsines: monomethylarsine (CH_3AsH_2), dimethylarsine, ($(CH_3)_2AsH$), and trimethylarsine ($(CH_3)_3As$). The various arsine compounds, including arsine (AsH_3), are volatile and tend to escape to the atmosphere (Gao, Tanji, and Goldberg, 1998). Methylated Se compounds are primarily the volatile species dimethylselenide (DMSe) and dimethyldiselenide(DMDSe) (Séby et al., 1998), although the nonvolatile dimethylselenonium ion ($DMSe^+$-R) has been detected in surface waters (Cooke and Bruland, 1987).

Historically, various organic arsenicals have been used as herbicides on cotton and other agricultural crops and on lawns, golf courses, and highway rights-of-way. These herbicides include monosodium methanearsonate (MSMA), disodium methanearsenonate (DSMA), calcium acid methanearsenate (CAMA), and cacodylic acid, which is dimethylarsenic acid. Due to toxicity concerns, the U.S. Environmental Protection Agency cancelled all agricultural uses of arsenical herbicides except MSMA as of September 30, 2009. Use of MSMA on golf courses, sod farms, and highway rights-of-way will be prohibited after December 31, 2013. Use of MSMA on cotton, where no alternative herbicide presently exists, will be permitted and reviewed beginning in 2013.

Speciation studies of As and Se in reservoir sediment suspensions under controlled redox and pH conditions found no evidence of organic arsenicals; however, methylated Se compounds were detected under oxidizing and moderately reducing conditions (Masscheleyn et al., 1991b). In soils, bacteria, fungi, and algae can convert organic and inorganic arsenicals to form volatile arsines. Microorganisms can also mineralize organic As forms to produce inorganic As via demethylation reactions. The relative rates of these transformations can control As cycling and accumulation in soils and are functions of As form, As concentration, soil moisture, soil temperature, and availability of organic carbon energy sources for the microbial population (Gao and Burau, 1997).

Organic and inorganic selenium forms can be methylated to the volatile forms, DMSe and DMDSe, by microbial action in soils under both aerobic and anaerobic conditions (Doran and Alexander, 1977). Addition of a supplemental carbon energy source resulted in gaseous release of indigenous Se from soils high in Se (Francis, Duxbury, and Alexander, 1974). Proteins were found to dramatically stimulate Se biomethylation in evaporation pond water, indicating that the process is protein peptide-limited rather than ammonium-, amino acid-, or carbon-limited (Thompson-Eagle and Frankenberger, 1990). The rate of gaseous Se evolution from soil is highly dependent on Se content, soil moisture, soil temperature, aeration status, and availability of organic carbon (Frankenberger and Karlson, 1992).

Precipitation–Dissolution Reactions

Precipitation and dissolution reactions of As and Se solid phases are processes that may influence the soil solution concentrations of these elements. Under oxidizing conditions, $Ca_3(AsO_4)_2$ and $Mn_3(AsO_4)_2$ are the most stable As minerals in alkaline soils, while in acid soils $Fe_3(AsO_4)_2$, $FeAsO_4$, and $AlAsO_4$ are the least soluble minerals (Sadiq, 1997). Soils that had been contaminated historically with lead arsenate were found to be supersaturated with $Pb_3(AsO_4)_2$ and $Mn_3(AsO_4)_2$ (Hess and Blanchar, 1976). In reduced, anoxic sediments, As(III) sulfides (i.e., As_2O_3 and FeAsS) are the most stable As minerals (Welch et al., 2000). Speciation calculations of As in reservoir sediment suspensions under controlled redox and pH conditions found that they were undersaturated by several orders of magnitude with respect to $FeAsO_4$, $AlAsO_4$, $Ca_3(AsO_4)_2$, and $Mn_3(AsO_4)_2$, and As solubility was not limited by the formation of As_2O_3 and FeAsS (Masscheleyn et al., 1991b). Chemical speciation calculations of As in shallow groundwater in the Carson Desert (Nevada) found undersaturation with respect to As minerals (Welch and Lico, 1998).

Selenate minerals are much too soluble to persist in aerated soils (Elrashidi et al., 1987). In acid soils, $MnSeO_3$ is the most stable selenite mineral, while $PbSeO_3$ is the least soluble mineral in alkaline soil (Elrashidi et al., 1987). Speciation calculations of Se in reservoir sediment suspensions under controlled redox and pH conditions found undersaturation with respect to metal selenates and selenites (Masscheleyn et al., 1991b). Selenium solubility is governed by precipitation reactions of metal selenides: FeSe and $FeSe_2$ and elemental Se under reducing conditions (Séby et al., 1998). Under reducing conditions, elemental Se is thermodynamically stable over a wide solution pH range (Masscheleyn et al., 1991b). Elemental Se exists as three allotropes: the red and black forms are amorphous and are more likely to occur in soils than the gray hexagonal crystalline form (Geering et al., 1968).

Oxidation–Reduction Reactions

Under oxic conditions, arsenate is the thermodynamically stable and dominant As species in soil solution, while arsenite is the thermodynamically stable and most abundant solution form under anoxic conditions. At intermediate pH and suboxic conditions, both As(III) and As(V) redox states coexist (Sadiq, 1997). Under highly reducing conditions, arsine (AsH_3) may be formed (Korte and Fernando, 1991). Under oxidizing conditions, selenate is the thermodynamically stable and dominant species of Se in soil solution, while selenite is the thermodynamically stable and most abundant form under reducing conditions (Séby et al., 1998). Under highly reducing conditions, the thermodynamically stable form of Se is selenide (Elrashidi et al., 1987).

Most oxidation–reduction reactions are mediated by microorganisms. Microbes act as kinetic reaction mediators but cannot promote thermodynamically unfavorable reactions (Séby et al., 1998). Oxidation of arsenite to arsenate was observed by bacteria isolated from soil such as *Alcaligenes* (Osborne and Ehrlich, 1976). Various bacteria isolated from river water have been shown to reduce arsenate to arsenite (Freeman, 1985). Reduction of arsenate to arsenite was also achieved by a *Clostridium* species isolated from an As-contaminated soil (Langner and Inskeep, 2000).

Under anaerobic conditions in sediments, bacteria can reduce selenate to selenite and elemental Se (Oremland et al., 1990). The capacity for selenate reduction was found to be widespread and to occur in diverse sediments, including those uncontaminated with Se (Steinberg and Oremland, 1990). Selenate-reducing bacterial communities exhibit broad phylogenetic diversity (Lucas and Hollibaugh, 2001). Oxidation of refractory Se(0) into soluble selenate upon reexposure to air was observed for contaminated sediments that had been deposited under strongly reducing conditions and/or previously ponded (Tokunaga, Pickering, and Brown, 1996; Zawislanski and Zavarin, 1996).

Oxidation of arsenite to arsenate in some aquatic systems can occur predominantly via abiotic processes (Oscarson, Huang, and Liaw, 1980). Removal of manganese oxides from lake sediments indicated that these minerals were responsible for As(III) oxidation (Oscarson, Huang, and Liaw, 1981). As(III) oxidation rates in aquifer samples increased with increasing manganese oxide content and were unaffected by anaerobic and sterile conditions (Amirbahman et al., 2006). Similarly, abiotic oxidation of selenite occurs on the surfaces of the synthetic manganese oxide, birnessite (Scott and Morgan, 1996). Abiotic reduction of Se(VI) to Se(IV) and Se(0) can occur in the presence of a mixed Fe(II,III) oxide called green rust that is present in many suboxic soils and sediments (Myneni, Tokunaga, and Brown, 1997).

Adsorption–Desorption Reactions

Adsorption is the accumulation of a chemical species at the solid–solution interface to form a two-dimensional molecular surface. Adsorption reactions on soil surfaces often control the concentration of As and Se in soil solution. Arsenic adsorption is significantly positively correlated with the clay, Al and Fe oxide, organic carbon, and inorganic carbon content of soils (Wauchope, 1975; Livesey and Huang, 1981; Elkhatib, Bennett, and Wright, 1984; Yang et al., 2002). Selenium adsorption is also significantly positively correlated with these same soil constituents (Lévesque, 1974; Elsokkary, 1980; Vuori et al., 1989, 1994).

For both As and Se, adsorption behavior in soils is determined by redox state. As(V) and As(III) both adsorb on soil surfaces. As(V) adsorption is greater than As(III) adsorption in the acid pH range, while at alkaline pH values, As(III) adsorption exceeds As(V) adsorption (Raven, Jain, and Loeppert, 1998; Goldberg and Johnston, 2001). Selenite adsorbs strongly on soil surfaces, while Se(VI) adsorbs weakly or not at all (Neal and Sposito, 1989). In contrast, Se(VI) adsorption at very high Se additions was consistently higher than Se(IV) adsorption on various soils (Singh, Singh, and Relan, 1981).

Specific adsorption of anions forms strong surface complexes that contain no water molecules between the adsorbing ion and the surface site and are called inner-sphere surface complexes. Outer-sphere surface complexes, on the other hand, contain at least one water molecule between the adsorbing anion and the surface site. Outer-sphere complexes are less strong than inner-sphere complexes. Arsenate has been observed to form inner-sphere complexes on goethite, maghemite, hematite, gibbsite, and amorphous Al and Fe oxides using various spectroscopic techniques (Waychunas et al., 1993; Fendorf et al., 1997; Goldberg and Johnston, 2001; Catalano et al., 2007; Morin et al., 2008). Similarly, arsenite was observed to form inner-sphere complexes on goethite, lepidocrocite, maghemite, hematite, and amorphous Fe oxide (Manning, Fendorf, and Goldberg, 1998; Ona-Nguema et al., 2005; Morin et al., 2008). A mixture of inner-sphere and outer-sphere surface complexes was observed for As(V) adsorption on hematite (Catalano et al., 2008) and As(III) adsorption on γ-Al_2O_3 (Arai, Elzinga, and Sparks, 2001).

Extended x-ray absorption fine structure spectroscopy investigations have observed selenate to form inner-sphere complexes on goethite (Manceau and Charlet, 1994) and hematite (Peak and Sparks, 2002). Various spectroscopic investigations have similarly observed selenite to form inner-sphere complexes on goethite, hematite, gibbsite, and amorphous Al and Fe oxide (Manceau and Charlet, 1994; Papelis et al., 1995; Peak, 2006; Catalano et al., 2006). A mixture of inner-sphere and outer-sphere surface complexes was observed for Se(IV) adsorption on corundum (Peak, 2006) and for Se(VI) adsorption on goethite, corundum, and amorphous Fe and Al oxide (Wijnja and Schulthess, 2000; Peak and Sparks, 2002; Peak, 2006). Predominantly outer-sphere surface complexes were observed for Se(VI) adsorption on γ-Al_2O_3 (Wijnja and Schulthess, 2000) and amorphous Al oxide (Peak, 2006).

Modeling of Adsorption by Soils: Empirical Models

Adsorption reactions by soils have historically been described using empirical adsorption isotherm equations. The most popular adsorption isotherms are the linear, Freundlich, and Langmuir equations. Typically, these equations are very good at describing experimental data despite their lack of theoretical basis. Their popularity stems in part from their simplicity and from the ease of estimation of their adjustable parameters.

The adsorption isotherm equation is a linear function and is written in terms of the distribution coefficient, K_d:

$$x = K_d c \tag{3.1}$$

where x is the amount of ion adsorbed per unit mass of soil and c is the equilibrium solution ion concentration. Because of the linear assumption, this equation usually describes adsorption data only over a very restricted solution concentration range. The linear adsorption isotherm equation has been used to describe As(V) adsorption by contaminated and uncontaminated Belgian soils (De Brouwere, Smolders, and Merckx, 2004) and Se(IV) adsorption by a calcareous soil from China (Wang and Liu, 2005).

The Freundlich adsorption equation is a nonlinear isotherm whose use implies heterogeneity of adsorption sites. The Freundlich isotherm equation is

$$x = Kc^{\beta} \tag{3.2}$$

where K is an affinity parameter and β is a heterogeneity parameter; the smaller the value of β, the greater the heterogeneity (Kinniburgh, 1985). When $\beta = 1$, the Freundlich equation reduces to a linear adsorption isotherm. Although the Freundlich equation is strictly valid only for ion adsorption at low solution ion concentration (Sposito, 1984), it has often been used to describe ion adsorption by soils over the entire concentration range investigated. The Freundlich adsorption isotherm equation has been used to describe As(III) (Elkhatib, Bennett, and Wright, 1984) and As(V) adsorption (Zhang and Selim, 2005) by soils and Se(IV) adsorption by alluvium, sediment, and basalt (Del Debbio, 1991).

The Langmuir adsorption isotherm equation was originally developed to describe gas adsorption onto clean surfaces. The Langmuir isotherm equation is

$$x = \frac{x_m Kc}{1 + Kc} \tag{3.3}$$

where x_m is the maximum ion adsorption per unit mass of soil and K is an affinity parameter related to the bonding energy of the ion to the surface. The Langmuir isotherm can be derived theoretically based on evaporation and condensation rates, assuming a finite number of uniform adsorption sites and the absence of any lateral interaction between adsorbed species

(Adamson, 1976). Despite the fact that these assumptions are violated in soils, the Langmuir isotherm has often been used to describe ion adsorption reactions by soils. In many studies, the Langmuir equation is only able to describe adsorption for low solution ion concentrations. The Langmuir adsorption isotherm equation has been used to describe As(V) adsorption by Chinese soils (Jiang et al., 2005) and Se(IV)and Se(IV) adsorption by diverse soils (Singh, Singh, and Relan, 1981).

Modeling of Adsorption by Soils: Constant Capacitance Model

The constant capacitance model is a chemical surface complexation model that was developed to describe ion adsorption at the oxide–solution interface (Schindler et al., 1976; Stumm, Kummert, and Sigg, 1980). As is characteristic of surface complexation models, this chemical model explicitly defines surface species, chemical reactions, equilibrium constants, mass balances, charge balance, and electrostatic potentials. The reactive surface site is defined as SOH, an average reactive surface hydroxyl ion bound to a metal, S, in the oxide mineral. The constant capacitance model has been extended to describe the adsorption of trace element anions on soil surfaces. The applications of the model to predict arsenate and selenite adsorption by soils will be presented (Goldberg et al., 2005, Goldberg, Lesch, and Suarez, 2007).

The constant capacitance model contains the following assumptions: (1) all surface complexes are inner-sphere; (2) anion adsorption occurs via ligand exchange with reactive surface hydroxyl groups; (3) no surface complexes are formed with background electrolyte ions; (4) the relationship between surface charge, σ ($mol_c.L^{-1}$), and surface potential, ψ (V), is linear and is given by

$$\sigma = \frac{CSa}{F}\psi \tag{3.4}$$

where C ($F.m^{-2}$) is the capacitance, S ($m^2.g^{-1}$) is the surface area, a ($g.L^{-1}$) is the solid concentration, and F ($C.mol_c^{-1}$) is the Faraday constant. The structure of the solid–solution interface for the constant capacitance model is depicted in Figure 3.1.

In the constant capacitance model, the protonation and dissociation reactions of the reactive surface site are

$$SOH + H^+ \leftrightarrow SOH_2^+ \tag{3.5}$$

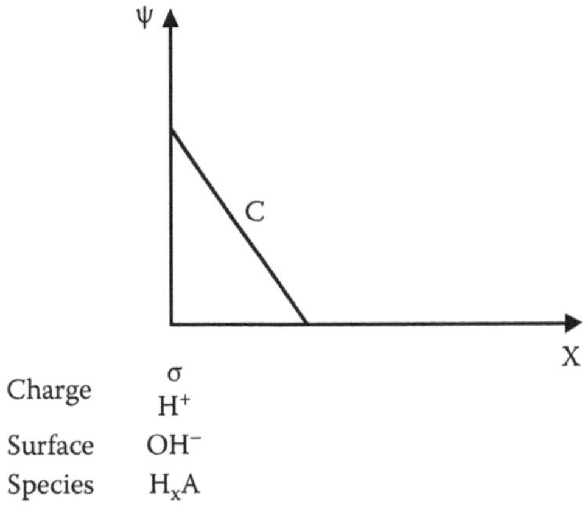

FIGURE 3.1
Structure of the solid–solution interface for the constant capacitance model.

$$SOH \leftrightarrow SO^- + H^+ \tag{3.6}$$

For soils, the SOH reactive site is generic, representing reactive hydroxyl groups including aluminol and silanol groups on the edges of clay mineral particles. The surface complexation reactions for the adsorption of arsenate and selenite are

$$SOH + H_3AsO_4 \leftrightarrow SH_2AsO_4 + H_2O \tag{3.7}$$

$$SOH + H_3AsO_4 \leftrightarrow SHAsO_4^- + H^+ + H_2O \tag{3.8}$$

$$SOH + H_3AsO_4 \leftrightarrow SAsO_4^{2-} + 2H^+ + H_2O \tag{3.9}$$

$$SOH + H_2SeO_3 \leftrightarrow SSeO_3^- + H^+ + H_2O \tag{3.10}$$

The equilibrium constants for the above reactions are

$$K_+(\text{int}) = \frac{[SOH_2^+]}{[SOH][H^+]}\exp(F\psi / RT) \tag{3.11}$$

$$K_-(\text{int}) = \frac{[SO^-][H^+]}{[SOH]}\exp(-F\psi / RT) \tag{3.12}$$

$$K_{As}^1(\text{int}) = \frac{[SH_2AsO_4]}{[SOH][H_3AsO_4]} \tag{3.13}$$

$$K_{As}^2(\text{int}) = \frac{[SHAsO_4^-][H^+]}{[SOH][H_3AsO_4]}\exp(-F\psi / RT) \tag{3.14}$$

$$K_{As}^3(\text{int}) = \frac{[SAsO_4^{2-}][H^+]^2}{[SOH][H_3AsO_4]}\exp(-2F\psi / RT) \tag{3.15}$$

$$K_{Se}(\text{int}) = \frac{[SSeO_3^-][H^+]}{[SOH][H_2SeO_3]}\exp(-F\psi / RT) \tag{3.16}$$

In the constant capacitance model, charged surface complexes create an average electric potential field at the solid surface. These coulombic forces provide the dominant contribution to the solid-phase activity coefficients because the contribution from other forces is considered equal for all surface complexes. In this manner, the exponential terms can be considered solid-phase activity coefficients that correct for the charges on the surface complexes (Sposito, 1983).

The mass balance of the surface reactive site for arsenate adsorption is

$$\begin{aligned}[SOH]_T &= [SOH] + [SOH_2{}^+] + [SO^-] + [SH_2AsO_4] \\ &\quad + [SHAsO_4{}^-] + [SAsO_4{}^{2-}]\end{aligned} \tag{3.17}$$

and for selenite adsorption is

$$[SOH]_T = [SOH] + [SOH_2{}^+] + [SO^-] + [SSeO_3{}^-] \tag{3.18}$$

where SOH_T is the total number of reactive sites. Charge balance for arsenate adsorption is

$$\sigma = [SOH_2^+] - [SO^-] - [SHAsO_4^-] - 2[SAsO_4^{2-}] \tag{3.19}$$

and for selenite adsorption it is

$$\sigma = [SOH_2^+] - [SO^-] - [SSeO_3^-] \tag{3.20}$$

The above systems of equations can be solved using a mathematical approach. The computer program FITEQL 3.2 (Herbelin and Westall, 1996) is an iterative nonlinear least squares optimization program that was used to fit equilibrium constants to experimental adsorption data using the constant capacitance model. The FITEQL 3.2 program was also used to predict chemical surface speciation using previously determined equilibrium constant values. The stoichiometry of the equilibrium problem for the application of the constant capacitance model to arsenate and selenite adsorption is shown in Table 3.1. The assumption that arsenate and selenite adsorptions take place on only one set of reactive surface sites is clearly a gross simplification because soils are complex multisite mixtures of a variety of reactive sites. Therefore, the surface complexation constants determined for soils will

TABLE 3.1

Stoichiometry of the Equilibrium Problem for the Constant Capacitance Model

	Components			
Species	**SOH**	**$e^{F\psi/RT}$**	**H_xA[a]**	**H^+**
H^+	0	0	0	1
OH^-	0	0	0	−1
SOH_2^+	1	1	0	1
SOH	1	0	0	0
SO^-	1	−1	0	−1
H_3AsO_4	0	0	1	0
$H_2AsO_4^-$	0	0	1	−1
$HAsO_4^{2-}$	0	0	1	−2
AsO_4^{3-}	0	0	1	−3
H_2SeO_3	0	0	1	0
$HSeO_3^-$	0	0	1	−1
SeO_3^{2-}	0	0	1	−2
SH_2AsO_4	1	0	1	0
$SHAsO_4^-$	1	−1	1	−1
$SAsO_4^{2-}$	1	−2	1	−2
$SSeO_3^-$	1	−1	1	−1

[a] A is an anion and x is the number of protons in the undissociated form of the acid.

be average composite values that include soil mineralogical characteristics and competing ion effects.

The total number of reactive surface sites, SOH_T (mol.L^{-1}), is an important input parameter in the constant capacitance model. It is related to the reactive surface site density, N_s (sites.nm^{-2}):

$$SOH_T = \frac{Sa10^{18}}{N_A} N_s \tag{3.21}$$

where N_A is Avogadro's number. Values of surface site density can be determined using a wide variety of experimental methods, including potentiometric titration and maximum ion adsorption. Site density results can vary by an order of magnitude between methods. The ability of the constant capacitance model to describe anion adsorption is dependent on the reactive site density (Goldberg, 1991). A surface site density value of 2.31 sites per nm^{-2} was recommended for natural materials by Davis and Kent (1990). This value has been used in the constant capacitance model to describe selenite adsorption by soils. A reactive surface site number of 21.0 μmol.L^{-1} was used to describe arsenate adsorption by soils.

For application of the constant capacitance model to soils, the capacitance value was chosen from the literature. To describe arsenate and selenite adsorption, the capacitance was set at 1.06 F.m^{-2}, considered optimum for aluminum oxide by Westall and Hohl (1980). For the development of self-consistent parameter databases, a constant value of capacitance is necessary.

Protonation and dissociation constant values were obtained from the literature for arsenate adsorption and by computer optimization for selenite adsorption. To describe arsenate adsorption, the protonation constant, $\log K_{+}$, was set to 7.35 and the dissociation constant, $\log K_{-}$ was set to −8.95. These values are averages of a literature compilation of protonation–dissociation constants for aluminum and iron oxides obtained by Goldberg and Sposito (1984).

Surface complexation constants for arsenate and selenite adsorption by soils were obtained using computer optimization. An evaluation of the goodness of model fit can be obtained from the overall variance *V* in *Y*:

$$V_Y = \frac{SOS}{DF} \tag{3.22}$$

where SOS is the weighted sum of squares of the residuals and *DF* represents the degrees of freedom.

Recently, general prediction models have been developed to obtain anion surface complexation constants from easily measured soil chemical properties—cation exchange capacity, surface area, organic carbon, inorganic carbon, Fe oxide, and Al oxide content—that correlate with soil adsorption capacity for trace element anions. This approach has been successfully applied to predict adsorption of arsenate (Goldberg et al., 2005) and selenite (Goldberg, Lesch, and Suarez, 2007) adsorption by soils. In this approach, arsenate and selenite adsorption behavior by soils is predicted independently of experimental adsorption measurements using the general prediction models.

Trace element adsorption was investigated using 53 surface and subsurface samples from soils belonging to 6 different soil orders chosen to provide a wide range of soil chemical properties. Soil chemical characteristics are listed in Table 3.2. Soils Altamont to Yolo constitute a set of 21 soil series from the southwestern Unites States, primarily California. This set of soils consists mainly of alfisols and entisols. Soils Bernow to Teller constitute a set of 17 soil series from the midwestern United States, primarily Oklahoma. This set of soils consists mainly of mollisols.

Soil pH values were measured in 1:5 soil:deionized water extracts (Thomas, 1996). Cation exchange capacities were obtained by sodium saturation and magnesium extraction (Rhoades, 1982). Ethylene glycol monoethyl ether adsorption was used to determine surface areas (Cihacek and Bremner, 1979). Free Fe and Al oxides were extracted with citrate buffer and hydrosulfite (Coffin, 1963); Al and Fe concentrations were measured using inductively coupled plasma optical emission spectrometry (ICP-OES). Carbon contents were obtained using a carbon coulometer. Organic C was calculated as the difference between total C measured by combustion at 950°C and inorganic C determined by acidification and heating (Goldberg et al., 2005, Goldberg, Lesch, and Suarez, 2007).

Adsorption experiments were carried out in batch systems to determine adsorption envelopes, the amount of ion adsorbed as a function of solution pH at fixed total ion mass. One-gram samples of soil were equilibrated with 25 mL of 0.1 M NaCl background electrolyte solution on a shaker for 2 hours. The equilibrating solution contained 20 μmol.L^{-1} of As(V) or Se(IV) and had been adjusted to the desired pH range of 2 to 10 using 1 M HCl or 1 M NaOH. After reaction, the samples were centrifuged, decanted, analyzed for pH, filtered, and analyzed for As or Se concentration using ICP-OES. Additional experimental details are provided in Goldberg et al. (2005) for As(V) and in Goldberg, Lesch, and Suarez (2007) for Se(IV) adsorption.

Arsenate adsorption envelopes were determined for 27 southwestern and 22 midwestern soil samples. Arsenate adsorption increased with increasing solution pH, exhibited an adsorption maximum in the pH range 6 to 7, and decreased with further increases in solution pH, as can be seen in Figure 3.2. Selenite adsorption envelopes were determined for 23 southwestern and 22 midwestern soil samples. Selenite adsorption decreased with increasing solution pH over the pH range 2 to 10, as can be seen in Figure 3.3.

The constant capacitance model was fit to the As(V) adsorption envelopes by optimizing the three As(V) surface complexation constants(log K^1_{As}, log K^2_{As}, and log K^3_{As}) simultaneously. The protonation constant was fixed at log

TABLE 3.2
Chemical Characteristics of Soils

Soil Series	Depth cm	pH	CEC $mmol_c$ kg^{-1}	S $m^2.g^{-1}$	IOC $g.kg^{-1}$	OC $g.kg^{-1}$	Fe $g.kg^{-1}$	Al $g.kg^{-1}$
Altamont	0–25	5.90	152	103	0.0099	9.6	7.7	0.58
	25–51	5.65	160	114	0.011	6.7	8.2	0.64
	0–23	6.20	179	109	0.12	30.8	9.2	0.88
Arlington	0–25	8.17	107	61.1	0.30	4.7	8.2	0.48
	25–51	7.80	190	103	0.16	2.8	10.1	0.60
Avon	0–15	6.91	183	60.1	0.083	30.8	4.3	0.78
Bonsall	0–25	5.88	54	15.7	0.13	4.9	9.3	0.45
	25–51	5.86	122	32.9	0.07	2.1	16.8	0.91
Chino	0–15	10.2	304	159	6.4	6.2	4.7	1.64
Diablo	0–15	7.58	301	190	0.26	19.8	7.1	1.02
	0–15	7.42	234	130	2.2	28.3	5.8	0.84
Fallbrook	0–25	6.79	112	68.3	0.023	3.5	6.9	0.36
	25–51	6.35	78	28.5	0.24	3.1	4.9	0.21
Fiander	0–15	9.60	248	92.5	6.9	4.0	9.2	1.06
Haines	20	9.05	80	59.5	15.8	14.9	1.7	0.18
Hanford	0–10	8.40	111	28.9	10.1	28.7	6.6	0.35
Holtville	61–76	8.93	58	43.0	16.4	2.1	4.9	0.27
Imperial	15–46	8.58	198	106	17.9	4.5	7.0	0.53
Nohili	0–23	8.03	467	286	2.7	21.3	49.0	3.7
Pachappa	0–25	6.78	39	15.1	0.026	3.8	7.6	0.67
	25–51	7.02	52	41.0	0.014	1.1	7.2	0.35
	0–20	8.98	122	85.8	0.87	3.5	5.6	0.86
Porterville	0–7.6	6.83	203	137	0.039	9.4	10.7	0.90
Ramona	0–25	5.89	66	27.9	0.02	4.4	4.5	0.42
	25–51	6.33	29	38.8	0.018	2.2	5.9	0.40
Reagan	Surface	8.39	98	58.8	18.3	10.1	4.6	0.45
Ryepatch	0–15	7.98	385	213	2.5	32.4	2.6	0.92
Sebree	0–13	5.99	27	21.2	0.0063	2.2	6.0	0.46
Wasco	0–5.1	5.01	71	30.9	0.009	4.7	2.4	0.42
Wyo		6.26	155	53.9	0.014	19.9	9.5	0.89
Yolo	0–15	8.43	177	73.0	0.23	11.5	15.6	1.13
Bernow	B	4.15	77.6	46.4	0.0028	3.8	8.1	1.1
Canisteo	A	8.06	195	152	14.8	34.3	1.7	0.44
Dennis	A	5.27	85.5	40.3	0.0014	18.6	12.9	1.7
	B	5.43	63.1	72.4	0.0010	5.2	30.0	4.1
Dougherty	A	4.98	3.67	241	0.0010	7.0	1.7	0.28
Hanlon	A	7.41	142	58.7	2.6	15.1	3.7	0.45
Kirkland	A	5.05	154	42.1	0.014	12.3	5.6	0.80
Luton	A	6.92	317	169	0.099	21.1	9.1	0.99
Mansic	A	8.32	142	42.2	16.7	10.1	2.7	0.40
	B	8.58	88.1	35.5	63.4	9.0	1.1	0.23

TABLE 3.2 (Continued)

Chemical Characteristics of Soils

Soil Series	Depth cm	pH	CEC $mmol_c$ kg^{-1}	S $m^2.g^{-1}$	IOC $g.kg^{-1}$	OC $g.kg^{-1}$	Fe $g.kg^{-1}$	Al $g.kg^{-1}$
Norge	A	3.86	62.1	21.9	0.0010	11.6	6.1	0.75
Osage	A	6.84	377	134	0.59	29.2	15.9	1.4
	B	6.24	384	143	0.0100	18.9	16.5	1.3
Pond Creek	A	4.94	141	35.4	0.0023	16.6	5.2	0.70
	B	6.78	106	59.6	0.016	5.0	5.1	0.81
Pratt	A	5.94	23.9	12.3	0.0026	4.2	1.2	0.18
	B	5.66	23.3	117	0.0007	2.1	0.92	0.13
Richfield	B	7.12	275	82.0	0.040	8.0	5.4	0.76
Summit	A	7.03	374	218	0.25	26.7	16.2	2.3
	B	6.23	384	169	0.0079	10.3	17.8	2.5
Taloka	A	4.88	47.4	87.0	0.0021	9.3	3.6	0.62
Teller	A	4.02	43.1	227	0.0008	6.8	3.2	0.53

Source: Adapted from Goldberg, S., S.M. Lesch, D.L. Suarez, and N.T. Basta. 2005. *Soil Sci. Soc. Am. J.* 69: 1389–1398; Goldberg, S., S.M. Lesch, and D.L. Suarez. 2007. *Geochim. Cosmochim. Acta* 71: 5750–5762.

$K_+ = 7.35$ and the dissociation constant was fixed at $\log K_- = -8.95$. The ability of the model to describe the As(V) adsorption data was very good. Figure 3.2 shows some model fits for surface horizons. Values of the optimized As(V) surface complexation constants are provided in Table 3.3. For the Bernow, Canisteo, Summit B, and Nohili soils, only two surface complexation constants were optimized because $\log K^2_{As}$ did not converge.

The constant capacitance model was fit to the Se(IV) adsorption envelopes by optimizing the Se(IV) surface complexation constant ($\log K^2_{Se}$) because $\log K^1_{Se}$ converged for only five of the soils. Subsequently, to improve the model fit, the protonation constant ($\log K_+$) and the dissociation constant ($\log K_-$) were simultaneously optimized with $\log K^2_{Se}$. Because initial optimizations indicated that the deprotonated surface species was present in only trace amounts ($\log K_- < -39$), it was omitted from the final optimizations. The ability of the model to describe the Se(IV) adsorption data was very good. Figure 3.3 shows some model fits for surface horizons. Values of the optimized Se(IV) surface complexation constants are provided in Table 3.4. Optimized constants are not listed for soils having >1% inorganic C because the model could not converge $\log K^2_{Se}$ and $\log K_+$ simultaneously.

A general regression modeling approach was used to relate the constant capacitance model As(V) and Se(IV) surface complexation constants to the following soil chemical properties: cation exchange capacity, surface area, inorganic carbon content, organic carbon content, iron oxide content, and aluminum oxide content. An exploratory data analysis revealed that the As(V) and Se(IV) surface complexation constants were linearly related to

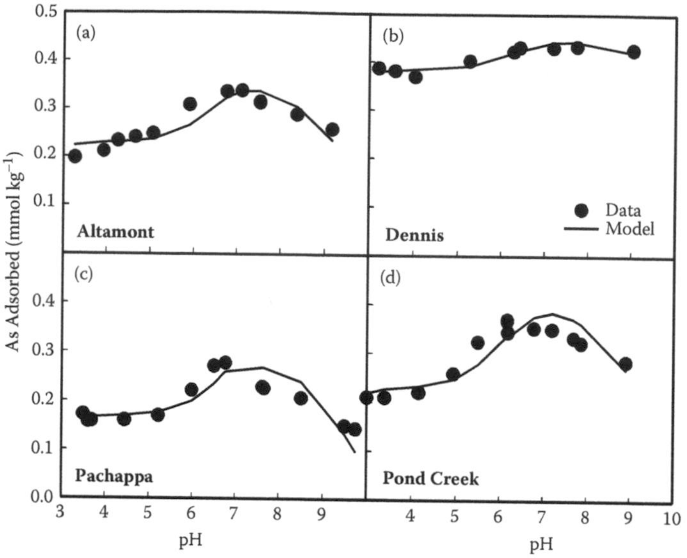

FIGURE 3.2
Fit of the constant capacitance model to As(V) adsorption on southwestern soils (a, c) and midwestern soils (b, d): (a) Altamont soil, (b) Dennis soil, (c) Pachappa soil, and (d) Pond Creek soil. Circles represent experimental data. Model fits are represented by solid lines. (*Source:* Adapted from Goldberg, S., S.M. Lesch, D.L. Suarez, and N.T. Basta. 2005. *Soil Sci. Soc. Am. J.* 69: 1389–1398.)

each of the log transformed chemical properties. Therefore, the following initial regression model was specified for each of the As(V) and Se(IV) surface complexation constants:

$$LogK_j = \beta_{0j} + \beta_{1j}(\ln CEC) + \beta_{2j}(\ln SA) + \beta_{3j}(\ln(IOC) + \beta_{4j}(\ln OC) + \beta_{5j}(\ln Fe) + \beta_{6j}(\ln Al) + \varepsilon \tag{3.23}$$

where the β_{ij} represent regression coefficients, ε represents the residual error component, and $j = 1, 2$ for southwestern and midwestern soils, respectively.

For As(IV) and Se(IV), an initial analysis of the regression model presented by Equation (3.23) yielded rather poor results when the two sets of soils were considered together. Additional statistical analyses revealed that the midwestern and southwestern soils represented two distinct populations exhibiting different soil property and surface complexation constant relationships. A multivariate analysis of covariance established a common intercept and common ln(CEC) term for the general regression prediction equations for As(V) adsorption. For Se(IV) adsorption, a multivariate analysis of covariance established a common ln(SA) term for log K^2_{Se} and a common ln(Fe) term for log K_+.

TABLE 3.3
Constant Capacitance Model Surface Complexation Constants for As(V) Adsorption

Soil Series	Depth (cm)	Optimized $LogK^1_{As}$	Optimized $LogK^2_{As}$	Optimized $LogK^3_{As}$	Predicted $LogK^1_{As}$	Predicted $LogK^2_{As}$	Predicted $LogK^3_{As}$
Altamont	0–25	9.99	3.92	−3.89	9.88	3.56	−4.10
	25–51	10.09	4.41	−3.69	9.98	3.56	−4.08
Arlington	0–25	9.57	2.04	−4.59	9.99	3.20	−4.15
	25–51	9.95	2.92	−4.24	10.20	3.33	−4.10
Avon	0–15	9.29	3.02	−4.52	9.38	3.22	−4.42
Bonsall	0–25	9.50	2.77	−4.64	9.92	3.20	−4.15
	25–51	10.90	3.64	−3.27	10.47	3.56	−3.82
Diablo	0–15	9.26	3.14	−4.45	9.90	3.51	−3.96
Fallbrook	0–25	10.02	3.00	−4.20	9.93	3.30	−4.27
	25–51	9.56	2.84	−4.62	9.68	2.85	−4.57
Fiander	0–15	10.08	2.10	-4.76	10.14	3.06	−4.14
Haines	20	9.43	2.39	−4.01	9.33	2.60	−4.45
Hanford	0–10	9.83	3.04	−4.16	9.53	2.92	−4.28
Holtville	61–76	10.32	3.66	−3.88	9.97	2.67	−4.26
Imperial	15–46	10.23	3.77	−3.77	10.10	2.98	−4.08
Nohili	0–23	12.82		−2.21	10.74	4.04	−3.17
Pachappa	0–25	9.67	3.55	−4.15	9.91	3.26	−4.15
	25–51	10.03	3.06	−4.44	10.05	3.16	−4.33
Porterville	0–7.6	10.36	3.89	−3.60	10.14	3.68	−3.84
Ramona	0–25	9.58	2.79	−4.37	9.55	3.02	−4.59
	25–51	9.96	2.99	−4.46	9.93	3.19	−4.18
Reagan	Surface	9.66	2.94	−4.07	9.74	2.84	−4.18
Ryepatch	0–15	9.40	3.07	−4.70	9.50	3.11	−4.26
Sebree	0–13	9.64	3.28	−4.70	9.76	3.11	−4.41
Wasco	0–5.1	9.65	3.31	−4.45	9.46	3.04	−4.59
Wyo		10.36	3.67	−3.80	9.79	3.62	−4.06
Yolo	0–15	10.00	3.96	−3.86	10.09	3.51	−3.93
Bernow	B	12.84		−1.78	11.21	5.29	−2.40
Canisteo	A	10.54		−3.70	9.39	2.21	−5.11
Dennis	A	10.99	5.02	−2.57	11.06	5.28	−2.77
	B	12.51	6.93	−0.73	12.50	6.97	−0.83
Dougherty	A	9.49	3.23	−4.21	9.69	3.21	−4.13
Hanlon	A	10.11	3.18	−4.17	10.35	3.61	−3.66
Kirkland	A	10.44	5.25	−3.14	10.42	4.26	−3.60
Luton	A	10.46	4.46	−3.31	10.96	4.66	−3.23
Mansic	A	9.71	3.05	−4.24	10.26	3.34	−3.65
	B	10.21	2.58	−3.65	9.47	2.19	−4.53
Norge	A	10.31	3.90	−3.99	10.37	4.46	−3.47
Osage	A	11.75	5.08	−2.63	11.55	5.27	−2.49
	B	12.26	5.97	−1.86	11.43	5.50	−2.66

—continued

TABLE 3.3 (Continued)

Constant Capacitance Model Surface Complexation Constants for As(V) Adsorption

Soil Series	Depth (cm)	Optimized $LogK^1_{As}$	Optimized $LogK^2_{As}$	Optimized $LogK^3_{As}$	Predicted $LogK^1_{As}$	Predicted $LogK^2_{As}$	Predicted $LogK^3_{As}$
Pond Creek	A	10.02	4.44	−3.86	10.09	4.04	−4.05
	B	10.85	4.98	−3.01	10.73	4.53	−3.09
Pratt	A	9.14	2.56	−4.78	9.09	2.73	−4.75
	B	9.26	2.55	−4.64	9.17	2.69	−4.87
Richfield	B	10.00	3.85	−4.17	10.62	4.34	−3.45
Summit	A	11.65		−2.57	11.58	5.34	−2.51
	B	13.14		−1.61	11.73	5.85	−2.24
Taloka	A	10.25	4.11	−3.89	10.09	3.88	−3.92
Teller	A	10.20	3.61	−4.34	10.10	3.87	−3.99

Source: Adapted from Goldberg, S., S.M. Lesch, D.L. Suarez, and N.T. Basta. 2005. *Soil Sci. Soc. Am. J.* 69: 1389–1398.

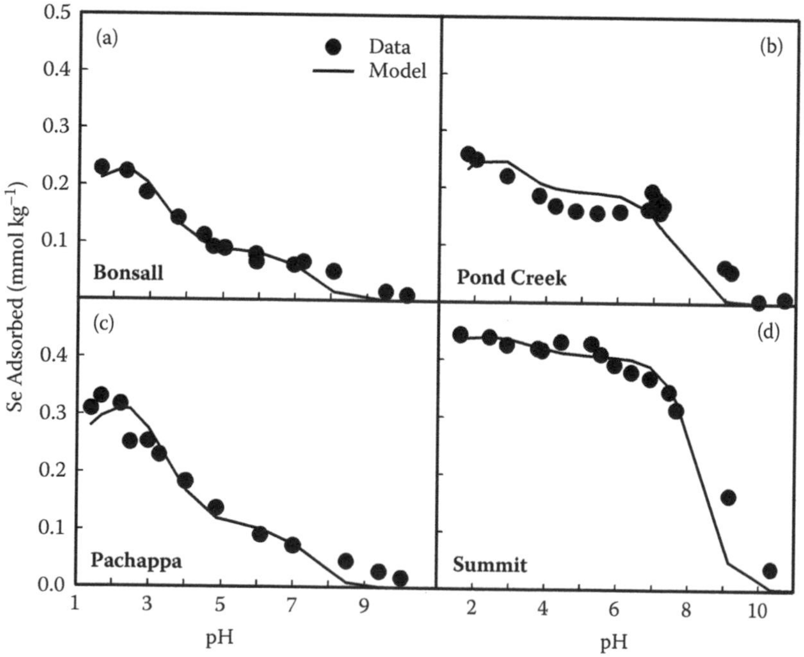

FIGURE 3.3

Fit of the constant capacitance model to Se(IV) adsorption on southwestern soils (a, c) and midwestern soils (b, d): (a) Bonsall soil; (b) Pond Creek soil; (c) Pachappa soil; (d) Summit soil. Circles represent experimental data. Model fits are represented by solid lines. (*Source:* Adapted from Goldberg, S., S.M. Lesch, and D.L. Suarez. 2007. *Geochim. Cosmochim. Acta* 71: 5750–5762.)

TABLE 3.4
Constant Capacitance Model Surface Complexation Constants for Se(IV) Adsorption

Soil Series	Depth (cm)	Optimized $LogK^2_{Se}$	Optimized $LogK_+$	Predicted $LogK^2_{Se}$	Predicted $LogK_+$
Altamont	0–20	−1.42	2.72	−1.23	2.79
Arlington	0–25	−1.04	2.82	−0.88	2.81
Avon	0–15	−1.24	2.27	−1.21	2.16
Bonsall	0–25	−0.90	2.80	−0.62	2.81
Chino	0–15	−1.57	2.96	−1.42	2.72
Diablo	0–15	−1.57	2.11	−1.47	2.68
	0–15	−1.75	2.85	−1.41	2.77
Fallbrook	25–51	−0.73	2.53	−0.70	2.38
Fiander	0–15	−1.01	3.20	−0.99	3.25
Haines	20			−1.40	2.04
Hanford	0–10	−0.53	2.68	−0.81	3.04
Holtville	61–76			−0.82	2.87
Imperial	15–46			−1.13	3.15
Nohili	0–23			−1.10	4.44
Pachappa	0–25		2.17	−0.69	2.47
	25–51	−0.53	2.50	−0.65	2.35
	0–20	−1.22	3.10	−1.09	2.63
Porterville	0–7.6	−1.04	2.86	−1.26	2.78
Reagan	Surface			−1.09	2.83
Sebree	0–13	−0.47	2.11	−0.53	2.12
Wasco	0–5.1	−0.93	1.28	−1.19	1.45
Wyo		−0.58	2.79	−1.05	2.57
Yolo	0–15	−0.79	3.28	−0.85	3.27
Bernow	B	0.62	2.85	0.59	2.24
Canisteo	A			1.77	1.77
Dennis	A	0.31	2.36	0.38	2.25
	B			1.63	2.19
Dougherty	A	−1.77	2.11	−1.94	2.14
Hanlon	A	−0.92	2.50	−0.96	2.37
Kirkland	A	−0.06	1.76	−0.31	2.21
Luton	A	−0.79	2.33	−0.59	2.41
Mansic	A			−0.97	2.21
	B			−1.78	1.96
Norge	A	−0.24	2.08	0.06	2.33
Osage	A	0.15	2.13	−0.08	2.57
	B	0.22	3.24	0.14	2.65
Pond Creek	A	-0.44	2.07	−0.46	2.26
	B	0.03	1.89	−0.11	2.13
Pratt	A	−0.99	2.56	−0.92	2.23
	B	−1.59	2.00	−1.73	2.29
Richfield	B	−0.77	2.22	−0.40	2.23
Summit	A	−0.14	2.66	−0.20	2.18
	B	0.49	1.54	0.44	2.18
Taloka	A	−0.88	2.38	−0.91	2.08
Teller	A	−1.46	2.15	−1.25	2.11

Source: Adapted from Goldberg, S., S.M. Lesch, and D.L. Suarez. 2007. *Geochim. Cosmochim. Acta* 71: 5750–5762.

The prediction equations for obtaining As(V) surface complexation constants to describe As(V) adsorption with the constant capacitance model are

$$LogK^1_{As} = 10.64 - 0.107\ln(CEC) + 0.094\ln(SA) + 0.078\ln(IOC) - 0.365\ln(OC) + 1.09\ln(Fe) \quad (3.24)$$

$$LogK^2_{As} = 3.39 - 0.083\ln(CEC) + 0.018\ln(SA) - 0.002\ln(IOC) - 0.400\ln(OC) + 1.36\ln(Fe) \quad (3.25)$$

$$LogK^3_{As} = -2.58 - 0.296\ln(CEC) - 0.004\ln(SA) + 0.115\ln(IOC) - 0.570\ln(OC) + 1.38\ln(Fe) \quad (3.26)$$

for midwestern soils,

$$LogK^1_{As} = 10.65 - 0.107\ln(CEC) + 0.256\ln(SA) + 0.022\ln(IOC) - 0.143\ln(OC) + 0.385\ln(Fe) \quad (3.27)$$

$$LogK^2_{As} = 3.39 - 0.083\ln(CEC) + 0.247\ln(SA) - 0.061\ln(IOC) - 0.104\ln(OC) + 0.313\ln(Fe) \quad (3.28)$$

$$LogK^3_{As} = -2.58 - 0.296\ln(CEC) - 0.376\ln(SA) + 0.024\ln(IOC) - 0.085\ln(OC) + 0.363\ln(Fe) \quad (3.29)$$

and for southwestern soils.

The prediction equations for obtaining Se(IV) surface complexation constants to describe Se(IV) adsorption with the constant capacitance model are

$$LogK^2_{Se} = 0.675 - 0.380\ln(SA) - 0.083\ln(OC) + 0.274\ln(Fe) \quad (3.30)$$

$$LogK_{+} = 3.36 + 0.115\ln(IOC) + 0.774\ln(Fe) \quad (3.31)$$

for midwestern soils

$$LogK_{Se}^{2} = 1.18 - 0.380\ln(SA) - 0.470\ln(OC) + 1.03\ln(Fe) \quad (3.32)$$

$$LogK_{+} = 0.613 + 0.774\ln(Fe) - 0.811\ln(Al) \quad (3.33)$$

for southwestern soils.

Surface complexation constants obtained with the prediction equations are provided in Table 3.3 for As(V) and Table 3.4 for Se(IV). A "jack-knifing" procedure was performed on Equations (3.24) to (3.33) to assess their predictive ability. Jack-knifing is a technique where each observation is sequentially set aside, the equation is reestimated without this observation, and the set-aside observation is then predicted from the remaining data using the reestimated equation. A jack-knifing procedure indicated good general agreement between ordinary predictions and jack-knife estimates for both As(V) and Se(IV) surface complexation constants, suggesting predictive capabilities. Additional details on statistical analyses are provided in Goldberg et al. (2005) for As(V) and Goldberg, Lesch, and Suarez (2007) for Se(IV).

The As(V) surface complexation constants were predicted using the prediction Equations (3.24) to (3.26) for midwestern soils and Equations (3.27) to (3.29) for southwestern soils. These surface complexation constants were then used in the constant capacitance model to predict As(V) adsorption on the soils. The ability of the model to predict As(V) adsorption is shown in Figure 3.4 for two soils not used to obtain the prediction equations. Prediction of As(V) adsorption by the Summit soil was good. Prediction of As(V) adsorption by the Nohili soil deviated from the experimental adsorption data by 30% or less, a reasonable result considering that the prediction was obtained without optimization of any adjustable parameters.

The Se(IV) surface complexation constants were predicted using the prediction Equations (3.30) and (3.31) for midwestern soils and Equations (3.32) and (3.33) for southwestern soils. These surface complexation constants were then used in the constant capacitance model to predict Se(IV) adsorption on the soils. The ability of the model to predict Se(IV) adsorption is shown in Figure 3.5 for four soils. For the Pond Creek and Summit soils, the prediction equations provide descriptions of the experimental data that are comparable in quality to the optimized fits (compare Figures 3.3b with 3.5b and 3.3d with 3.5d). For the Bonsall and Pachappa soils, the prediction equations describe the experimental data less closely than the optimized fits (compare Figure 3.3a with 3.5a and Figure 3.3c with 3.5c). However, the model always correctly predicts the shape of the adsorption envelope as a function of solution pH.

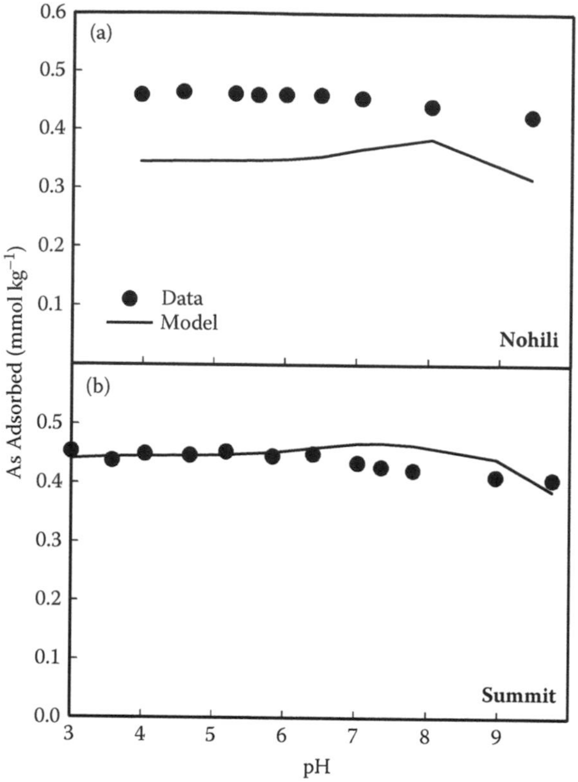

FIGURE 3.4
Constant capacitance model prediction of As(V) adsorption by soils not used to obtain the prediction equations: (a) Nohili soil and (b) Summit soil. Experimental data are represented by circles. Model predictions are represented by solid lines. (*Source:* Adapted from Goldberg, S., S.M. Lesch, D.L. Suarez, and N.T. Basta. 2005. *Soil Sci. Soc. Am. J.* 69: 1389–1398.)

Predictions of As(V) and Se(IV) adsorption were reasonable for most of the soils. These predictions were obtained independently of any experimental measurement of As(V) or Se(IV) adsorption on these soils, that is, solely from values of a few easily measured chemical properties. Incorporation of these prediction equations into chemical speciation transport models should allow simulation of As(V) and Se(IV) concentrations in soil solution under diverse environmental conditions.

Summary

The chemical processes and soil factors that affect the concentrations of As and Se in soil solution were discussed. Both elements occur in two redox

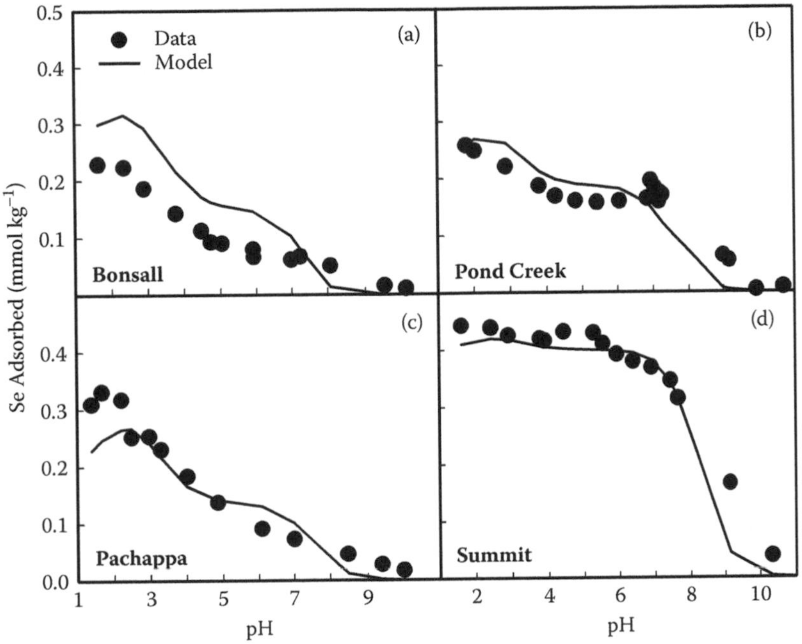

FIGURE 3.5
Constant capacitance model prediction of Se(IV) adsorption by southwestern soils (a, c) and midwestern soils (b, d): (a) Bonsall soil, (b) Pond Creek soil, (c) Pachappa soil, and (d) Summit soil. Experimental data are represented by circles. Model predictions are represented by solid lines. (*Source:* Adapted from Goldberg, S., S.M. Lesch, and D.L. Suarez. 2007. *Geochim. Cosmochim. Acta* 71: 5750–5762.)

states differing in toxicity and reactivity. Methylation and volatilization reactions occur in soils and can act as detoxification pathways. Precipitation-dissolution reactions control As and Se concentrations only in highly contaminated situations. Oxidation-reduction reactions determine which redox state is thermodynamically stable, although more than one redox state may be present in soils. Adsorption-desorption reactions most often control the dissolved As and Se concentrations. Empirical and chemical models of adsorption were discussed. The constant capacitance model, a chemical surface complexation model, has been used to predict As and Se adsorption under changing conditions of solution pH and equilibrium ion concentration.

References

Adamson, A.W. 1976. *Physical chemistry of surfaces. 3rd ed.* John Wiley & Sons, New York, NY.

Adriano, D.C. 1986. *Trace elements in the terrestrial environment.* Springer-Verlag, New York, NY.

Amirbahman, A., D.B. Kent, G.P. Curtis, and J.A. Davis. 2006. Kinetics of sorption and abiotic oxidation of arsenic(III) by aquifer materials. *Geochim. Cosmochim. Acta* 70: 533–547.

Arai, Y., E.J. Elzinga, and D.L. Sparks. 2001. X-ray absorption spectroscopic investigation of arsenite and arsenate adsorption at the aluminum oxide-water interface. *J. Colloid Interface Sci.* 235: 80–88.

Catalano, J.G., Z. Zhang, P. Fenter, and M.J. Bedzyk. 2006. Inner-sphere adsorption geometry of Se(IV) at the hematite (100)-water interface. *J. Colloid Interface Sci.* 297: 665–671.

Catalano, J.G., Z. Zhang, C. Park, P. Fenter, and M.J. Bedzyk. 2007. Bridging arsenate surface complexes on the hematite (012) surface. *Geochim. Cosmochim. Acta* 71: 1883–1897.

Catalano, J.G., C. Park, P. Fenter, and Z. Zhang. 2008. Simultaneous inner- and outer-sphere arsenate adsorption on corundum and hematite. *Geochim. Cosmochim. Acta* 72: 1986–2004.

Cihacek, L.J., and J.M. Bremner. 1979. A simplified ethylene glycol monoethyl ether procedure for assessing soil surface area. *Soil Sci. Soc. Am. J.* 43: 821–822.

Coffin, D.E. 1963. A method for the determination of free iron oxide in soils and clays. *Can. J. Soil Sci.* 43: 1–17.

Cooke, T.D., and K.W. Bruland. 1987. Aquatic chemistry of selenium: Evidence of biomethylation. *Environ. Sci. Technol.* 21: 1214–1219.

Cullen, W.R., and K.J. Reimer. 1989. Arsenic in the environment. *Chem. Rev.* 89: 713–714.

Davis, J.A., and D.B. Kent. 1990. Surface complexation modeling and aqueous geochemistry. *Rev. Mineral. 23*:177–260.

De Brouwere, K., E. Smolders, and R. Merckx. 2004. Soil properties affecting solid-liquid distribution of As(V) in soils. *Eur. J. Soil Sci.* 55: 165–173.

Del Debbio, J.A. 1991. Sorption of strontium, selenium, cadmium, and mercury in soil. *Radiochim. Acta* 52/53: 181–186.

Doran, J.W., and M. Alexander. 1977. Microbial formation of volatile selenium compounds in soil. *Soil Sci. Soc. Am. J.* 41: 70–73.

Elkhatib, E.A., O.L. Bennett, and R.J. Wright. 1984. Arsenite sorption and desorption in soils. *Soil Sci. Soc. Am. J.* 48: 1025–1030.

Elrashidi, M.A., D.C. Adriano, S.M. Workman, and W.L. Lindsay. 1987. Chemical equilibria of selenium in soils: A theoretical development. *Soil Sci.* 144: 141–152.

Elsokkary, I.H. 1980. Selenium distribution, chemical fractionation and adsorption in some Egyptian alluvial and lacustrine soils. *Z. Pflanzenernaehr. Bodenk.* 143: 74–83.

Fendorf, S., M.J. Eick, P. Grossl, and D.L. Sparks. 1997. Arsenate and chromate retention mechanisms of goethite. 1. Surface structure. *Environ. Sci. Technol.* 31: 315–320.

Francis, A.J., J.M. Duxbury, and M. Alexander. 1974. Evolution of dimethylselenide from soils. *Appl. Microbiol.* 28: 248–250.

Frankenberger, W.T., and U. Karlson. 1992. Dissipation of soil selenium by microbial volatilization. pp. 365–381. In D.C. Adriano (Ed.) *Biogeochemistry of trace metals.* Lewis Publishers, Boca Raton, FL.

Freeman, M.C. 1985. The reduction of arsenate to arsenite by an *Anabeana*-bacteria assemblage isolated from the Waikato River. *New Zealand J. Marine Freshwater Res.* 19: 277–282.

Gao, S., and R.G. Burau. 1997. Environmental factors affecting rates of arsine evolution from and mineralization of arsenicals in soil. *J. Environ. Qual.* 26: 753–763.

Gao, S., K.K. Tanji, and S. Goldberg. 1998. Reactivity and transformations of arsenic. pp. 17–38. In L.M. Dudley and J.C. Guitjens (Eds.) *Agroecosystems and the environment. Sources, control, and remediation of potentially toxic, trace element oxyanions.* Pacific Division, AAAS, San Francisco, CA.

Garbarino, J.R., A.J. Bednar, D.W. Rutherford, R.S. Beyer, and R.L. Wershaw. 2003. Environmental fate of roxarsone in poultry litter. I. Degradation of roxarsone during composting. *Environ. Sci. Technol.* 37:1509–1514.

Geering, H.R., E.E. Cary, L.H.P. Jones, and W.H. Allaway. 1968. Solubility and redox criteria for the possible forms of selenium in soils. *Soil Sci. Soc. Am. Proc.* 32: 35–40.

Goldberg, S. 1991. Sensitivity of surface complexation modeling to the surface site density parameter. *J. Colloid Interface Sci.* 145: 1–9.

Goldberg, S., and C.T. Johnston. 2001. Mechanisms of arsenic adsorption on amorphous oxides evaluated using macroscopic measurements, vibrational spectroscopy, and surface complexation modeling. *J. Colloid Interface Sci.* 234: 204–216.

Goldberg, S., and G. Sposito. 1984. A chemical model of phosphate adsorption by soils. I. Reference oxide minerals. *Soil Sci. Soc. Am. J.* 48: 772–778.

Goldberg, S., S.M. Lesch, D.L. Suarez, and N.T. Basta. 2005. Predicting arsenate adsorption by soils using soil chemical parameters in the constant capacitance model. *Soil Sci. Soc. Am. J.* 69: 1389–1398.

Goldberg, S., S.M. Lesch, and D.L. Suarez. 2007. Predicting selenite adsorption by soils using soil chemical parameters in the constant capacitance model. *Geochim. Cosmochim. Acta* 71: 5750–5762.

Harr, J.R. 1978. Biological effects of selenium. pp. 393–426. In F.W. Oehme (Ed.) *Toxicity of heavy metals in the environment, Part 1,* Marcel Dekker, New York, NY.

Herbelin, A.L., and J.C. Westall. 1996. A computer program for determination of chemical equilibrium constants from experimental data. Rep. 96-01, Version 3.2, Dept. of Chemistry, Oregon State Univ., Corvallis, OR.

Hess, R.E., and R.W. Blanchar. 1976. Arsenic stability in contaminated soils. *Soil Sci. Soc. Am. J.* 40: 847–852.

Jiang, W., S. Zheng, X. Shan, M. Feng, Y.-G. Zhu, and R.G. McLaren. 2005. Adsorption of arsenate on soils. Part 1: Laboratory batch experiments using 16 Chinese soils with different physiochemical properties. *Environ. Pollut.* 138: 278–284.

Kim, M.-J., J. Nriagu, and S. Haack. 2000. Carbonate ions and arsenic dissolution by groundwater. *Environ. Sci. Technol.* 34: 3094–3100.

Kinniburgh, D.G. 1985. ISOTHERM. A computer Program for Analyzing Adsorption Data. Report WD/ST/85/02. Version 2.2. British Geological Survey, Wallingford, UK.

Korte, N.E., and Q. Fernando. 1991. A review of arsenic (III) in groundwater. *CRC Crit. Rev. Environ. Control* 21: 1–39.

Lakin, W.H. 1961. *Selenium content of soils. USDA-ARS agricultural handbook.* 200. U.S. Government Printing Office, Washington, DC.

Langner, H.W., and W.P. Inskeep. 2000. Microbial reduction of arsenate in the presence of ferrihydrite. *Environ. Sci. Technol.* 34: 3131–3136.

Lee, J.S., and J.O. Nriagu. 2007. Stability constants for metal arsenates. *Environ. Chem.* 4: 123–133.

Lévesque, M. 1974. Selenium distribution in Canadian soil profiles. *Can. J. Soil Sci.* 54:63–68.

Livesey, N.T., and P.M. Huang. 1981. Adsorption of arsenate by soils and its relation to selected chemical properties and anions. *Soil Sci.* 131: 88–94.

Lucas, F.S., and Hollibaugh. 2001. Response of sediment bacterial assemblages to selenate and acetate amendments. *Environ. Sci. Technol.* 35: 528–534.

Manceau, A., and L. Charlet. 1994. The mechanism of selenate adsorption on goethite and hydrous ferric oxide. *J. Colloid Interface Sci.* 168: 87–93.

Manning, B.A., S.E. Fendorf, and S. Goldberg. 1998. Surface structures and stability of arsenic(III) on goethite: Spectroscopic evidence for inner-sphere complexes. *Environ. Sci. Technol.* 32: 2383–2388.

Masscheleyn, P.H., R.D. Delaune, and W.H. Patrick. 1990. Transformations of selenium as affected by sediment oxidation-reduction potential and pH. *Environ. Sci. Technol.* 24: 91–96.

Masscheleyn, P.H., R.D. Delaune, and W.H. Patrick. 1991a. Effect of redox potential and pH on arsenic speciation and solubility in a contaminated soil. *Environ. Sci. Technol.* 25:1414–1419.

Masscheleyn, P.H., R.D. Delaune, and W.H. Patrick. 1991b. Arsenic and selenium chemistry as affected by sediment redox potential and pH. *J. Environ. Qual.* 20: 522–527.

Morin, G., G. Ona-Nguema, Y. Wang, N. Menguy, F. Juillot, O. Proux, F. Guyot, G. Calas, and G.E. Brown. 2008. Extended x-ray absorption fine structure analysis of arsenite and arsenate adsorption on maghemite. *Environ. Sci. Technol.* 42: 2361–2366.

Myneni, S.C.B., T.K. Tokunaga, and G.E. Brown. 1997. Abiotic selenium redox transformations in the presence of Fe(II,III) oxides. *Science* 278: 1106–1109.

Neal, R.H., and G. Sposito. 1989. Selenate adsorption on alluvial soils. *Soil Sci. Soc. Am. J.* 53: 70–74.

Neuberger, C.S., and G.R. Helz. 2005. Arsenic(III) carbonate complexing. *Appl. Geochem.* 20: 1218–1225.

Ona-Nguema, G., G. Morin, F. Juillot, G. Calas, and G.E. Brown. 2005. EXAFS analysis of arsenite adsorption onto two-line ferrihydrite, hematite, goethite, and lepidocrocite. *Environ. Sci. Technol.* 39: 9147–9155.

Oremland, R.S., N.S. Steinberg, A.S. Maest, L.G. Miller, and J.T. Hollibaugh. 1990. Measurement of in situ rates of selenate removal by dissimilatory bacterial reduction in sediments. *Environ. Sci. Technol.* 24: 1157–1164.

Osborne, F.H., and H.L. Ehrlich. 1976. Oxidation of arsenite by a soil isolate of *Alcaligenes*. *J. Appl. Microbiol.* 41: 295–305.

Oscarson, D.W., P.M. Huang, and W.K. Liaw. 1980. The oxidation of arsenite by aquatic sediments. *J. Environ. Qual.* 9: 700–703.

Oscarson, D.W., P.M. Huang, and W.K. Liaw. 1981. Role of manganese in the oxidation of arsenite by freshwater lake sediments. *Clays Clay Miner.* 29: 219–224.

Papelis, C., G.E. Brown, G.A. Parks, and J.O. Leckie. 1995. X-ray absorption spectroscopic studies of cadmium and selenite adsorption on aluminum oxides. *Langmuir* 11: 2041–2048.

Peak, D. 2006. Adsorption mechanisms of selenium oxyanions at the aluminum oxide/water interface. *J. Colloid Interface Sci.* 303: 337–345.

Peak, D., and D.L. Sparks. 2002. Mechanisms of selenate adsorption on iron oxides and hydroxides. *Environ. Sci. Technol.* 36: 1460–1466.

Penrose, W.R. 1974. Arsenic in the marine and aquatic environments: Analysis, occurrence, and significance. *CRC Crit. Rev. Environ. Control* 4: 465–482.

Raven, K.P., A. Jain, and R.H. Loeppert. 1998. Arsenite and arsenate adsorption on ferrihydrite: Kinetics, equilibrium, and adsorption envelopes. *Environ. Sci. Technol.* 32: 344–349.

Rhoades, J.D. 1982. Cation exchange capacity. pp. 149–157. In A.L. Page et al. (Eds.) *Methods of soil analysis. Part 2. 2nd ed.* Agron. Monogr. 9, ASA, Madison, WI.

Sadiq, M. 1997. Arsenic chemistry in soils: An overview of thermodynamic predictions and field observations. *Water, Air, Soil Pollut.* 93: 117–136.

Schindler, P.W., R. Fürst, R. Dick, and U. Wolf. 1976. Ligand properties of surface silanol groups. I. Surface complex formation with Fe^{3+}, Cu^{2+}, Cd^{2+}, and Pb^{2+}. *J. Colloid Interface Sci.* 55:469–475.

Scott, M.J., and J.J. Morgan. 1996. Reactions at oxide surfaces. 2. Oxidation of Se(IV) by synthetic birnessite. *Environ. Sci. Technol.* 30: 1990–1996.

Séby, F., M. Potin-Gautier, E. Giffaut, and O.F.X. Donard. 1998. Assessing the speciation and the biogeochemical processes affecting the mobility of selenium from a geological repository of radioactive wastes to the biosphere. *Analusis* 26: 193–198.

Singh, M., N. Singh, and P.S. Relan. 1981. Adsorption and desorption of selenite and selenate selenium on different soils. *Soil Sci.* 132: 134–141.

Sposito, G. 1983. Foundations of surface complexation models of the oxide-solution interface. *J. Colloid Interface Sci.* 91: 329–340.

Sposito, G. 1984. *The surface chemistry of soils.* Oxford University Press, Oxford, UK.

Steinberg, N.A., and R.S. Oremland. 1990. Dissimilatory selenate reduction potentials in a diversity of sediment types. *Appl. Environ. Microbiol.* 56: 3550–3557.

Stumm, W., R. Kummert, and L. Sigg. 1980. A ligand exchange model for the adsorption of inorganic and organic ligands at hydrous oxide interfaces. *Croat. Chem. Acta* 53: 291–312.

Thomas, G.W. 1996. Soil pH and acidity. pp. 475–490. In D.L. Sparks et al. (Eds.) *Methods of soil analysis. Part 3.* SSSA Book Series 5, SSSA, Madison, WI.

Thompson-Eagle, E.T., and W.T. Frankenberger. 1990. Protein-mediated selenium biomethylation in evaporation pond water. *Environ. Toxicol. Chem.* 9: 1453–1462.

Tokunaga, T.K., I.J. Pickering, and G.E. Brown. 1996. Selenium transformations in ponded sediments. *Soil Sci. Soc. Am. J.* 60: 781–790.

Vuori, E., J. Vääriskoski, H. Hartikainen, P. Vakkilainen, J. Kumpulainen, and K. Niinivaara. 1989. Sorption of selenate by Finnish agricultural soils. *Agric. Ecosystems Environ.* 25: 111–118.

Vuori, E., J. Vääriskoski, H. Hartikainen, J. Kumpulainen, T. Aarnio, and K. Niinivaara. 1994. A long-term study of selenate sorption in Finnish cultivated soils. *Agric. Ecosystems Environ.* 48: 91–98.

Wang, X., and X. Liu. 2005. Sorption and desorption of radioselenium on calcareous soil and its components studied by batch and column experiments. *Appl. Radiat. Isot.* 62: 1–9.

Wauchope, R.D. 1975. Fixation of arsenical herbicides, phosphate, and arsenate in alluvial soils. *J. Environ. Qual.* 4: 355–358.

Wauchope, R.D. 1983. Uptake, translocation and phytotoxicity of arsenic in plants. pp. 348–377. In W.H. Lederer and R.J. Fensterheim (Eds.) *Arsenic: Industrial, biomedical, environmental perspectives.* Van Nostrand Reinhold, New York, NY.

Waychunas, G.A., B.A. Rea, C.C. Fuller, and J.A. Davis. 1993. Surface chemistry of ferrihydrite. 1. EXAFS studies of the geometry of coprecipitated and adsorbed arsenate. *Geochim. Cosmochim. Acta* 57: 2251–2269.

Welch, A.H., D.B. Westjohn, D.R. Helsel, and R.B. Wanty. 2000. Arsenic in ground water of the United States: Occurrence and geochemistry. *Ground Water* 38: 589–604.

Welch, A.H., and M.S. Lico. 1998. Factors controlling As and U in shallow ground water, southern Carson Desert, Nevada. *Appl. Geochem.* 13: 521–539.

Westall, J., and H. Hohl. 1980. A comparison of electrostatic models for the oxide/solution interface. *Adv. Colloid Interface Sci.* 12: 265–294.

Wijnja, H., and C.P. Schulthess. 2000. Vibrational spectroscopy study of selenate and sulfate adsorption mechanisms on Fe and Al (hydr)oxide surfaces. *J. Colloid Interface Sci.* 229: 286–297.

Yang, J.-K., M.O. Barnett, P.M. Jardine, N.T. Basta, and S.W. Casteel. 2002. Adsorption, sequestration, and bioaccessibility of As(V) in soils. *Environ. Sci. Technol.* 36: 4562–4569.

Zawislanski, P.T., and M. Zavarin. 1996. Nature and rates of selenium transformations: A laboratory study of Kesterson Reservoir soils. *Soil Sci. Soc. Am. J.* 60: 791–800.

Zhang, H., and H.M. Selim. 2005. Kinetics of arsenate adsorption-desorption in soils. *Environ. Sci. Technol.* 39: 6101–6108.

4

Heavy Metal and Selenium Distribution and Bioavailability in Contaminated Sites: A Tool for Phytoremediation

Beatrice Pezzarossa, F. Gorini, and G. Petruzzelli

CONTENTS

Metal Dynamics in Soil

Various metals, including heavy metals and metalloids, are common contaminants in soil and can reach critical levels in terms of human health, food safety, soil fertility, and ecological risks (Sharma and Agrawal, 2005). The presence of metals in soil is due to natural processes, for example the formation of soil, and anthropogenic activities such as the use of sludge or municipal compost, pesticides, and fertilizers, industrial manufacturing processes, residues from metalliferous mines and smelting industries, car exhaust, and emissions from municipal waste incinerators (Yadav et al., 2009).

Heavy metals cannot be degraded or destroyed. Some are important essential elements for the nutrition of higher plants and for animal nutrition (e.g., Cu, Fe, Mn, Zn, and Co), if present in optimal concentration ranges, while others (e.g., Hg, Pb, and Cd) are potentially toxic (Järup, 2003; Tchounwou et al., 2003).

The heavy metal content in plants varies according to the metal concentration and chemical species in soils, and the soluble fraction of heavy metals in soil is recognized as significantly affecting the uptake of metals by plants. The effects of heavy metals on plant development vary according to different soil characteristics, the type of plant, and the metal (Guala, Vegaa, and Covelo, 2010).

Heavy metals bioaccumulate and thus their concentration in organisms increases over time compared to the levels measured in the environment. This is because the absorption rate is higher in organisms than the excretion rate (Sridhara Chary, Kamala, and Suman Raj, 2008).

The Case of Selenium

Selenium (Se), whose chemical and physical properties are intermediate between metals and nonmetals, has been recognized as an essential nutrient for animals, humans, and microorganisms. However, it has also been found to be toxic at levels just above those required for health, and the range between its essentiality and its toxicity is quite narrow (Wilber, 1980).

In nature, Se is widely, but unevenly, distributed in the earth's crust. Growing health and environmental concerns about Se exposure are due to its increasing use in several industrial fields such as mining; electronics and photography; glass production and pigment formulation; copper, aluminum, steel, and other metal manufacturing activities; coal- or oil-burning power plants; metal recycling; and sewage sludge disposal (Cappon, 1991; Fishbein, 1983; Shamberger, 1981). The main releases of Se into the environment, as a consequence of human activities, result from the combustion of fossil fuels. Furthermore, critical Se accumulation can also occur in agricultural drainage water as an effect of the repeated irrigation of seleniferous soils that characterizes geographical areas such as the western United States (Mayland et al., 1989; Weres, Jaouni, and Tsao, 1989; Stephens and Waddell, 1998) and in several countries, including Australia, Canada, China, Colombia, India, Ireland, and Mexico (Dhillon and Dhillon, 2001). The primary exposure pathways through which Se reaches humans are food, water, and air, in order of decreasing importance.

Although Se is considered nonessential to higher plants, it may have beneficial biological functions. At low concentrations, it acts as an antioxidant and stimulates plant growth; whereas at higher concentrations, it acts as a pro-oxidant, thus reducing the yield (Hartikainen, Xue, and Piironen, 2000).

The Se content in plants varies according to Se concentrations and chemical species in soils, and the soluble fraction of soil Se is recognized as significantly affecting the uptake of Se by plants (Mikkelsen, Page, and Bingham, 1989).

The dynamics of Se in soil resembles the dynamics of heavy metals. The determination of different species by selective chemical extraction is essential in order to understand the chemistry of Se and its interaction with other soil components, and to assess its mobility and availability to plants (Pyrzynska, 2002). In fact, the mobility, bioavailability, and toxicity of Se in water and soil depend on its chemical speciation and partitioning. The prominent factor determining the fate of Se in the environment is its oxidation state and related water solubility.

Selenium is present in soil as selenides (Se^{2-}), elemental selenium (Se^{0}), selenite (Se^{4+}), and selenate (Se^{6+}). Selenate and selenite are the main inorganic forms that are readily absorbed by plants and can be converted to organic Se, but selenate is the most predominant inorganic form found in both animal and plant tissues. Selenite appears to be less mobile in soil than selenate as it is more strongly adsorbed by minerals and thus less available for plant uptake than selenate (Wang and Chen, 2003; Pezzarossa et al., 2007).

Se toxicity is strictly related to the intrinsic bioavailability of its chemical forms; thus, the oxyanions selenite and selenate are the most toxic chemical species occurring in oxidative conditions.

The fate of Se in contaminated soils has long been linked to microbial activity, which is thought to be the primary means by which soluble, biologically active forms of selenium can be reduced to an almost inert elemental state (Oremland and Stolz, 2000; Stolz and Oremland, 1999; Stolz, Basu, and Oremland, 2002). Thus, an accurate knowledge of the speciation and biogeochemical transformation of selenium is needed in order to characterize Se-contaminated sites and, if necessary, to design appropriate remediation protocols (Ellis et al., 2003). In this context, a promising strategy for reducing the impact of naturally occurring or anthropogenic Se in soils is by phytoremediation (Banuelos et al., 1997). Phyoremediation is a plant-based, low-cost technology relying on plant species that have the ability to absorb and accumulate bioavailable selenium forms and to convert them into volatile forms, which are then (possibly) released into the atmosphere (Raskin, Smith, and Salt, 1997). Recent attention has focused on the use of plants and their associated microbes for environmental cleanup operations (Pilon-Smits, 2005). Rhizosphere microorganisms can strongly affect availability, plant uptake, translocation, and the volatilization of pollutants, as well as chelation and immobilization of contaminants in root-explored soil.

The concentration and availability of Se forms depend on soil properties such as pH, organic matter content, texture, microbiological activity, and the presence of competitive ions and organic compounds. The organic matter associated with silt and clay fractions significantly contributes to the association with Se, thus affecting its bioavailability in soil (Tam, Chow, and Hadley, 1995; Wang and Gao, 2001)

Bioavailability of Heavy Metals

The total metal content of contaminated soil can indicate the extent of contamination but is not a good indicator of metal availability to organisms or their phytotoxicity. To predict the extent of the adverse effects of metals in soil and in organisms and the probability with which adverse effects occur, it is necessary to determine the fraction of contaminants that are active in organisms.

Bioavailability can be defined as the fraction of a substance that will exert an effect on an organism and which is environmentally available and appraised chemically.

The bioavailability of metals depends on biotic factors, such as being organism specific, and abiotic factors, which determine in which form the metals are present in the solid and solution phases of the soil system.

The determination of the different species by selective chemical extraction is essential in order to understand the chemistry of metals and their interactions with other soil components, and to assess their mobility and availability to plants. The mobility, bioavailability, and toxicity of metals in water and soil depend, in fact, on their chemical speciation and partitioning.

The bioavailability of metals in soils is considered in terms of a series of single-extraction methods, including the use of ethylenediaminetetraacetic acid (EDTA), acetic acid, diethylenetriaminepentaacetic acid (DTPA), ammonium nitrate, calcium chloride, and sodium nitrate.

The distribution of metals between solid phases and solutions in soil is considered the key factor when assessing the environmental consequences of the accumulation of metals in soil (Adriano, 2001; He, Yang, and Stoffella, 2005).

The retention and release process of metals in the solid phase of the soil includes precipitation and decomposition, ionic exchanges, and adsorption and desorption. The precipitation and release reactions may involve discrete solid phases or solid phases that are absorbed onto the soil surface. The ion-exchange reactions derive from an exchange between an ionic species in the soil solution and an ionic species retained in sites with permanent charges on the soil surface. The adsorption and desorption processes can affect all ionic or molecular species and generally affect absorbent sites with a pH-dependent charge. These surfaces are iron, aluminum, and manganese oxides and hydroxides, clay minerals, and humic substances.

Bioavailability can vary, depending on the soil characteristics, the contaminant properties, and the target living organisms, which in turn are sometimes able to modify the bioavailability of some pollutants.

Bioavailability of Selenium

The concentration and speciation of Se in soil are governed by several physical and chemical factors, including pH, redox status, mineralogical composition, and adsorbing surface (Dhillon and Dhillon, 1999). In poorly aerated, acidic, organic-rich soils under strong reducing conditions, Se primarily occurs as insoluble elemental (Se^0) and selenide (Se^{2-}) forms, while in oxidizing environments such as aerobic soils, Se is present as soluble selenite (Se^{4+}) and selenate (Se^{6+}) species, the latter being predominant at alkaline pH (Fordyce, 2007; Munier-Lamy et al., 2007). Usually, Se^0 is considered to have little toxicological significance for most organisms (Combs et al., 1996; Schlekat et al., 2000), although biological activity has been observed for elemental Se nanoparticles (Zhang et al., 2005)

Under strongly reducing conditions, Se^{2-} forms several metastable anions such as selenosulfate (SO_3Se^{2-}) (Ball and Milne, 1995), which is largely used as a source of Se for the formation of metallic selenides in industrial processes (Riveros et al., 2003). Selenite is a weak acid that can exist as H_2SeO_3, $HSeO_3^-$, or SeO_3^{2-}, depending upon the solution pH ($pK_{a1} = 2.70$ and $pK_{a2} = 8.54$) (Seby et al., 2001). In the moderate redox potential range, selenite is the major species in soil. Its mobility is mainly controlled by sorption–desorption processes on metal oxyhydroxides (Papelis et al., 1995; Saeki, Matsumoto, and Tatsukawa, 1995); clays (Bar-Yosef and Meek, 1987); and organic matter (Gustafsson and Johnsson, 1994; Tam, Chow, and Hadley, 1995). Selenate, the predominant soluble species with a high redox potential, is characterized by low adsorption and precipitation (Fordyce, 2007). In aqueous environments, Se^{6+} exists as biselenate ($HSeO_4^-$) or selenate (SeO_4^{2-}), with $pK_{a2} = 1.8$ (Seby et al., 2001).

There are conflicting reports on the relative toxicity of Se^{4+} and Se^{6+}. For example, Buhl and Hamilton (1996) showed that Se^{4+} is more toxic than Se^{6+}, while Reddy (1999) reported the opposite. The distribution of selenite and selenate between the solid phase and the solution phase depends on the pH and the mineral phase composition. Both species are present in the upper layer of the soil and are reduced to elementary selenium in the deeper layer through biotic and abiotic reduction (Vodyanitskii, 2010). Selenate adsorption by soil was found to be insignificant within the pH range of 5.5 to 9 (Neal and Sposito, 1987). Selenite is well adsorbed by iron hydroxides at pH 7, whereas its adsorption is minimal at pH > 8. Therefore, neutral and alkaline soils appear to be the most dangerous, due to increased selenite availability for plants (Balistrieri and Choa, 1987; Vodyanitskii, 2010).

Se mobility in soil is mainly controlled by geochemical processes such as adsorption and desorption, oxidation and reduction, precipitation, co-precipitation, and complexation with other minerals (Lim and Goh, 2005). Se interacts with Fe, Al, and Mn oxide and hydroxides, carbonates, and organic matter (Selim and Sparks, 2001). The association of Se with humic and fulvic acids occurs both in aquatic and terrestrial environments, and is related to the bioavailability of the element to organisms (Zhang and Moore, 1996). In fact, a decrease in Se uptake by plant roots has been observed due to interactions between selenite and fulvic substances. This effect seems to govern the bioavailability of Se to animals (Wang, Xu, and Peng, 1996).

A very interesting example of the importance of organic matter in Se speciation and bioavailability in soil is the case of Se deficiency in the Zhangijakou District in China. In this geographic area, a heart pathology known as Keshan disease affects a section of the population. However, the reported Se deficiency is not correlated to a lack of Se in the soil (it ranges from 0.1 to 0.3 $mg.kg^{-1}$), but rather to the presence of Se in a not-available chemical form for plants. Specifically, Se is immobilized on organic matter either through direct adsorption or through a reduction reaction of selenate to selenite, and the consequent adsorption of the latter into iron and aluminum oxyhydroxides (Férnandez-Martínez and Charlet, 2009). At low pH values, both selenite and selenate species adsorb onto the positive surfaces of Fe and Al oxides whose affinity for selenite is generally higher than that for selenate (Zhang and Sparks, 1990; Lo and Chen, 1997; Yu, 1997; Su and Suarez, 2000).

The Se chemical forms differ widely in their short-term bioavailability. They are operationally defined using sequential extraction procedures (SEPs), which are useful in obtaining information on the fractionation of an element in soil, before the determination of the species using chromatographic techniques (Tokunaga et al., 1991; Dauchy et al., 1994; Zhang and Moore, 1996). While most SEPs were developed for heavy metal cations, the anionic nature of Se requires different sequential extraction schemes (Bujdos, Kubová, and Streško, 2000). Because soil selenate is the most prevalent form of soluble Se, it is also the most bioavailable. It is usually extracted from soil using KCl (Dhillon and Dhillon, 1999; Gao et al., 2000). The determination of selenite, which is probably specifically adsorbed to anion exchange sites, occurs by a subsequent extraction step, with KH_2PO_4 (Dhillon and Dhillon, 1999; Goodson et al., 2003) or K_2HPO_4 (Dhillon and Dhillon, 1999; Gao et al., 2000). Elemental Se is solubilized by Na_2SO_3 (Zhang and Moore, 1996; Gao et al., 2000; Zhang and Frankenberger, 2003), whereas organo-bound Se can be extracted by NaOCl (Zhang and Moore, 1997); $K_2S_2O_8$ (Martens and Suarez, 1997; Bujdos, Kubová, and Streško, 2000); or NaOH (Martens and Suarez, 1997; Heninger et al., 1997). The use of 0.5 M HCl permits the extraction of a relatively uniform fraction of Se (Bujdos et al., 2005).

Se speciation, mobility, bioavailability, and uptake in soil–plant systems are very much affected by the presence of microorganisms in the environment through the control of Se oxidation states (Fernández-Martinez and Charlet, 2009). Bacteria can use selenate and selenite as terminal electron acceptors in energy metabolism, or they can reduce and incorporate Se into organic compounds. The dissimilatory process by which the selenium oxyanions are used during the respiration of organic carbon has been shown in several bacteria strains (Macy, Lawson, and Demolldecker, 1993; Zhang, Zahir, and Frankenberger, 2003). Some produce elemental Se nanospheres that accumulate inside and outside the bacteria cell wall (Oremland et al., 2004). Most of the gaseous Se is in a dimethylselenide form, which is the product of soil microbial activity (Siddique et al., 2006). Soil factors that influence the volatilization of Se through the direct control of microorganism methylation reactions are pH, temperature, redox status, and soil chemical composition (Frankenberger and Karlson, 1994; Zhang and Frankenberger, 1999). The addition of organic amendments to Se-contaminated soils favors the dimethylselenide production rate through stimulation of the methylation process (Karlson and Frankenberger, 1989).

As reported above, whereas oxidized forms of Se are highly soluble under a wide range of geochemical conditions, elemental Se and organic Se^{2-} compounds are characterized by a very low solubility (Seby et al., 2001). Hence, the biomethylation of microorganisms is considered a selenium remediation process for soils that present a very high Se content due to both natural and anthropic events (Chasteen and Bentley, 2003; Banuelos and Lin, 2007). This process consists of subsequent reactions that lead to a reduction in Se^{6+} or Se^{4+} to Se^{2-}. The ensuing methylation reaction produces one or two methyl groups in the formed selenide species (Chasteen and Bentley, 2003). The reduction of these oxyanions into insoluble selenium precipitate by soil bacteria occurs under both aerobic and anaerobic conditions (Losi and Frankenberger, 1997a; Stolz and Oremland, 1999). Other relevant processes are alkylation, dealkylation, and oxidation (Fernández-Martinez and Charlet, 2009).

Factors Affecting Bioavailability

Several abiotic and biotic factors can affect the chemical speciation and availability of metals in the environment. These factors include soil properties such as pH, clay and organic matter content, cation exchange capacity, redox potential, and iron and manganese oxides found in the soil. Soil properties control cation exchange, specific adsorption, precipitation, and complexation, which are the main processes governing the partition of metals between the solid and solution phases of soils.

pH

Soil acidity (pH) is the most important parameter governing the concentrations of metals in soil solutions that regulate precipitation–dissolution phenomena. Metal solubility tends to decrease at a higher pH. In alkaline conditions, the precipitation of solid phases diminishes the concentration of metal ions in solutions and the reverse happens with a lower pH. pH values also regulate specific adsorption and complexation processes. The sorption of metals is often directly proportional to soil pH due to the competition of H^+ (and Al^{3+}) ions for adsorption sites; however, this competition may be reduced by specific adsorption. Metal hydrolysis at higher pH values also promotes the adsorption of the resulting metal hydroxo complexes, which beyond a threshold pH level (which is specific for each metal) drastically reduce the concentration of metal ions in the soil solution. At low pH levels, on the other hand, sorption processes are reduced due to the acid-catalyzed dissolution of oxides and their sorption sites, whereas the complexation by organic matter tends to decrease with increasing acidity. As far as Se is concerned, several studies have reported that selenite is sorbed more than selenate in a wider range of pH (Balistrieri and Choa, 1987; Saeki, Matsumoto, and Tatsukawa, 1995), although both species show a higher sorption onto Fe oxides at low pH values.

Clay Content

Ion exchange and specific adsorption are the mechanisms by which clay minerals adsorb metal ions. This is done through the adsorption of hydroxyl ions followed by the attachment of the metal ion to the clay by linking it to the adsorbed hydroxyl ions or directly to sites created by proton removal. Highly selective sorption occurs at the mineral edges. However, notable differences exist among clay minerals in terms of their ability to retain heavy metals, which are more strongly adsorbed by kaolinite than montmorillonite. This is probably due to a higher amount of weakly acidic edge sites on kaolinite surfaces. In expandable clays (vermiculite and smectite), the sorption processes essentially involve the inter-layer spaces, and are greater than in nonexpandable clays such as kaolinite. The importance of clay minerals, and of soil texture, in determining the distribution of heavy metals between the solid and the liquid phases of soil has a direct consequence on the metal bioavailability of plants. For the same total concentration, heavy metals are more soluble and available for plant uptake in sandy soil than in clay soil.

As far as Se is concerned, due to the negative charge of clay, it can only be sorbed on the edges of clay minerals, particularly kaolinite (Bar-Yosef, 1987). In fact, broken edges of kaolinite sorb both H^+ and OH^- ions by developing surfaces with variable charges depending on pH.

Organic Matter Content

The organic matter content of soils is often small compared to clay. However, the organic fraction has a great influence on metal mobility and bioavailability due to the tendency of metals to bind with humic compounds in both the solid and solution phases in soil. The formation of soluble complexes with organic matter, in particular the fulvic fraction, is responsible for increasing the metal content of soil solutions. However, higher molecular weight humic acids can greatly reduce heavy metal bioavailability due to the strength of the linkages. Both complexation and adsorption mechanisms are involved in the linking of metals by organic matter, thus including inner-sphere reactions and ion exchange. Negatively charged functional groups (phenol, carboxyl, amino groups, etc.) are essential in metals retained by organic matter. The increase in these functional groups during humification produces an increase in the stability of metal organic complexes, which also show a greater stability at higher pH values.

Organic matter significantly contributes to selenium mobility in soil (Singh, Singh, and Relan, 1981; Johnsson, 1989; Gustafsson and Johnsson, 1992; Tam, Chow, and Hadley, 1995; Wang and Gao, 2001). Several organic compounds of high molecular weight present in soil interact with selenium and may act as active binding agents, thus affecting its availability and, consequently, selenium uptake by plants (Ferri et al., 2003).

Cation Exchange Capacity

The density of negative charges on the surfaces of soil colloids defines the cation exchange capacity of soil. This capacity is governed by the type of clay and the amount of organic colloids present in the soil. Montmorillonitic-type clays have a higher net electrical charge than kaolinitic-type clays; consequently, they have a higher cation exchange capacity. Soils containing a high percentage of organic matter also tend to have high cation exchange capacities. The surface negative charges may be pH dependent or permanent; and to maintain electroneutrality, they are reversibly balanced by equal amounts of cations from the soil solution. Weak electrostatic bonds link cations to soil surfaces, and heavy metals can easily substitute alkaline cations on these surfaces by exchange reactions. Moreover, specific adsorption promotes the retention of heavy metals, also by partially covalent bonds, although major alkaline cations are present in soil solutions at much greater concentrations.

Redox Potential

Reduction–oxidation reactions in soils are controlled by the aqueous free electron activity pE, which is often expressed as the E_h redox potential. High levels of E_h are encountered in dry, well-aerated soils, while soils with a high content of organic matter or subject to waterlogging tend to have low E_h values. Low E_h values generally promote the solubility of heavy metals. This can be ascribed to the dissolution of Fe–Mn oxyhydroxides under reducing conditions, thus resulting in the release of adsorbed metals. However, under anaerobic conditions, the solubility of metals could decrease when sulfides are formed from sulfates. Differences in individual metal behavior and soil characteristics result in conflicting reports regarding the effects of redox conditions on metal solubility. Redox conditions, together with pH, influence Se speciation in the soil. Selenate, for example, is present when the pH ranges from 3 to 10 and E_h is less than 1 volt (Mayland et al., 1989).

Iron and Manganese Oxides

Hydrous Fe and Mn oxides are particularly effective in influencing metal solubility in relatively oxidizing conditions. They are important in reducing metal concentrations in soil solutions by both specific adsorption reactions and precipitation. Although Mn oxides are typically less abundant in soils than Fe oxides, they are particularly involved in sorption reactions with heavy metals. Mn oxides also adsorb heavy metals more strongly, thus reducing their mobility. This action is particularly important in contaminated soils. The specific adsorption of metals by hydrous oxides follows the preferential order: Pb > Cu >> Zn > Cd.

In alkaline soils, the interactions between selenite and iron oxides are particularly important in controlling Se solubility (Neal, 1995). Because Mn oxides have a lower zero charge point than Fe oxides, a lower Se sorption has even been recorded (Balistrieri and Chao, 1990).

Other Factors

Several other factors could affect the solubility of metals in soils. Temperature, which influences the decomposition of organic matter, can modify the mobilization of organo-metal complexes and, consequently, plant uptake. Balistrieri and Chao (1990) reported that an increase of 10°C in temperature decreases the selenite sorption. An increase in the ionic strength of soil solutions reduces the sorption of heavy metals by soil surfaces due to the increased competition from alkaline metals. Similar effects also derive from the simultaneous presence in soil solutions of many heavy metals, which compete for the same sorption sites. Competitive anions, such as phosphate, affect the desorption process of Se, especially when the concentration of phosphate is much higher than selenate (Cowan, Zachara, and Resch, 1990).

This results in an increase in Se mobility in contaminated soils due to the saturation of adsorption sites.

The living phase of soil is also of great importance in determining metal solubility, which is dependent to some extent on both microbial and root activity. In the rhizosphere, plants can increase metal mobility by increasing their solubility. This happens following a release in the exudates of both protons, which increase the acidity, and organic substances, which act as complexing agents. Microbial biomass may promote the removal of heavy metals from soil solutions by precipitation such as sulfides and by sorption processes on newly available surfaces characterized by organic functional groups.

Measuring Bioavailability

One of the main difficulties in the practical application of bioavailability in reclamation lies in the lack of clear consensus concerning the appropriate methodology for measuring bioavailability. Using a series of tests to assess bioavailability rather than looking for a universal method that is valid under all conditions is one possible strategy to overcome this obstacle. The only direct way of measuring the bioavailability of a contaminant for a certain organism would be to use the organism itself. The number of organisms in the natural ecosystem is obviously too great to use this approach, and we need to resort to approximations in the laboratory with "test organisms" that supply, by means of biological tests, some indications as to the real bioavailability. On the other hand, if the limiting factor in the absorption from soil of a chemical compound by an organism is linked to a series of chemical processes in the soil, indications regarding bioavailability can also come from a chemical test that identifies the nature of the bonds with which a certain substance is held by the surfaces of the soil (e.g., using an extracting solution).

According to Harmsen, Rulkens, and Eijsackers (2005), information on the bioavailability of a certain organism derives from laboratory tests in which bioavailability is determined both biologically and chemically:

$$\text{Bioavailability} = f_1\,(\text{Bioavailability}_{\text{bio}}) + f_2\,(\text{Bioavailability}_{\text{chem}}) \qquad (4.1)$$

where f_1 and f_2 are mathematical functions. Chemical and biological tests alone do not define the bioavailability, but both must be considered tools that provide information on bioavailability.

If, for example, in the environmental sector we consider a metal as "contaminated soil-vegetation," then the processes that determine bioavailability are the release from a solid phase into a liquid phase of the soil, and the uptake of the element in soluble form by the root system of the plant. Thus, bioavailability tests need to consider two distinct aspects: a physical, chemically driven solubilization process and a physiologically driven uptake process. Soil characteristics and plant characteristics determine bioavailability.

From a chemical point of view, it is possible to determine the amount of pollutant in the soil solution and/or the amount that can be most easily released from the solid phase (e.g., metals retained with electrostatic bonds). For this purpose we can use either direct sampling of the solution present in the pore system or make use of bland extractants, such as water or alkaline saline solutions that have the specific characteristic of not modifying the soil particle surfaces. Alternatively, more energetic extracting agents can be used, such as complexing agents (EDTA, DTPA), whose action, however, is more aggressive than that of a plant.

The biological test can be carried out using growth trials of some plant species on the polluted soil in question, for example, climatic cell trials in controlled temperature and humidity conditions (Alexander, 2000).

Plant growth trials on contaminated soils are very interesting in terms of risk assessment because they enable us to directly examine the retention of metals by the soil associated with aging, which tends to reduce the bioavailability of these elements (Alexander, 2000). At the end of the growing period, the metal content in plants will provide information on the bioavailable fraction.

It is clear that a chemical test, for example, cannot take into account those antagonistic or synergic effects that may take place in the case of a range of pollutants, even if they are of the same class. On the other hand, a very important aspect, which is often underestimated, is the fact that negative results (e.g., the stunted growth of a plant) can also occur in the absence of effects deriving from the contaminants. These can be due to other causes (e.g., the poor physical characteristics of a soil).

This multiple approach for determining bioavailability is an essential aspect of the recent ISO proposal on bioavailability and agrees well with the latest EPA recommendations (ISO, 2008), which suggest using the criterion of "weight of evidence" (WOE) as a support tool for decisions to evaluate the contributions from different tests.

Regarding bioavailability trials, some of the most important aspects that give a particular methodological approach greater validity are as follows:

- Relevance in terms of soil chemistry
- Relevance in terms of the final receptor
- Relevance in terms of the environmental path considered
- Acceptability and validation of the method

Chemical and biological tests provide different lines of evidence regarding bioavailability in a specific site, and the WOE is used to refer to how the results of various combined tests should be used.

This dual approach can combine specific empirical measurements of contaminant concentration in soil extracts with toxicity and accumulation measurements, in addition to other parameters (i.e., seedling emergence, survival, root length, shoot height, root and shoot mass) in plants. The procedure can be greatly improved by using results from several chemical extractions instead of one or more plant species. In this way the combination of chemical and biological tests is particularly important in overcoming the various disadvantages of each individual empirical method. It also provides information that will enable us to leave the wrong approximation that the total amount of contaminant at a site must be considered as bioavailable. Together with the combined use of chemical, biological, and toxicological methods, we also need an in-depth understanding of the characteristics of the site, which will enable us to accurately identify the most important processes involving the contaminants. This entails knowing the soil characteristics of the polluted site and the parameters that influence the chemical behavior of the contaminants.

Bioavailability as a Tool in Remediation Strategies

Our knowledge of the processes of bioavailability can be used in the selection and application of remediation strategies. Bioavailability, in fact, is an essential step in a feasability test. It also improves the risk assessment and provides information to select appropriate remediation technology.

In recent years there has been a great deal of development as regards the remediation technologies that use previous knowledge and experience regarding the bioavailability of contaminants in the soil. These developments reflect two different strategies—both the reduction and increase in bioavailability.

Techniques that effectively reduce bioavailability focus on preventing the movement of pollutants from the soil to living organisms, essentially by

- Removing the labile phase of the contaminant (i.e., the fraction that is intrinsic to bioavailability);
- Converting the labile fraction into a stable fraction (e.g., the precipitation of metals), or modification of the redox state towards insoluble forms; and
- Increasing the resistance of contaminants to mass transfer (e.g., inertization).

There are also other procedures that are aimed at increasing the bioavailability of pollutants, which can be used with technologies that remove or destroy the solubilized contaminants.

Physical and chemical methods for cleaning up soils of heavy metals and metalloids involve chemical extraction with acids or chelating agents, and size separation of the fraction containing the highest content of contaminants (Iskandar and Adriano, 1997; Page and Page, 2002). These techniques can increase the mass transfer from the absorbed phase in three main ways: (1) by sieving in order to decrease diffusion, (2) by increasing the temperature, or (3) through chemical additives, such as surface-active or chelating agents. Naturally, these operations must be applied together with techniques that monitor, remove, or destroy the released contaminants. However, these techniques have high costs and significantly alter soil properties (Kortba et al., 2009).

For inorganic pollutants, it is common to use chelating agents, which can transfer metals into the liquid phase of the soil. A typical case is phytoremediation where the mobilization of metals with limited mobility, such as Pb, can be triggered by the addition of EDTA. These operations naturally involve monitoring the side effects of the treatment, such as increased leaching. The capacity of higher plants to clean up agricultural and industrial wastewaters and soil has been comprehensively reviewed in recent years. Phytoremediation is a low-cost, effective, and sustainable technology to remove heavy metals (such as Cd, Pb, and Hg) and metalloids (such as As and Se) from the environment and convert them to less toxic forms (Terry, Sambukumar, and LeDuc, 2003). Phytoremediation includes different strategies.

Phytoextraction makes use of a small number of plant species that have the inherent ability to accumulate extremely high concentrations of metals in all tissues in their natural habitat (Reeves, 1992). To date, in the world there are approximately 400 plants known as "hyperaccumulators" (Reeves and Baker, 2000) and the last few years have seen a steady expansion of the list (Zhao and McGrath, 2009). These include plants belonging to Brassicaceae, such as *Alyssum* species and *Brassica juncea,* Violaceae such as *Viola calaminaria,* and Leguminosae such as *Astragalus racemosus* (Reeves and Baker, 2000). The currently accepted concentration limits in shoot tissues of hyperaccumulators on a dry weight basis are 0.1 wt.% for most metals, except, for example, for Zn (1 wt.%), Cd (0.01 wt.%), or Au (0.0001 wt.%) (Baker et al., 2000). Although hyperaccumulators have limited use in phytoremediation due to their slow growth rate and low biomass, they represent an important source of genes, through which it is possible to combine the ability of a hyperaccumulator with fast-growing plant species in order to enhance phytoremediation (Terry, Sambukumar, and LeDuc, 2003). Ideal plants for genetic engineering for phytoremediation have a high biomass; are easy to cultivate and harvest; have a widely distributed, highly branched root system; and are suitable for genetic transformation (Kortba et al., 2009). Many of these are crop plants and when used for phytoremediation, they should not be consumed by humans and animals (Eapen and Souza, 2005). Rhizofiltration uses plant roots to absorb, concentrate, and/or precipitate pollutants. In phytostabilization, the contaminants are transformed into a harmless form by plants. Phytovolatilization is a process through which the accumulated pollutants in plant leaves are converted into volatile forms (Cherian and Oliveira, 2005; Eapen and D'Souza, 2005; Doty, 2008; Macek et al., 2008)

It is precisely in biological techniques like phytoremediation that an understanding of the bioavailability of pollutants plays a primary role in the assessment of the applicability and efficiency of a technology. For the technique to function as a process of phytoextraction, the contaminants must be in the soil in a mobile form and available for absorption by plant root systems. On the other hand, if phytostabilization is needed, the pollutants must be present in insoluble forms that are not bioavailable for plant uptake.

Knowledge of the bioavailability, or at least of the chemical forms, of metals in the soil is particularly important for "soil washing," which is essential for defining the amount of metals in soluble form. It is performed in order to choose the type of liquid phase needed to carry out the washing process. In fact, in the case of heavy metals in prevalently insoluble forms, it is advisable to use water as a washing fluid due its low cost, its easy recovery, its limited corrosive action on machinery, and its overall ease of use (Petruzzelli et al., 2004). In contrast, when metals are in soluble forms, it is advisable to modify the washing fluid, for example, using pH adjustments, which facilitate the complete solubilization of the contaminants (which will be later recovered from the aqueous phase). This leaves the solid phase cleaner, although in a different chemical situation from the original. Finally, some techniques can have unforeseen effects on bioavailability. The most evident case is the excavation and removal of soil, which is widespread in Italy, where landfilling is one of the most commonly applied techniques. During the excavation and removal of contaminated soil, the exposure levels of workers are much higher than when the soil is maintained in situ. The finest particles can enter the atmosphere, where even the most volatile contaminants can be dispersed. Furthermore, the accumulation of differently contaminated materials in the same landfill can cause modifications in the original chemical and physical conditions of the landfill, with an increase in the leaching of some pollutants, for example, due to changed redox conditions.

The potential increase in the leaching of contaminants in landfill is of increasing concern in many industrialized countries. This is due to the presence of many uncontrollable reactions, and it is generally considered less hazardous to keep the contaminated soil in situ rather than move it to landfill.

Plant-Based Remediation of Selenium

Phytoremediation is an inexpensive and environmentally friendly plant-based technology that can be applied in the cleanup of Se-contaminated soils. Plants, in fact, are able to take up and accumulate Se and to convert inorganic Se into volatile forms that are released in the atmosphere (Pilon-Smits and LeDuc, 2009). Research has focused on two broad groups of plants, which often grow in soils naturally enriched in Se and can accumulate high Se levels in their leaves: primary accumulators (hyperaccumulators), which include members of the Fabaceae such as certain species of *Astragalus* and a member of the Brassicaceae, *Stanleya pinnata* (Feist and Parker, 2001); and secondary accumulators, which include species of *Aster, Atriplex,* and *Melilotus* and, above all, *Brassica juncea* (Indian mustard) (Banuelos and Meek, 1990; Guo and Wu, 1998).

Primary accumulators grow in seleniferous soils, accumulate Se (from selenate) to leaf/shoot concentrations in the range of thousands of milligrams per kilograms (Mayland et al., 1989), and can also volatilize great quantities of Se (Parker et al., 2003). Secondary accumulators, which grow in soils containing a low-to-medium Se content, take up the available Se in the soil and the Se concentration in their tissue ranges in the hundreds of milligrams per kilogram (Bell, Parker, and Page, 1992).

Selenate accumulates in plant cells by active transport against its electrochemical potential gradient (Brown and Shrift, 1982). Owing to their similar chemical properties, the uptake and assimilation of Se proceed through sulfur transporters and metabolic pathways. Unlike selenate, there is no evidence that selenite uptake is mediated by membrane transporters (Shrift and Ulrich, 1976; Asher, Butler, and Peterson, 1977; Abrams et al., 1990; Arvy, 1993). After its absorption into the root, selenate is translocated to the leaves through the xylem (DeSouza et al., 1998; Zayed, Lytle, and Terry, 1998).

In Se-hyperaccumulating taxa, once inside the leaf, selenate enters chloroplasts and is converted by ATP sulfurylase into adenosine 5′-phosphoselenate (APSe), which is then reduced to selenite by adenosine 5′-phosphosulfate (APS) reductase (De Souza et al., 2000). Transgenic plants overexpressing the ATP sulfurylase gene (APS) have a fourfold higher APS enzymatic activity and accumulate three times as much Se per plant than the wild-type (Pilon-Smits et al., 1999). Hence, the reduction of selenate to selenite seems to be the rate-limiting step for the assimilation of selenate into organic Se (Pilon-Smits and LeDuc, 2009).

However, there are conflicting reports regarding the effect of APS overexpression in Se accumulation and tolerance. In fact, while Pilon-Smits et al. (1999) showed an increased Se accumulation in the APS-transgenic *Brassica juncea,* other authors observed the opposite phenomenon in the APS-transgenic *Astragalus thaliana* (Sors et al., 2005). Selenite is converted into selenocysteine (SeCys) and selenomethionine (SeMet).

Plants have evolved several mechanisms to avoid the toxic misincorporation of seleno amino acids into proteins (Brown and Shrift, 1981). Methylation of SeCys and SeMet leads to the production of methyl-SeMet (MetSeMet), which accumulates safely in the cell but cannot be incorporated into proteins. This strategy is used by Sehyperaccumulators, such as *Astragalus bisulcatus,* which thrive on seleniferous soils and accumulate up to 1.5% Se in their tissues (15,000 mg Se kg^{-1} dry weight). The enzyme responsible for the methylation of SeCys to MetSeCys is SeCys methyltransferase (SMT). When the *A. bisulcatus* SMT enzyme has been overexpressed into two different host plants, *A. thaliana* and *Brassica juncea,* SMT-expressing *A. thaliana* was observed to accumulate up to 1000 mg Se.kg^{-1} dry weight and SMT-expressing *B. juncea* accumulated almost 4000 mg Se kg^{-1} dry weight (LeDuc et al., 2004). Moreover, SMT *B. juncea* exhibited a threefold higher content of foliar MetSeCys than the wild-type control (LeDuc et al., 2004). The limitation rate represented by the reduction reaction of selenate to selenite can be overcome through the creation of double-transgenic plants that overexpress both APS and SMT. The double transgenics accumulated up to nine times higher Se content than wild-type (LeDuc et al., 2006).

In addition, MetSeCys or MetSeMet can be metabolized to nontoxic volatile dimethylselenide (DMSe, in non-hyperaccumulators) or dimethyldiselenide (DMDSe, in hyperaccumulators) (Zayed, Lytle, and Terry, 1966; Terry and Zayed, 1994). The conversion of MetSeMet to DMSe is probably carried out by S-methymethionine hydrolase (SMM) (Mudd and Datko, 1990). Alternatively, DMSe can be produced by the preliminary conversion of MetSeMet to dimethylselenopropionate (DMSeP) in a reaction catalyzed by two enzymes, a transaminase/carboxylase and a dehydrogenase (Kocsis et al., 1998). Thus, the dimethylsulfoniopropionate (DMSP) lyase is probably responsible for the conversion of DMSeP to DMSe (De Souza et al., 2000). The rate-limiting step in the DMSe formation pathway is the conversion of SeCys to selenocystathionine, which is catalyzed by the cystathionine-γ-synthase (CSG) enzyme (Terry et al., 2000; Van Huysen et al., 2003). The constitutive expression of *CSG1* of *Astragalus thaliana* in *Brassica juncea* determined an increased DMSe formation and evaporation when compared to the wild-type control (Van Huysen et al., 2003).

The production of methyl selenide has been proposed as a means to optimize the effectiveness of phytoremediation (Banuelos et al., 1997) because it has the advantage of not producing highly Se-enriched plant material that would otherwise require special disposal (Terry et al., 2000). In the non-accumulators and secondary accumulators such as *Brassica juncea* , the root is the primary site of Se volatilization (Terry et al., 2000). The question as to whether or not rhizosphere microorganisms are involved in this process remains to be answered (Zayed, Lytle, and Terry, 1998). It is probable that bacteria take part in selenite volatilization by producing organo-Se compounds such as SeMet, which is more volatilizable than selenite or selenate (Zayed, Lytle, and Terry, 1998). In hyperaccumulating plants, volatilization mainly occurs in the leaf, although the significance of this process in these plants has not been completely explained (Parker et al., 2003).

Another mechanism involved in the process of Se tolerance enhancement consists of overexpressing SeCys lyase (SL). The enzyme catalyzes the breakage of SeCys into alanine and elemental Se, which can be incorporated into essential selenoproteins (Pilon-Smits et al., 2002; Pilon et al., 2003). The SL transgenic *Brassica juncea* accumulated approximately twofold more Se than the wild-type when grown in Se-contaminated soil (Banuelos et al., 2007).

Astragalus bisulcatus is a well-known Sehyperaccumulator that grows in soils containing high Se concentrations in southwestern United States. When these plants are grown hydroponically in the presence of selenate, while the older leaves contain mainly inorganic Se, Se in young leaves and shoots is up to 90% to 95% inorganic (Pickering et al., 2003). MetSeCys and SeCys represent 5% to 10% and 20% to 30%, respectively, of the total amino acid pool in the young shoots (Ellis and Salt, 2003).

The biosynthesis of MetSeCys catalyzed by SMT in hyperaccumulating plants determines the inactivation of SeMet production from SeCys. SMT-transgenic *Brassica juncea* produced approximately 80% of the organic Se as MetSeCys. The reduced production of SeMet (compared to wild-type plants) has a decreased volatilization of DMSe as a direct consequence (Montes-Bayon et al., 2002; Meija et al., 2002). The SMT-transformed plants also showed a greater volatilization of DMDSe than the wild-type (Terry et al., 2000), suggesting that the DMDSe is produced by MetSeCys, as occurs in *Astragalus racemosus*, which is known to hyperaccumulate MetSeCys and produce large quantities of DMDSe (Evens, Asher, and Johnson, 1968; Meija et al., 2002).

The translocation of Se from roots to shoots depends on the Se species present in soil Selenate is transported much more easily than selenite or organic Se, such as SeMet (Terry et al., 2000). In addition in accumulator plants, Se is accumulated in young leaves during the early vegetative stage of growth, whereas during the reproductive stage, Se is mostly found in the seeds and decreases in leaves (Ernst, 1982).

As already observed, inorganic Se is more toxic than the organo-Se species. In plants, selenate and selenite toxicity is due to the ready adsorption and the consequent assimilation into organo-Se compounds. Some studies indicate that selenite is more toxic than selenate because of its faster conversion to seleno amino acids, which can be incorporated into plant proteins instead of sulfur (Zayed, Lytle, and Terry, 1998). Other authors have reported that selenate may interfere with GSH (glutathione) production, and the decreased GSH content in the cells may diminish plant defenses against hydroxyl radicals and oxidative stress (De Kok and Kuiper, 1986; Bosma et al., 1991).

Remediation of Selenium through Phytoextraction by *Brassica juncea*

Experiments were conducted to determine the possible mutual effects in plants and the microbial community of soil on Se removal, distribution, and speciation.

Brassica juncea was tested in pots filled with soil whose main physical and chemical properties were as follows: pH 8.1; silt 21.1%, sand 68.9%, clay 10.0%, organic matter 2.23%, and CEC 18.7 meq.100g^{-1}. Before seed sowing, the soil was amended with 1 or 2.5 mg Se kg-1 (on a dry matter basis) as sodium selenite (Na_2SeO_3) and sodium selenate (Na_2SeO_4).

Pots prepared with pristine soil were vegetated, while nonvegetated pots were considered as controls. Twelve pots were set up for each treatment. Five plants for each treatment were randomly collected from vegetated pots 60 days (at blooming) and 82 days (at maturity) after sowing. Soil sampling was carried out twice in all pots: after selenium addition (t_0) and at the end of the cultivation trials (t_1). Plants collected at blooming were separated into roots and shoots, while those harvested at maturity were separated into roots, shoots, and pods. Fresh (FW) and dry (DW) weights of the different portions of plant biomass were recorded. Total selenium content was quantified according to Zasoski and Burau (1977) both in the fine fraction (<2 mm) of air-dried soil and in plant tissues after drying in an oven at 50°C until a constant weight was achieved.

A sequential extraction procedure was used to investigate the different Se oxidation states and assess the availability of Se in soil, following the method proposed by Martens and Suarez (1997). All digests were analyzed using a hydride generation atomic absorption spectrophotometer (Varian VGA 77).

Microbial total counts were performed in pristine and selenium-spiked soil, sampled in vegetated or nonvegetated pots. A total of 10 g soil from each pot were incubated in 90 mL of 0.9% NaCl (wt.vol^{-1}) for 2 hours at 28°C on an orbital shaker. Serial dilutions were plated on Petri dishes prepared with nutrient agar.

Enrichment cultures for bacterial isolation were inoculated with soil samples from the rhizosphere of *Brassica juncea* grown for 82 days in pots amended with 2.5 mg Se kg^{-1} soil as Se^{4+} or Se^{6+}. Bacterial isolates were clustered in different Operational Taxonomic Units (OTUs) by amplified ribosomal DNA restriction analysis (ARDRA) (Weisburg et al., 1991). ARDRA was performed using F8 and R11 primers (Weisburg et al., 1991) to amplify the 16S rRNA genes of the isolates and by digesting the amplification products with AluI, HinfI, and HhaI (New England Biolabs). The gene encoding for the 16S rRNA of one microorganism for each OTU was amplified, sequenced on both strands, aligned to the sequence databases using BLASTN (Altshul et al., 1997), and analyzed using ARB software (Ludwig et al., 2004).

Plant Growth

The chemical form as well as the concentration of Se added to the soil affected plant growth. In 2.5 mg Se kg^{-1} soil, Se^{6+} induced a significant reduction in both fresh and dry biomass production (Table 4.1). No significant differences in fresh and dry matter production were detected in plants grown in the presence of Se^{4+}. Plants grown in the presence of Se^{6+} accumulated higher

TABLE 4.1

Biomass Production by *Brassica juncea* at Maturity

Se Added	Fresh Matter (g)				Dry Matter (g)			
(mg Se kg^{-1} soil)	Roots	Leaves	Pods	Total Plant	Roots	Leaves	Pods	Total Plant
Control 0	2.79 ± 0.34c	34.0 ± 5.58c	28.1 ± 4.23ab	64.9 ± 5.79b	0.39 ± 0.04a	4.1 ± 0.39abc	3.9 ± 0.43ab	8.39 ± 0.23b
Se^{6+} 1	1.71 ± 0.38a	22.3 ± 3.7ab	27.7 ± 2.21ab	51.7 ± 2.44ab	0.32 ± 0.05a	3.3 ± 0.38ab	5.3 ± 0.30ab	8.92 ± 0.31b
2.5	2.25 ± 0.27b	24.5 ± 2.96ab	21.7 ± 1.91ab	48.4 ± 1.12ab	0.36 ± 0.07a	3.4 ± 0.47abc	3.7 ± 0.33ab	7.46 ± 0.63a
Se^{4+} 1	2.67 ± 0.54c	33.6 ± 3.76c	33.7 ± 5.68b	70.0 ± 7.73b	0.48 ± 0.09b	5.04 ± 0.47c	6.4 ± 0.38b	11.92 ± 0.65b
2.5	2.72 ± 0.36c	32.9 ± 2.67bc	27.3 ± 1.95ab	62.9 ± 4.43b	0.49 ± 0.05b	4.90 ± 0.57bc	5.2 ± 0.39ab	10.59 ± 0.90b
Significance								
Chemical form (A)	n.s.	n.s.	n.s.	n.s.	n.s.	**	n.s.	n.s.
Concentration (B)	n.s.	n.s.	n.s.	n.s.	n.s.	n.s.	n.s.	n.s.
A x B	n.s.	n.s.	n.s.	n.s.	n.s.	n.s.	n.s.	n.s.

Note: Numbers followed by different letters in the same column differ significantly at the 5% level by L.S.D. test; n.s. = not significant; * = significant at 5% level ($P< 0.05$); ** = significant at 1% level ($P < 0.01$).

amounts of Se in the shoots in comparison to plants grown in the presence of Se^{4+} (Figure 4.1). This may explain the negative effects that selenate showed toward *Brassica juncea* growth. Over the past two decades, Se^{6+} has been observed to be more easily absorbed than Se^{4+} by plant roots and to be readily transported to the shoots via the xylem (Asher et al., 1977; Smith and Watkinson, 1984; Wu, Huang, and Burau, 1988; Arvy, 1993; De Souza et al., 1998; Zayed et al., 1998; Hopper and Parker, 1999).

Selenium Distribution in Vegetated Pots

Immediately after Se addition to the soil, 98% of the metalloid was recovered in the same oxidation state as the amended selenium salts (Table 4.2). During the experimental trials, a reduction in Se concentration initially supplied to the soil was detected, along with a redistribution of Se within chemical forms of different oxidation states (Table 4.2). At the same time, a portion of the initial amended Se was recovered in the plant biomass (Figure 4.1 and Table 4.3). The amended Se^{6+} was taken up by plant roots and then

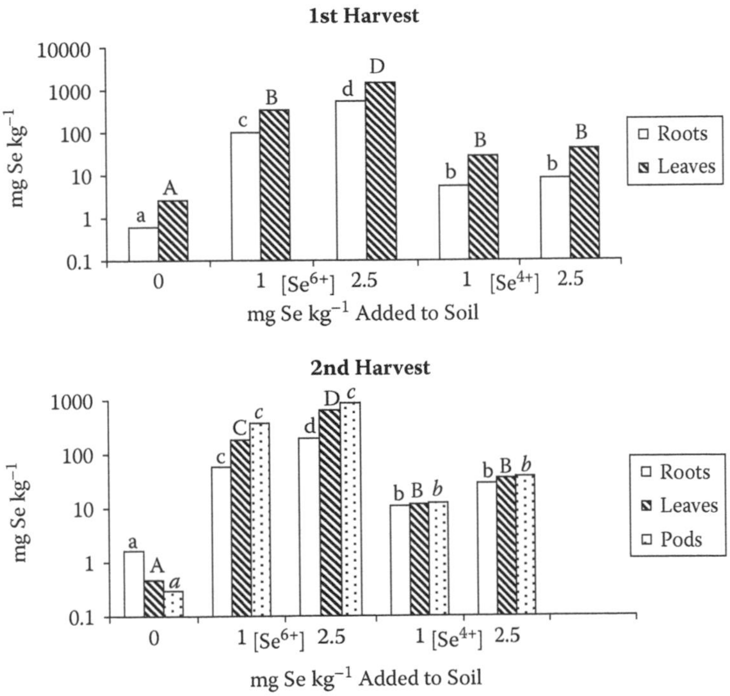

FIGURE 4.1
Total Se concentration (mg.kg^{-1} dry matter) in different portions of *Brassica juncea* at both plant blooming (first harvest) and maturity (second harvest). Different letters in the same plant part differ significantly at the 5% level by L.S.D. test.

TABLE 4.2

Selenium Speciation in Soil Amended with Different Amounts of Se^{4+} and Se^{6+} after Harvesting of *Brassica juncea* Plants

Se Added	Total Se (t_0)		0.1 M K_2HPO_4-KH_2PO_4 (t_1)			0.1M $K_2S_2O_8$ (t_1)	17M HNO_3 (t_1) $HClO_4$- HNO_3		
(mg Se kg^{-1} soil)	Se^{4+}	Se^{6+}	Se^{4+}	Se^{6+}	Se^{2-}	$Se(^{4+};^{2-})$	Se^0	Total Se	ΣSe (t_1)
Control 0	0.01 ± 0.00a	0.01 ± 0.00a	0.01 ± 0.00a	0.00 ± 0.00a	0.06 ± 0.01b	0.07 ± 0.01a	0.03 ± 0.00a	0.18 ± 0.02a	0.17 ±0.03a
Se^{6+} 1	0.01 ± 0.00a	0.98 ± 0.07b	0.06 ± 0.01ab	0.47 ± 0.04c	0.02 ± 0.00a	0.21 ± 0.02b	0.14 ± 0.02b	0.70 ± 0.11b	0.90 ±0.04b
2.5	0.01 ± 0.00a	2.45 ± 0.29b	0.10 ± 0.02b	0.51 ± 0.07c	0.06 ± 0.01b	0.20 ± 0.03b	0.10 ± 0.01b	0.89 ± 0.08d	0.97 ±0.08b
Se^{4+} 1	0.96 ± 0.12b	0.01 ± 0.00a	0.16 ± 0.03b	0.02 ± 0.00b	0.15 ± 0.01c	0.31 ± 0.05c	0.23 ± 0.02c	0.81 ± 0.04c	0.87 ±0.07b
2.5	2.30 ± 0.31c	0.01 ± 0.00a	0.50 ± 0.04c	0.04 ± 0.01b	0.12 ± 0.02c	0.54 ± 0.03d	0.39 ± 0.06d	1.52 ± 0.05e	1.58 ±0.10c
Significance									
Chemical form (A)	**	**	**	**	**	**	**	**	**
Concentration (B)	**	**	**	**	**	**	**	**	**
A x B	**	**	**	**	**	**	**	**	**

Note: Numbers followed by different letters in the same column differ significantly at the 5% level by LSD test; n.s. = not significant; * = significant at 5% level ($P < 0.05$); ** = significant at 1% level ($P < 0.01$).

TABLE 4.3

Total Se Content (mg) in Soil, Either Soon After Se Addition (t_0) or at the End of Cultivation Trials (t_1), and in *Brassica juncea* at Plant Maturity

Se Added (mg Se kg^{-1} soil)	**Soil**		*Brassica juncea*			
	t_0	t_1	**Roots**	**Leaves**	**Pods**	**Whole Plant**
Control 0	1.50 ± 0.18a	1.62 ± 0.22a	0.001 ± 0.00a	0.002 ± 0.00a	0.001 ± 0.00a	0.004 ± 0.00a
Se^{6+} 1	9.12 ± 1.25b	6.30 ± 0.96b	0.020 ± 0.00b	0.61 ± 0.03c	1.99 ± 0.22e	2.62 ± 0.23d
2.5	22.05 ± 2.67c	8.01 ± 0.70c	0.071 ± 0.017b	2.24 ± 0.40d	3.27 ± 0.34d	5.58 ± 0.52e
Se^{4+} 1	8.82 ± 1.27b	7.29 ± 0.33bc	0.005 ± 0.00a	0.06 ± 0.01a	0.06 ± 0.01a	0.145 ± 0.02b
2.5	22.05 ± 3.08c	13.68 ± 0.49d	0.014 ± 0.00b	0.18 ± 0.03b	0.20 ± 0.04b	0.39 ± 0.05c
Significance						
Chemical form (A)	n.s.	**	**	**	**	**
Concentration (B)	**	**	**	**	**	**
A x B	n.s.	**	**	*	*	**

Note: Numbers followed by different letters in the same column differ significantly at the 5% level by L.S.D. test; n.s. = not significant; * = significant at 5% level ($P < 0.05$); ** = significant at 1% level ($P < 0.01$)

translocated to the plant shoots, throughout the plant's life cycle from blooming to maturity (Figure 4.1). Se^{4+} was also absorbed by plant roots, although in lower amounts.

Multifactorial analysis revealed that Se concentrations in shoots and pods were statistically higher in plants treated with selenate than in plants treated with selenite (Figure 4.1). An evaluation of the mass balance of Se at plant maturity (Table 4.3) indicated that an aliquot of the metalloid added to the soil as both selenate and selenite seemed to have volatilized. This fraction increased with the amount of Se added to the soil. This evidence is consistent with previous findings demonstrating that *Brassica juncea* efficiently accumulates and volatilizes selenium (Terry et al., 1992; Terry and Zayed, 1994; Terry et al., 2000). As far as the redox state of the Se fraction still present in the soil is concerned (Table 4.2), a general trend of transforming Se oxyanions to more reduced forms was observed. It is worth noting that, with reference to rhizosphere-assisted phytoremediation, the transformation of soluble Se oxyanions into insoluble forms in soil, elicited by the presence of *B. juncea,* tends to restrain Se bioavailability and consequently its toxicity. The biotic reduction of Se oxyanions in soil has been previously observed and attributed to the metabolism of bacteria that have to cope with the toxicity of the oxidized species of Se (Losi and Frankenberger, 1997b). No evidence of the plant's capacity to catalyze the same reactions in soil was observed. Furthermore, the lack of transformation where microorganisms were nearly absent sustained the direct involvement of the microbial community of the soil in the reduction of Se oxidized chemical forms. Therefore, when Se^{4+} was added to the soil, Se^{2-} and Se^0 were recovered as a consequence of microbial detoxification. Interestingly, the presence of reduced Se and insoluble forms was higher in Se^{4+}-amended pots than in those added with Se^{6+}. This might be related to the toxicity exerted by oxyanions toward microbes.

Selenium Distribution in Nonvegetated Pots

When Se was amended as selenate at 1 and 2.5 mg Se kg^{-1}, the Se fraction presumably lost through volatilization accounted for 0.8% and 1.2% of the total Se added, respectively. In the soil added with selenite at 1 and 2.5 mg Se kg^{-1}, the Se fraction presumably lost through volatilization accounted for 1.3% and 2% of the total Se added, respectively. A gradual change in the oxidation state of spiked Se oxyanions was still observed.

Soil Bacterial Community

Total counts of aerobic bacteria indicated that Se had a slight effect on the number of cultivable bacteria. PCR-DGGE (PCR-denaturing gradient gel electrophoresis) analysis of the 16S rDNAs was applied to pristine soil and to the rhizosphere soil of plants grown in increasing Se^{4+} and Se^{6+} concentrations.

The results suggest that the presence of *Brassica juncea* caused a significant stimulation of the putative microbial capacity to reduce Se oxyanions. A more detailed description of the taxonomic and phylogenetic characterization of the soil's microbial community colonizing the rhizosphere of *B. juncea* is reported by Di Gregorio et al. (2006).

The higher capacity of the microbial cenosis to reduce Se oxyanions in the presence of *Brassica juncea* was probably related to the increase in its metabolic activity rather than to changes in its taxonomic structure. In fact, both variety and the amount of organic compounds released by the root system into the surrounding soil contribute to the so-called rhizosphere effect, which thus stimulates microbial growth and activity to a higher extent than in bulk soil (Anderson, Guthrie, and Walton, 1993).

The addition of Se as selenite or selenate to soil was found to proceed through partial volatilization and the transformation of toxic oxyanions into less toxic reduced forms. Both processes were amplified by the presence of *Brassica juncea,* which was also responsible for Se phytoextraction from the soil. The results concerning the capacity of *B. juncea* to phytoextract and phytovolatilize selenium are consistent with previous findings by other authors (Terry et al., 1992; Terry and Zayed, 1994).

However, if plants can elicit different mechanisms of Se detoxification in soil microbes, even a direct microbial Se volatilization cannot be ruled out as a further mechanism of microbial Se detoxification (Frankenberger and Karlson, 1994; Lin et al., 2000; Lin and Terry, 2003).

Confirmation of the direct involvement of the soil's microbial community in reducing soil content into Se toxic forms is also due to the capacity of the cultivable fraction of soil microorganisms to reduce Se oxyanions to elemental Se. Rhizobacteria were isolated and taxonomically divided into different genera. Their capacity to reduce Se^{4+} and Se^{6+} to Se^0 was tested in vitro. The rhizosphere of *Brassicsa juncea* was an effective source and carrier of microorganisms capable of in vitro reduction and, thus, to precipitate toxic Se^{4+} and Se^{6+} to nontoxic elemental Se (data not shown). The information gained thus far on these strains suggests that they could be exploited in Se precipitation from selenite- or selenate-polluted matrices using a microbe–plant interaction-based phytoremediation protocol.

Concluding Remarks

We confirmed that selenium volatilization by *Brassica juncea* is a relevant mechanism of Se detoxification in contaminated soil. In fact, *B. juncea* cultivation in the soil tested resulted in a higher volatilization rate compared with nonvegetated soil. However, the presence of *B. juncea* in soil amended with selenium as selenite or selenate promoted significant Se precipitation by eliciting rhizobacteria capable of reducing the metalloid oxyanions to Se^0. In this context, the possibility of relying on plant–rhizospheric microbe systems leads to new and interesting perspectives in phytoremediation. In fact, the capacity of certain rhizobacteria to precipitate Se oxyanions, thus reducing their toxicity in contaminated matrices, could be seen as an alternative option to Se phytoextraction and phytovolatilization, which are currently the principal methods used for removing toxic metalloids from soil.

Incorporating the concept of bioavailability into remediation schemes by starting with a risk assessment is particularly appropriate when this leads to a different view of the issues associated with pollution and brings about a different way of carrying out remediation. There are four main factors that can determine whether or not bioavailability should be taken into account:

1. When only some chemical forms of the contaminants are sources of risk for the site
2. When default assumptions regarding bioavailability are not suitable because of the site's specific characteristics
3. When there is a substantial difference in remediation goals when the bioavailability of the pollutants is taken into account
4. When the final destination of the site is not likely to be modified at least in the near future

A typical assumption often made is that the total concentration of a contaminant in the solid phase is bioavailable, with negative effects on the receptors. For heavy metals, this is scientifically inaccurate, and an accurate assessment of the transfer routes of the pollutants and of the processes of bioavailability carried out in the preliminary phases of the remediation can help to save a considerable amount of time and resources. Naturally, abandoning the default values must be done with caution, in the light of our knowledge of the chemical forms of the contaminants and their distribution among the various soil phases and the specific characteristics of the soil itself within a contaminated site. However, inserting site-specific parameters in the risk assessment is essential in defining the most appropriate remediation operation and in assessing the real hazards that derive from the contaminants.

On the basis of information regarding bioavailability, it is also possible to significantly modify the target objectives of the remediation, especially if these target levels are not far from the concentrations present in the soil. Particularly important results can be obtained if the quantity of soil to be decontaminated is high. Finally, taking bioavailability into account can increase the number of applicable technologies, especially regarding the option of soil excavation and transport to a landfill, and where this choice could damage neighboring ecosystems. On the basis of current scientific knowledge, there is a precise understanding of the mechanisms and reactions involved in many processes of bioavailability, while for others there remains uncertainty and further studies and in-depth analyses are required. The scientific community is making a great effort to broaden our understanding of bioavailability processes, in order to make them more explicitly used in risk assessment procedures, and using the following fundamental guidelines:

- Select appropriate measurement tools for determining bioavailability.
- Understand, assess, and, when possible, reduce the uncertainty concerning the parameters and models used for the particular processes of bioavailability.
- Develop a coordinated monitoring plan to assess the possible, but unlikely, variations in bioavailability over time.

Regarding the last point, there is little likelihood of a variation in bioavailability. In fact, variations tend to happen only due to an unforeseen sudden geochemical variation, a change in the use of the site, thus changing the targets of bioavailability, or the introduction of a new species into the area. In addition, although long-term studies on bioavailability are still needed, a monitoring program is part of many reclamation strategies and is certainly not a conceptually complex aspect.

To promote bioavailability as a practical tool in soil remediation, there is a need for methods that can combine chemical measures, which define bioaccessibility in terms of a potential release of a metal bound to the solid phases, with biological and toxicological measures, which consider the fraction of the contaminant absorbed by the target organism as being bioavailable. Legislative aspects and public acceptance regarding bioavailability are very important from the practical viewpoint of application, but they do not fall within the scientific context. However, if particularly innovative remediation strategies are used, it is clearly essential to involve all interested parties from the beginning of the planning stage.

References

Abrams M.M., Shennan C., Zazoski J., and Burau R.G. 1990. Selenomethionine uptake by wheat seedlings. *Agron. J.*, 82: 1127–1130.

Adriano D C. 2001. *Trace elements in the terrestrial environment.* Springer, New York

Anderson T.A, Guthrie E.A., and Walton B.T., 1993. Bioremediation in the rhizosphere. *Environ. Sci. Technol.*, 27: 2630–2636.

Alexander M. 2000. Aging, bioavailability and overestimation of risk from environmental pollutants. *Environ. Sci. Technol.* 34: 4259–4265.

Altshul, S.F., Madden, T.L., Shaffer, A.A., Zhang, J., Zhang, Z., Miller, W., and Lipman, D.J. 1997. Gapped BLAST and PSI-BLAST: A new generation of protein database search programs. *Nucleic Acids Res.*, 25: 3389–3402.

Arvy, M.P. 1993. Selenate and selenite uptake and translocation in bean plants (*Phaseolus vulgaris*). *J .Exp. Bot.*, 44: 1083–1087.

Asher, C.J., Butler, G.W., and Peterson, P.J. 1977. Selenium transport in root systems of tomato. *J. Exp. Bot.*, 28: 279–291.

Baker, A., McGrath, S., Reeves, R., and Smith, J. 2000. Metal hyperaccumulator plants: A review of the ecology and physiology of a biological resource for phytoremediation of metal polluted soils. In Terry, N., and Banuelos, G.S., (Eds.) *Phytoremediation of contaminated soil and water*, CRC Press, Boca Raton, FL. pp. 85–107.

Balistrieri, L.S., and Choa, T.Y. 1987. Selenium adsorption by goethite. *Soil Sci. Soc. Am. J.*, 51: 1145–1151.

Balistrieri, L.S., and Chao, T.T. 1990. Adsorption of selenium by amorphous iron oxyhydroxide and manganese dioxide. *Geochem. Cosmochim. Acta*, 54: 739–751.

Ball, S., and Milne, J. 1995. Studies on the interaction of selenite and selenium with sulfur donors. 3. Sulfite. *Can. J. Chem-Revue Canadienne de Chimie*, 73: 716–724.

Banuelos, G.S., and Meek, D.W. 1990. Accumulation of selenium in plants grown on selenium-treated soil. *J. Environ. Qual.*, 19: 772–777.

Banuelos, G.S., Ajwa, H.A., Terry, N., and Zayed, A. 1997. Phytoremediation of selenium laden soil: a new technology. *J. Soil Water Conserv.*, 52: 426–430.

Banuelos, G.S., Leduc, D.L., Pilon-Smits, E.A.H., and Terry, N. 2007. Transgenic Indian mustard overexpressing selenocysteine lyase or selenocysteine methytransferase exhibit enhanced potential for selenium phytoremediation under field conditions. *Environ. Sci. Technol.*, 41: 599–605.

Banuelos, G.S., and Lin, Z.Q. 2007. Acceleration of selenium volatilization in seleniferous drainage sediments amended with methionine and casein. *Environ. Pollut.*, 150: 306–312.

Bar-Yosef, B. 1987. Selenium desorption from Ca-kaolinite. *Comm. Soil Sci. Plant Anal.*, 18: 771–779.

Bar-Yosef, B., and Meek, D. 1987. Selenium sorption by kaolinite and montmorillonite. *Soil Sci.*, 144: 11–19.

Bell, P.F., Parker, D.R., and Page, A.L. 1992. Contrasting selenate sulfate interactions in selenium accumulating and nonaccumulating plant species. *Soil Sci. Soc. Am. J.*, 56: 1818–1824.

Bosma, W., Schupp, R., De Kok, L.J., and Rennenberg, H. 1991. Effect of selenate on assimilatory sulfate reduction and thiol content spruce needles. *Plant Physiol. Biochem.*, 29: 131–138.

Brown, T., and Shrift, A. 1981. Exclusion of selenium from proteins in selenium-tolerant Astragalus species. *Plant Physiol.*, 67: 1951–1953.

Brown, T.A., and Shrift, A. 1982. Selenium: toxicity and tolerance in higher plants. *Biol. Rev.*, 57: 59–84.

Buhl, K.J., and Hamilton, S.J. 1996. Toxicity of inorganic contaminants, individually in environmental mixtures to three endangered fishes (Colorado Squawfish, Bonytail and Razorback Sucker). *Arch. Environ. Contam. Toxicol.*, 30: 84–92.

Bujdos, M., Kubová, J., and Streško, V. 2000. Problems of selenium fractionation in soils rich in organic matter. *Anal. Chim. Acta*, 408: 103–109.

Bujdos, M., Mul'ová, A., Kubová, J., and Medved', J. 2005. Selenium fractionation and speciation in rocks, soil, waters and plants in polluted surface mine environment. *Environ. Geol.*, 47: 353–360.

Cappon, C.J. 1991. Sewage sludge as a source of environmental selenium. *Sci. Total Environ.*, 100: 177–205.

Chasteen, T.G., and Bentley, R. 2003. Biomethylation of selenium and tellurium: Microorganisms and plants. *Chem. Rev.*, 103: 1–25.

Cherian, S., and Oliveira, M. 2005. Transgenic plants in phytoremediation: Recent advances and new possibilities. *Environ. Sci. Technol.*, 39: 9377–9390.

Combs, G.F., Garbisu, C., Yee, B.C., Yee, A., Carlson, D.E., Smith, N.R., et al. 1996. Bioavailability of selenium accumulated by selenite-reducing bacteria. *Biol. Trace Elem. Res.*, 52: 209–225.

Cowan, E.C., Zachara, J.M., and Resch, C.T. 1990. Solution ion effects on the surface exchange of selenite on calcite. *Geochim. Cosmochim. Acta*, 54: 2223–2234.

Dauchy, X., Potin-Gautier, M., Astruc, A., and Astruc, M. 1994. Analytical methods for the speciation of selenium compounds: A review. *Fresenius J. Anal. Chem.*, 348: 792–805.

De Kok, L.J., and Kuiper, P.J.C. 1986. Effect of short-term dark incubation with sulfate, chloride and selenate on the glutathione content of spinach [*Spinacia oleracea* cultivar *Estivato*] leaf discs. *Physiol. Plant.*, 68: 477–482.

De Souza, M.P., Pilon-Smits, E.A.H., Lytle, C.M., Hwang, S., Tai, J., Honma, T.S.U., Yeh, L., and Terry, N. 1998. Rate limiting steps in Se assimilation and volatilization by *Brassica juncea*. *Plant Physiol.*, 117: 1487–1494.

De Souza, M.P., Lytle, C.M., Mulholland, M.M., Otte, M.L., and Terry, N. 2000. Selenium assimilation and volatilization from dimethylselenipropionate by Indian mustard. *Plant Physiol.*, 122: 1281–1288.

De Souza, M.P., Pilon-Smits, E.A.H., and Terry, N. 2000. The physiology and biochemistry of selenium volatilization by plants. In Raskin, I., and Ensley, B.D. (Eds.) *Phytoremediation of toxic metals: using plants to clean up the environment*. Wiley and Sons, New York. pp. 171–188.

Dhillon, K.S., and Dhillon, S.K. 1999. Adsorption-desorption reactions of selenium in some soils of India. *Geoderma*, 93: 19–31.

Dhillon, K.S., and Dhillon, S.K. 2001. Restoration of selenium contaminated soils. In Iskandar, I.K. (Ed.) *Environmental restoration of metals-contaminated soils*. Lewis Publishers, Boca Raton FL. pp. 199–227.

Di Gregorio, S., Lampis, S., Malorgio, F., Petruzzelli, G., Pezzarossa, B., and Vallini, G. 2006. *Brassica juncea* can improve selenite and selenate abatement in selenium contaminated soil through the aid of its rhizospheric bacterial population. *Plant and Soil*, 285: 223–244.

Doty, S.L. 2008. Enhancing phytoremediation through the use of transgenics and endophytes. *New. Phytol.*, 179: 318–333.

Eapen, S., and D'Souza, S.F. 2005. Prospects of genetic engineering of plants for phytoremediation of toxic metals. *Biotechnol. Adv.*, 23: 97–114.

Ellis, A.S., Johnson, T.M., Herbel, M.J., and Bullen, T.D. 2003. Stable isotope fractionation of selenium by natural microbial consortia. *Chem. Geol.*, 195: 119–129.

Ellis, D.R. and Salt, D.E. 2003. Plants, selenium and human health. *Current Opinion in Plant Biology*, 6:273–279.

Ernst, W.H.O. 1982. Seelenpflanzen (Selenophyten). In *Pflanzenokologie and Mineralstoffwechsel*, Kinzel H. (Ed.) pp. 511–519. Stuttgart: Verlag Eugen Ulmer.

Evens, C.D., Asher, A.J., and Johnson, C.M. 1968. Isolation of dimethylselenide and other volatile selenium compounds from *Astragalus racemosus* (Pursh.). *Aust. J. Biol. Sci.*, 21: 13–20.

Feist, L.J., and Parker, D.R. 2001. Ecotypic variation in selenium accumulation among population of *Stanley pinnata*. *New Phytol.*, 149: 61–69.

Férnández-Martínez, A., and Charlet, L. 2009. Selenium environmental cycling and bioavailability: a structural chemist point of view. *Rev. Environ. Sci. Biotechnol.*, 8: 81–110.

Ferri, T., Petruzzelli, G., Pezzarossa, B., Santaroni, P., Brunori, C., and Morabito, R. 2003. Study of the influence of carboxymethylcellulose on the absorption of selenium (and selected metals) in a target plant. *Microchem. J.*, 74 : 257–265.

Fishbein, L. 1983. Environmental selenium and its significance. *Fundam. Appl. Toxicol.*, 3: 411–419.

Fordyce, F. 2007. Selenium geochemistry and health. *Ambio*, 36: 94–97.

Frankenberger, W.T. Jr., and Karlson, U. 1994. Microbial volatilization of selenium from soils and sediments. In Frankenberger W.T. Jr., and Bensoc S. (Eds.) *Selenium in the environment*. Marcel Dekker Inc., New York, NY. pp. 369–387.

Gao, S., Tanji, K.K., Peters, D.W., and Hebel, M.J. 2000. Water selenium speciation and sediment fractionation in a California flow-through wetland system. *J. Environ. Qual.*, 29: 1275–1283.

Goodson, C.C., Parker, D.R., Amrhein, C., and Zhang, Y. 2003. Soil selenium uptake and root system development in plant taxa differing in Se-accumulating capability. *New Phytologist*, 159: 391–401.

Guala, S.D., Vegaa, F.A., and Covelo, E.F. 2010. The dynamics of heavy metals in plant–soil interactions. *Ecolog. Model.*, 221: 1148–1152.

Guo, X., and Wu, L. 1998. Distribution of free seleno-aminoacids in plant tissue of *Melilotus indica* L. grown in selenium laden soils. *Ecotoxicol. Environ. Saf.*, 39: 207–214.

Gustafsson, J.P., and Johnsson, L. 1992. Selenium retention in the organic matter of Swedish forest soils. *J. Soil Sci.*, 43: 461–472.

Gustafsson, J.P., and Johnsson, L. 1994. The association between selenium and humic substances in forested ecosystems-laboratory evidence. *Appl. Organomet. Chem.*, 8: 141–147.

Harmsen, J., Rulkens, W., and Eijsackers, H. 2005. Bioavailability: concept for understanding or tool for prediting. *Land Contam. Reclam.*, 13: 161–171.

Hartikainen, H., Xue, T., and Piironen, V. 2000. Selenium as an antioxidant and prooxidant in ryegrass. *Plant and Soil*, 225:193–200.

He, Z.L., Yang, X.E., and Stoffella, P.J. 2005. Trace elements in agroecosystems and impacts on the environment. *J. Trace Elem. Med. Biol.*, 19; 125–140.

Heninger, I., Potin-Gautier, M., Astruc, M., Snidaro, D., Vignier, V., and Manem, J. 1997. Speciation of selenium and organotin compounds in sewage sludge applied to land. *Chem. Spec. Bioavailab.*, 10: 1–10.

Hopper, J.L., and Parker, D.R. 1999. Plant availability of selenite and selenate as influenced by the competing ions phosphate and sulfate. *Plant and Soil*, 210: 199–207.

Järup, L. 2003. Hazards of heavy metal contamination. *Br. Med. Bull.*, 68: 167–182.

Johnsson, L., 1989. Se levels in the mor layer of Swedish forest soils. *J. Soil Sci.*, 43: 461–472.

Iskandar, I.K., and Adriano, D.C. 1997. Remediation of soils contaminated with metals—A review of current practices in the USA. In Iskandar, I.K., and Adriano, D.C. (Eds.) *Remediation of soils contaminated with metals*. Northwood: Science Reviews. pp. 1–16.

ISO 17402. 2008. Soil quality-requirements and guidance for the selection of methods for the assessment of bioavailability of contaminants and soil materials.

Karlson, U., and Frankenberger, W.T. 1989. Accelerated rates of selenium volatilization from California soils. *Soil Sci. Soc. Am. J.*, 53: 749–753.

Kocsis, M.G., Nolte, K.D., Rhoads, D., Shen, T.L., Gage, D.A., and Hanson, A.D. 1998. Dimethysulfonopriopionate biosynthesis in *Spartina alterniflora*. *Plant Physiol.*, 117: 273–281.

Kortba, P., Naymanova, J., Macek, T., Ruml, T., and Mackova, M. 2009. Genetically modified plants in phytoremediation of heavy metals and metalloid soil and sediment pollution. *Biotechnol. Adv.*, 27: 799–810.

LeDuc, D.L., Tarun, A.S., Montes-Bayón, M., Meija, J., Malit, M.F., Wu, C.P., AbdelSamie, M., Chiang, C.Y., Tagmount, A., De Souza, M.P., et al. 2004. Overexpression of selenocysteine methytransferase in Arabidopsis and Indian mustard increases selenium tolerance and accumulation. *Plant Physiol.*, 135: 377–383.

LeDuc, D.L., AbdelSamie, M., Montes-Bayón, M., Wu, C.P., Reisenger, S.J., and Terry, N. 2006. Overexpression both ATP sulfurylase and selenocysteine methytransferase enhances selenium phytoremediation traits in Indian mustard. *Environ. Pollut.*, 144: 70–76.

Lewis, B., Johnson, C., and Delwiche, C. 1966. Release of volatile selenium compounds by plants: Collection procedures and preliminary observations. *J. Agric. Food Chem.*, 14: 638–664.

Lim, T.T., and Goh, K.H. 2005. Selenium extractability from a contaminated fine soil fraction: Implication on soil cleanup. *Chemosphere*, 58: 91–101.

Lin, Z.Q., Schemenauer, R.S., Cervinka, V., Zayed, A., and Terry, N. 2000. Selenium volatilization from a soil-plant system for the remediation of contaminated water and soil in the San Joaquin Valley. *J. Environ. Qual.*, 29:1048–1056.

Lin, Z.Q., and Terry, N. 2003. Selenium removal by constructed wetlands: Quantitative importance of biological volatilization in the treatment of selenium-laden agricultural drainage water. *Environ. Sci. Technol.*, 37: 606–615.

Lo S.L., and Chen, T.Y. 1997. Adsorption of Se(IV) and Se(VI) on an iron-coated sand from water. *Chemosphere*, 35: 919–930.

Losi, M.E., and Frankenberger Jr. W.T. 1997a. Bioremediation of selenium in soil and water. *Soil Sci.*, 162: 692–702.

Losi, M.E., and Frankenberger, Jr., W.T. 1997b. Reduction of selenium oxyanions by *Enterobacter cloacae* SLD1a-1: Isolation and growth of the bacterium and its expulsion of selenium particles. *Appl. Environ. Microbiol.*, 63: 3079–3084.

Ludwig, W., Strunk, O., Westram, R., Richter, L., Meier, H., Kumar, Y., et al. 2004. ARB: A software environment for sequence data. *Nucleic Acids Res.*, 32: 1363–1371.

Macek, T., Kortba, P., Svatos, A., Novakova, M., Demnerova, K., and Mackova, M. 2008. Novel roles for genetically modified plants in environmental protection. *Trends Biotechnol.*, 26: 146–152.

Macy, J.M., Lawson, S., and Demolldecker, H. 1993. Bioremediation of selenium oxyanions in San Joaquin drainage water using *Thauera selenatis* in a biological reactor system. *Appl. Microbiol. Bioctechnol.*, 40: 588–594.

Martens, D.A., and Suarez, D.L. 1997. Selenium speciation of soil/sediment determined with sequential extractions and hydride generation atomic absorption spectrophotometry. *Environ. Sci. Technol.*, 31: 171–177.

Mayland, H.F., James, L.F., Panter, K.E., and Sonderreger, J.L. 1989. Selenium in seleniferous environments. In Jacobs, L.W. (Ed.) *Selenium in agriculture and the environment*. American Society of Agronomy, Inc., Soil Science Society of America, Inc., Madison, WI. SSSA. Special Publication Number 23, pp. 15–50.

Meija, J., Montes-Bayon, M., LeDuc, D.L., Terry, N., and Caruso, J.A. 2002. Simultaneous monitoring of volatile selenium and sulphur species from Se accumulating plants (wild type and genetically modified) by GC/MS and GC/ICPMS using solid-phase microextraction for sample induction. *Anal. Chem.*, 74: 5837–5844.

Mikkelsen, R.L., Page, A.L., and Bingham, F.T. 1989. Factors affecting selenium accumulation by agricultural crops. In Jacobs, L.W. (Ed.) *Selenium in agriculture and the environment*. Spec. Pubb. SSSA, Madison, WI. pp. 65–94.

Montes-Bayon, M., LeDuc, D.L., Terry, N., and Caruso J.A. 2002. Selenium speciation in wildtype and genetically modified Se accumulating plants with HPLC separation and ICP-MS/ES-MS detection. *J. Anal. Atomic Spectrom.*, 17: 872–879.

Mudd, S.H., and Datko, A.H. 1990. The S-methylmethionine cycle in *Lemna paucicostata*. *Plant Physiol.*, 93: 623–630.

Munier-Lamy, C., Deneux-Mustin, S., Mustin, C., Merlet, D., Berthelin, J., and Leyval, C. 2007. Selenium bioavailability and uptake as affected by four different plants in a loamy clay soil with particular attention to mycorrhizae inoculated ryegrass. *J. Environ. Radioactivity*, 97: 148–158.

Neal, R.H., and Sposito, G. 1987. Selenate adsorption on alluvial soils. *Soil Sci. Soc. Am. J.*, 53: 79–74.

Neal, R.H. 1995. Selenium. In *Heavy metals in soil*. Alloway, B. (Ed.) Blackwell Publishers, Oxford, UK.

Oremland, R.S., and Stolz, J.F. 2000. Dissimilatory reduction of selenate and arsenate in nature. In Lovley, D.R. (Ed.) *Environmental metal-microbe interaction*. Amer. Soc. Microbiology Press, Washington, DC. pp. 199–224.

Oremland, R.S., Herbel, M.J., Blum, J.S., Langley, S., Beveridge, T.J., Ajayan, P.M., Sutto, T., et al. 2004. Structural and spectral features of selenium nanospheres produced by Se-respiring bacteria. *Appl. Environ. Microbiol.*, 70: 52–60.

Page, M., and Page, C. 2002. Electroremediation of contaminated soils. *J. Environ. Eng.*, 128: 208–219.

Papelis, C., Brown, G.E., Parks, G.A., and Leckie, K.O. 1995. X-ray-absorption spectroscopic studies of cadmium and selenite adsorption on aluminium-oxides. *Lagmuir*, 11: 2041–2048.

Parker, D.R., Feist, L.J., Varvel, T.W., Thomason, D.N., and Zhang, Y. 2003. Selenium phytoremediation potential of *Stanleya pinnata*. *Plant and Soil*, 249: 157–165.

Petruzzelli, G., Barbafieri, M., Bonomo, L., Saponaro, S., Milani, A., and Pedron, F. 2004. Bench scale evaluation of soil washing for heavy metal contaminated soil at a former manufactured gas plant site. *Bull. Environ. Contam. Toxicol.*, 73: 38–44.

Pezzarossa, B., Petruzzelli, G., Petacco, F., Malorgio, F., and Ferri, T. 2007. Absorption of selenium by *Lactuca sativa* as affected by carboxymethylcellulose. *Chemosphere*, 67: 322–329.

Pickering, I.J., Wright, C., Bubner, B., Ellis, D., Persans, M.W., Yu, E.Y., George, G.N., Prince, R.C., and Salt, D.E. 2003. Chemical form and distribution of selenium and sulphur in the selenium hyperaccumulator *Astragalus bisulcatus*. *Plant Physol.*, 131: 1460–1467.

Pilon, M., Owen, J.D., Garifullina, G.F., Kurihara, T., Mihara, H., Esaki, N., and Pilon-Smits, E.A.H. 2003. Enhanced selenium tolerance and accumulation in transgenic *Arabidopsis* expressing a mouse selenocysteine lyase. *Plant Physiol.*, 131: 1250–1257.

Pilon-Smits, E.A.H., Hwang, S., Mel Lytel, C., Zhu, Y., Tai, J.C., Bravo, R.C., et al. 1999. Overexpression of ATP-sulfurylase in Indian mustard leads to increased selenate uptake, reduction and tolerance. *Plant Physiol.*, 119: 123–132.

Pilon-Smits, E.A.H., Garifullina, G.F., Abdel-Ghany, S.E., Kato, S.I., Mihara, H., Hale, K.L., Burkhead, J.L., Esaki, N., Kurihara, T., and Pilon, M. 2002. Characterization of a NifS-like chloroplast protein from *Arabidopsis thaliana*—Implications for its role in sulphur and selenium metabolism. *Plant Physiol.*, 130: 1309–1318.

Pilon-Smits, E.A.H. 2005. Phytoremediation. *Annu. Rev. Plant Biol.*, 56: 15–39.

Pilon-Smits, E.A.H., and LeDuc, D.L. 2009. Phytoremediation of selenium using transgenic plants. *Curr. Opin. Biotechnol.*, 20: 207–212.

Pyrzynska, K. 2002. Determination of selenium species in environmental samples. *Microchim. Acta*, 140: 55–62.

Raskin, I., Smith, R.D., and Salt, D.E. 1997. Phytoremediation of metals using plants to remove pollutants from the environment. *Curr. Opin. Biotechnol.*, 8: 221–226.

Reddy, K.J. 1999. Selenium speciation in soil water: experimental and model predictions. In Selim, H.M., and Iskandar, I.K. (Eds.) *Fate and transport of heavy metals in the vadose zone*. Lewis Publishers, Boca Raton, FL. pp. 147–157.

Reeves, R.D. 1992. Hyperaccumulation of nickel by serpentine plants. In Proctor, J. (Ed.) *The vegetative of ultramafic (serpentine) soils*. J. Intercept, Andover: UK. pp. 253–277.

Reeves, R.D., and Baker, A.J.H. 2000. Metal accumulating plants. In Raskin, I., and Ensley, E.D. (Eds.) *Phytoremediation of toxic metals: Using plants to clean up the environment*. Wiley, New York. pp. 193–229.

Riveros, G., Loncot, D., Guillemoles, J.F., Henriquez, R., Schrebler, R., Cordova, R., and Gomez, H. 2003. Redox and solution chemistry of the $SeSO_3^{2-}$-Zn-$EDTA^{2-}$ system and electrodeposition behavior of ZnSe from alkaline solutions. *J. Electroanal. Chem.*, 558: 9–17.

Saeki, K., Matsumoto, S., and Tatsukawa, R., 1995. Selenite sorption by manganese oxides. *Soil Sci.*, 160: 265–272.

Schlekat, C.E., Dowdle, P.R., Lee, B.G., Luoma, S.N., and Oremland, R.S. 2000. Bioavailability of particle-associated Se to the bivalve *Potamocorbula amurensis*. *Environ. Sci. Technol.*, 34: 4504–4510.

Seby, F., Potin-Gautier, M., Giffaut, E., Borge, G., and Donard, O.F.X. 2001. A critical review of thermodynamic data for selenium species at 25°C. *Chem. Geol.*, 171: 173–194.

Selim, H.M., and Sparks, D. 2001. *Heavy metals release in soils*. CRC Press, Boca Raton, FL.

Shamberger, R.J. 1981. Selenium in the environment. *Sci. Total Environ.*, 17: 59–74.

Sharma, R.K., and Agrawal, M. 2005. Biological effects of heavy metals: an overview. *J. Environ. Biol.*, 26: 301–313.

Shrift, A., and Ulrich, J.M. 1976. Transport of selenate and selenite into *Astragalus* roots. *Plant Physiol.*, 44: 893–896.

Siddique, T., Zhang, Y., Okeke, B.C., and Frankenberger, Jr., W.T. 2006. Characterization of sediment bacteria involved in selenium reduction. *Bioresour. Technol.*, 97: 1041–1049.

Singh, M., Singh, N., and Relan, P.S. 1981. Adsorption and desorption of selenite and selenate selenium on different soils. *Soil Sci.*, 132: 134–141.

Smith, G.S., and Watkinson, J.H. 1984. Selenium toxicity in perennial ryegrass and white clover. *New Phytol.*, 97: 557–564.

Sors, T.G., Ellis, D.R., Na, G.N., Lahner, B., Lee, S., Leustek, T., Pickering, I.J., and Salt, D.E. 2005. Analysis of sulfur and selenium assimilation in *Astragalus* plants with varying capacities to accumulate selenium. *Plant J.*, 42: 785–797.

Sridhara Chary, N., Kamala, C.T., and Suman Raj, S.D. 2008. Assessing risk of heavy metals from consuming food grown on sewage irrigated soils and food chain transfer. *Ecotoxicol. Food Saf.*, 69: 513–524.

Stephens, D.W., and Waddell, B. 1998. Field screening of water quality, bottom sediment, and biota associated with irrigation on the Uintah and Ouray Indian Reservation, eastern Utah. U. S. Geological Survey, Report: WRI 98–4161, 45.

Stolz, J.F., and Oremland, R.S. 1999. Bacterial respiration of arsenic and selenium. *FEMS Microbiol. Rev.*, 23: 615–627.

Stolz, J.F., Basu, P., and Oremland, R.S. 2002. Microbial transformation of elements: The case of arsenic and selenium. *Int. Microbiol.*, 5: 201–207.

Su, C.M., and Suarez, D.L. 2000. Selenate and selenite sorption on iron oxides: An infrared and electrophoretic study. *Soil Sci. Soc. Am. J.*, 64: 101–111.

Tam, S.C., Chow, A., and Hadley, D. 1995. Effects of organic-component on the immobilization of selenium on iron oxyhydroxide. Sci. Total. Environ., 164: 1–7.

Tchounwou, P.B., Ayensu, W.K., Ninashvili, N., and Sutton, D. 2003. Environmental exposure to mercury and its toxipathologic implications for human health. *Environ. Toxicol.*, 18: 149–175.

Terry, N. Karlson, C., Raab, T.K., and Zayed, A.M. 1992. Rates of selenium volatilization among crop species. *J .Environ. Qual.*, 21: 341–344.

Terry, N., and Zayed, A.M. 1994. Selenium volatilization by plants. In Frankenberger W.T. Jr., Benson S (Eds.) *Selenium in the environment*. Marcel Dekker Inc., New York. p. 343–369.

Terry, N., Zayed, A.M., de Souza, M.P., and Tarun, A.S. 2000. Selenium in higher plants. *Annu. Rev. Plant Physiol. Plant Molec. Biol.*, 51: 401–432.

Terry, N., Sambukumar, S.V., and LeDuc, D.L. 2003. Biotechnological approaches for enhancing phytoremediation of heavy metals and metalloids. *Acta Biotechnol.*, 23: 281–288.

Tokunaga, T.K., Lipton, D.S., Benson, S.M., Yee, A.W., Oldfather, J.M., Duckart, E.C., Johannis, P.W., and Halvorsen, K.E. 1991. Soil selenium fractionation depth profiles and time rends in a vegetated site at Kesterson Reservoir. *Water, Air and Soil Pollut.*, 57–58: 38–41.

Van Huysen, T., Abdel-Ghany, S., Hale, K.L., LeDuc, D., Terry, N., and Pilon-Smits, E.A.H. 2003. Overexpression of cystathionine-γ-synthase enhances selenium volatilization in *Brassica juncea*. *Planta*, 218: 71–78.

Vodyanitskii, Y.N. 2010. Status and behavior of natural and technogenic forms of As, Sb, Se and Te in ore tailings and contaminated soils: a review. *Eurasian Soil Sci.*, 43: 30–38.

Wang, M.C., and Chen, H.M. 2003. Forms and distribution of selenium at different depths and among particles size fractions of three Taiwan soils. *Chemosphere*, 52: 585–593.

Wang, Z., and Gao, Y. 2001. Biogeochemical cycling of selenium in Chinese environments. *Appl. Geochem.*, 16, 1345–1351.
Wang, Z.J., Xu, Y., and Peng, A. 1996. Influences of fulvic acid on bioavailability and toxicity. *Biol. Trace Elem. Res.*, 55: 147–162.
Weisburg, W.G., Barns, S.M., Pelletier, D.A., and Lane, D.J. 1991. 16S Ribosomal DNA amplification for phylogenetic study. *J. Bacteriol.*, 173: 697–703.
Weres, O., Jaouni, A., and Tsao, L. 1989. The distribution, speciation, and geochemical cycling of selenium in a sedimentary environment, Kesterson Reservoir, California, U.S.A. *Appl. Geochem.*, 4: 543–563.
Wilber, C.G. 1980. Toxicology of selenium: A review. *Clin. Toxicol.*, 17, 171–230.
Wu, L., Huang, Z.Z., and Burau, R.G. 1988. Selenium accumulation and selenium-salt co-tolerance in five grass species. *Crop Sci.*, 28: 517–522.
Yadav, S.K., Juwarkar, A.A., Kumar, G.P., Thawale, P.R., Singh, S.K., and Chakrabarti, T. 2009. Bioaccumulation and phyto-translocation of arsenic, chromium and zinc by *Jatropha curcas* L.: Impact of dairy sludge and biofertilizer. *Bioresour. Technol.*, 100(20): 4616–4622.
Yu, T.R. 1997. *Chemistry of variable charge soils.* Oxford University Press, New York
Zasoski, R.J., and Burau, R.G. 1977. A rapid nitric-perchloric acid digestion procedure for multi-element tissue analyses. *Commun. Soil Sci. Plant Anal.*, 8: 425–436.
Zayed, A., Lytle, C.M., and Terry, N. 1998. Accumulation and volatilization of different chemical species of selenium by plants. *Planta*, 206: 284–292.
Zhang, P.C., and Sparks, D.L. 1990. Kinetics of selenate and selenite adsorption desorption at the goethite water interface. *Environ. Sci. Technol.*, 24: 1848–1856.
Zhang, Y.Q., and Moore, J.N. 1996. Selenium fractionation and speciation in a wetland system. *Environ. Sci. Technol.*, 33: 1652–1656.
Zhang, Y.Q., and Moore, J.N. 1997. Interaction of selenate with a wetland sediment. *Appl. Geochem.*, 12: 685–691.
Zhang, Y.Q., and Frankenberger, W.T. 1999. Effects of soil moisture, depth, and organic amendments on selenium volatilization. *J. Environ. Qual.*, 28: 1321–1326.
Zhang, Y.Q., and Frankenberger, W.T. 2003. Determination of selenium fractionation and speciation in wetlands sediments by parallel extraction. *Int. J. Environ. Anal. Chem.*, 83: 315–326.
Zhang, Y.Q., Zahir, Z.A., and Frankenberger, W.T. 2003. Factors affecting reduction of selenate to elemental selenium, in agricultural drainage water by *Enterobacter taylorae*. *J. Agric. Food Chem.*, 51: 3609–3613.
Zhang, E.S., Wang, H.L., Yan, X.X., and Zhang, L.D. 2005. Comparison of short-term toxicity between nano-Se and selenite in mice. *Life Sci.*, 76: 1099–1109.
Zhao, F.J., and McGrath, S.P. 2009. Biofortification and phytoremediation. *Curr. Opin. Plant Biol.*, 12: 373–380.

5

Colloid-Associated Transport and Metal Speciation at Reclaimed Mine Sites Following Biosolid Application

A.D. Karathanasis and J. O. Miller

CONTENTS

Introduction

The *unsaturated soil zone* (McCarthy and Zachara, 1989) is often assumed to act as a buffer for groundwater contaminants due to potential immobilization by sorption onto the soil matrix (Levin et al., 2002). However, water-dispersible colloid particles released from the soil matrix during rainstorm events can carry mobile and previously immobilized pollutants beyond this barrier (Seta and Karathanasis, 1997; De Jonge, Kjaergaard, and Moldrup, 2004). Reclaimed mine sites can be a source of heavy metals, derived either from the original unweathered spoil material or from industrial wastes, fertilizers, fly ash, or biosolids applied during various reclamation stages (Haigh, 1995). Unoxidized spoil materials can contain copper (Cu), lead (Pb), or zinc (Zn) sulfides (Geidel and Caruccio, 2000), while rock phosphate fertilizers can contain cadmium (Cd), chromium (Cr), iron (Fe), manganese (Mn), and lead (Pb) (Haigh, 1995).

Metal mobility within soil matrices is controlled by hydraulic properties, pH, mineralogy, surface adsorption (Konig, Baccini, and Ulrich, 1986), or complexation by organic compounds (Pohlman and McColl, 1986). Mass balance studies of soils receiving biosolid applications have indicated that up to 95% of biosolid-associated metals (McGrath and Lane, 1989; Sukkariyah et al., 2005) could be adsorbed in the upper 15 to 30 cm of the soil matrix, thereby reducing their mobility (Streck and Richter, 1997; Gove et al., 2001). More recent studies, however, have revealed significant metal transport in association with dispersed colloidal material to greater soil depths (Grolimund et al., 1996).

Following coal mining, reclaimed soils may become a source of mineral colloids due to the disturbance of the original soil matrix. Loss of aggregation due to mechanical disturbance by mining equipment, or loss of binding agents such as organic matter and carbonates, can increase the possibility of mineral colloid suspension. The amount of clay and its mineralogical composition, ionic strength, pH, soil moisture, and soil management are all factors that affect colloid mobilization (De Jonge, Kjaergaard, and Moldrup, 2004). Application of biosolids to promote structural formation and revegetation (Haering, Daniels, and Feagley, 2000) may be an additional source of organic colloids (Karathanasis and Ming, 2002; Karathanasis and Johnson, 2006), the transport of which can be facilitated through pseudo-karst channels that are common in reclaimed mine sites (Skousen, Sexstone, and Ziemkiewicz, 2000; Geidel and Carcuccio, 2000).

Although there are no data available for metal mobility in reclaimed mine sites, the high surface area and charge density of colloids generated from the disturbance and management of these sites could be an important vector in metal transport (Karathanasis, 1999; Bertsch and Seaman, 1999). Elevated concentrations of Cd, Cu, and Zn within the dispersible clay fraction have been documented in soil sites receiving increasing rates of biosolid application (Sukkariyah et al., 2005). Colloid-facilitated transport of Cu, Cr, Ni, Pb, and Zn has also frequently been observed in packed and undisturbed soil columns (De Jonge, Kjaergaard, and Moldrup, 2004). Biosolid-induced metal transport varies with soil properties, with greater mobility observed in coarse-textured soils than fine-textured soils (Stehouwer, Day, and Macneal, 2006). Clay loam soils receiving biosolid applications retained 90% of Cu, Ni, and Zn within the upper 25 cm 17 years after the original application (Sukkariyah et al., 2005). However, biosolid-derived colloids applied to undisturbed soil monoliths considerably enhanced the transport of Cu,

Zn, Pb, Cd, Cr, and Mo even in fine-textured soils (Karathanasis, Johnson, and Matocha, 2005; Karathanasis and Johnson, 2006). Organic functional groups and humic substances associated with biosolid colloids can promote metal binding and mobility (Sopper, 1993). Even lime-stabilized biosolids with high pH that reduce metal solubility (Brown and Chaney, 1997; Haering et al., 2000) may increase the dispersivity of organic colloids and increase the likelihood of transportability through the soil matrix (Karathanasis and Ming, 2002).

Case Study

In this case study, the goals were to (1) assess the mobility of Cd, Cr, Cu, Ni, Pb, and Zn within reclaimed mine sites of different ages with or without spoil materials following biosolid application; and (2) evaluate colloid, soil, and reclamation practices enhancing or inhibiting metal mobilization and transport.

Soil Monolith Preparation

Intact soil monoliths and spoil materials were obtained from the Powell River Project (PRP), near Wise, Virginia, in the southern Appalachian Mountains to represent 30-year-old reclaimed mine sites, and from Robinson Forest, near Jackson, Kentucky, to represent recently reclaimed (<5 years old) mine sites.

Monoliths were removed by digging a pedestal approximately 50 × 50 × 40 cm, then trimming them with knives and soil picks to fit within a polyvinyl chloride (PVC) tube of 20-cm internal diameter and 30-cm height. The 1-cm gap between the PVC and the soil was sealed with expandable Poly-U-Foam (Kardol, Lebanon, OH, 1-800-252-7365) to stabilize the monoliths and prevent preferential flow along the walls.

- Treatment 1 included two replicated, reclaimed mine sites.
- Treatment 2 was the same as Treatment 1 but with a 10-cm spoil layer attached in the bottom of the monolith.
- Treatment 3 was the same as Treatment 2 but receiving biosolid application.

For the Kentucky sites receiving biosolid application, Treatment 2 did not include the spoil material.

The lime-stabilized biosolid material used in the study was obtained from a municipal wastewater treatment facility in Winchester (Clark County), Kentucky. It was dried and applied to the surface of the soil at a rate of 20 $T.ha^{-1}$.

Bulk Soil Analysis

The reclaimed, spoil, and biosolid materials used in the experiments were air-dried and passed through a 2-mm sieve before analysis. Environmentally available Cd, Cr, Cu, Ni, Pb, and Zn levels were extracted with HNO_3 and HCl at 95°C using EPA method 3050b (USEPA, 1996). Metal levels in the extractants were determined by inductively coupled plasma-mass spectrometry (ICP-MS). The pH and electrical conductivity (EC) were measured on a Denver Instrument Model 250 pH*ISE* conductivity meter. Ammonium acetate extracts were used to determine cation exchange capacity (CEC) and total exchangeable bases (TEB). The affinity of materials for Cd, Cr, Cu, Ni, Pb, and Zn was assessed with adsorption isotherms. Duplicate 1-g soil, spoil, and biosolid samples were added to 50-mL centrifuge tubes with 20 mL of 0 to 5 $mg.L^{-1}$ metal concentrations. Samples were shaken on a reciprocating shaker for 24 h at room temperature and centrifuged for 1 h at 3,500 rpm. Supernatants were collected and analyzed for Cd, Cr, Cu, Ni, Pb, and Zn via inductively coupled plasma (ICP) spectrophotometry. Freundlich isotherms fitted on log-scale by linear regression were used to describe the experimental adsorption data.

Colloid Fractionation and Characterization

Water-dispersible colloids were fractionated from bulk samples of soil, spoil, and biosolid materials using 50-g samples mixed in a 1-L centrifuge bottle filled with deionized water. The slurry was shaken for 1 h and centrifuged at 750 rpm for 3.5 min. The suspended colloid particles were decanted and the procedure was repeated twice on the same 50-g sample. Metal affinity for the colloid particles was assessed with adsorption isotherms using 100 mg dried colloid samples following the procedure described for the bulk materials.

In Situ Colloid Elution

In situ colloid generation and elution from monoliths was assessed with leaching experiments. A rainfall simulator was set up to apply deionized water at a rate of 250 $mL.h^{-1}$ (1.0 $cm.h^{-1}$) to the surface of each monolith. The application rate was controlled with a peristaltic pump. The upper boundary condition of the monolith was 0 cm and the lower boundary was kept at −10 cm using a Mariotte device. To keep the lower boundary at −10 cm, the monolith was placed in a large funnel and sealed around the edges with a silicone gel. A tube attached at the bottom of the funnel allowed the leachate to drip into a sealed 2-L flask. This collection flask was connected to the Mariotte device in order to maintain the lower boundary condition at the desired level. A tensimeter inserted at the bottom of the monolith was used to monitor the pressure within the funnel, which was adjusted to −10 cm through the Mariotte device.

The leaching of each monolith was conducted in six cycles, corresponding to two to three pore volumes (PV) of elution. Each cycle consisted of 2 L of water elution at 24-h intervals. Leachate was collected at the bottom of the monolith every hour for a total of 8 h. Suspension concentrations were determined gravimetrically by taking a 20-mL aliquot from each hourly sample and drying it at 105°C in a preweighed aluminum tin for 24 h. The pH and EC were determined for each hourly elution. Eluted dissolved organic carbon (DOC) was measured in H_2SO_4-acidified eluents with a Shimadzu TOC 500DA carbon analyzer. Colloid particle size was measured with a Beckman Coulter N5 submicron particle size analyzer.

Eluents were also analyzed for dissolved metals following filtration through a 0.2-μm filter to remove colloidal material. The filtered material was analyzed for dissolved metals by ICP. Colloid materials remaining on the filters were washed with 1 M HCl/HNO_3 to extract colloid-bound metals. The HCl/HNO_3 filtrates were also analyzed for metals by ICP. Selected samples were assessed for anionic eluent composition (F, NO_2, NO_3, Br, PO_4, and SO_4) for metal speciation purposes with a Metrohm 792 basic ion chromatograph (IC) following filtration through a 0.2-μm filter. The concentrations of DOC, metals, and anions were entered into a Visual Minteq version 2.23 software program to estimate dissolved metal speciation.

Findings

Metal Concentrations in Bulk Samples

All metals except Cd were detected in all reclaimed, spoil, and biosolid materials using the EPA total recoverable metals digestion (Table 5.1). Zinc was highest

TABLE 5.1

Extractable Levels of Cd, Cr, Cu, Ni, Pb, and Zn (in mg.L^{-1}) by HCl/HNO_3 in Soils, Spoil, and Biosolids

	Virginia		Kentucky		
	Reclaimed	Spoil	Reclaimed	Spoil	Biosolids
			(mg.L^{-1})		
Cd	nd	nd	nd	nd	nd
Cr	0.06	0.11	0.03	0.15	0.06
Cu	0.12	0.32	0.07	0.64	0.60
Ni	0.13	0.20	0.08	0.31	0.04
Pb	0.08	0.13	0.15	0.18	0.06
Zn	0.66	0.46	0.49	1.40	1.70

Note: nd = none detected.

in all samples, ranging from 0.46 to 1.70 mg/L, followed by Cu (0.07 to 0.64 ppm). Chromium (0.03 to 0.15 mg/L) typically had the lowest concentrations. The Kentucky spoil materials had the highest Cr (0.15 ppm), Cu (0.64 mg/L), Ni (0.3 mg/L), and Pb (0.18 mg/L) concentrations, and the second highest Zn (1.4 mg/L) levels of all the treatment materials. Although this spoil material was minimally weathered, the similarity in metal content and affinity suggested limited potential for significant differences in metal elution across treatments.

Adsorption Isotherms

The linear Freundlich isotherm had the best fit for the metal sorption data, with r^2 values usually above 0.8. Whole biosolid materials had the highest overall affinity for Cd, Pb, and Zn (Table 5.2). There were no significant differences between Virginia (old) and Kentucky (young) reclaimed bulk soil materials, with Cu consistently showing the highest affinity, and Ni and Zn showing the lowest. The Virginia spoil bulk materials generally showed an elevated affinity for all metals except Cr. Surprisingly, colloid material affinity for most metals was in the same range as the respective bulk materials, except for Cu in the Virginia spoil and Pb in the Kentucky reclaimed materials, which were nearly twofold higher. The biosolid colloids also showed a slight increase for Cu and Pb preference over their respective bulk materials. The general absence of higher affinity of the colloid particles for individual metals compared to their bulk counterparts suggests that surface area may not be the only factor controlling metal sorption processes. Oxide and organic coatings may have played a role in this behavior. Overall, there were no drastic sorption affinity differences across metals for soils or colloids, suggesting that any of these metals had an equal probability of being transported by mobile particles through the matrix.

TABLE 5.2

Adsorption Isotherm Constants K_d (L.kg^{-1}) for Cd, Cr, Cu, Ni, Pb, and Zn in Virginia and Kentucky Bulk Soil and Colloid Samples

	Bulk Materials					Colloids				
	VR	VS	KR	KS	B	VR	VS	KR	KS	B
					(L.kg^{-1})					
Cd	1.43	2.15	1.88	1.17	3.11	1.52	1.57	1.58	1.56	3.26
Cr	1.29	1.58	1.02	2.17	1.49	1.95	2.16	1.92	1.89	1.63
Cu	1.76	2.48	2.08	2.02	1.15	1.67	2.28	4.54	1.15	1.72
Ni	1.42	1.92	1.67	0.97	1.01	1.53	1.77	2.03	1.16	1.64
Pb	1.92	2.73	1.58	2.65	2.74	2.07	5.57	2.52	2.39	2.53
Zn	1.26	1.69	1.58	1.49	3.39	1.09	1.50	1.88	1.12	2.96

Note: VR = Virginia Reclaimed; VS = Virginia Spoil; KR = Kentucky Reclaimed; KS = Kentucky Spoil; B = Biosolids.

Metal Elution in Virginia Monoliths

Reclaimed Monoliths

The Virginia reclaimed monoliths (Figure 5.1) eluted only traces (<0.05 mg) of soluble Cd, Cr, Cu, and Pb, but considerably higher amounts of Ni and Zn. Total metal loads transported throughout the leaching cycle were much higher for the first four metals, suggesting a significant contribution by mobile colloids (Table 5.3). Colloid-bound metal transport increased for Cr, Cd, Cu, and Pb by 85% to 92%, but only 14% for Ni. Both Cd and Cu displayed a pattern similar to colloid elution at 0.75 PV, while Cd, Cr, Cu, and Ni release

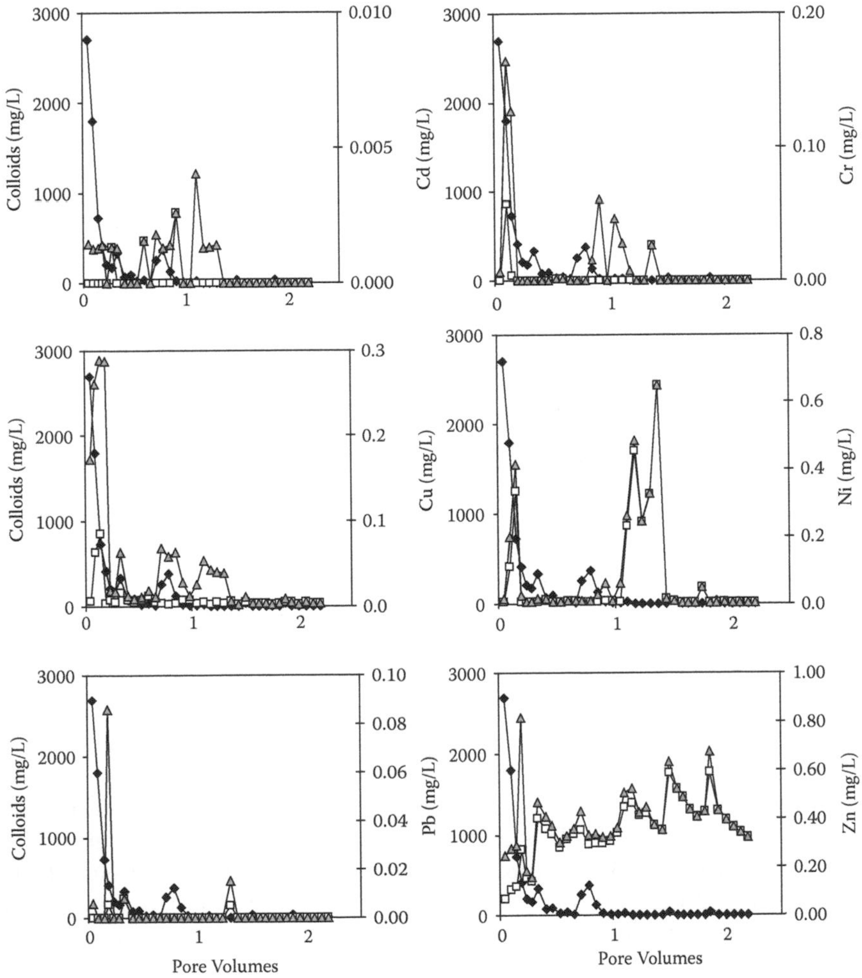

FIGURE 5.1
Elution results for Cd, Cr, Cu, Ni, Pb, and Zn from Virginia reclaimed monoliths.

matched colloid elution patterns after 1 PV. However, only Ni was dominantly transported as a solute. Zn was released over the entire leaching cycle, but only 11% (Table 5.3) was colloid bound. Total Zn was much higher and the percent colloid bound was lower due to its consistent release as a solute after colloids ceased to elute from the monoliths. Zn release resembled the leaching cycle pattern, with peaks occurring at the beginning of each cycle and descending as the 8-h cycle continued. The Virginia reclaimed monolith colloids carried the highest fraction of metals of any treatment (Table 5.3). This could be attributed to the higher total colloid load released from these monoliths, which was influenced by the higher pH and smaller colloid particle diameter (Table 5.4). The reclaimed colloids also had a higher percentage of more reactive 2:1 minerals and gibbsite to bind the metals.

Spoil Monoliths

When a spoil layer was present in the bottom of the reclaimed monoliths, colloid transport was significantly inhibited (Figure 5.2, Table 5.4), thus reducing the chance of colloid-mediated transport of metals. Trace amounts of Cd and Cu were detected in some eluents, with less than 13% being colloid bound (Table 5.3). Only Zn and Ni had a consistent release throughout the leaching cycle. More total Zn was released from the reclaimed soil monoliths when spoil material was present, with less than 1% being colloid bound. The Zn elution pattern also resembled the leaching cycle of the reclaimed monoliths discussed above.

Biosolid Monoliths

When biosolids were applied to the reclaimed spoil combination monoliths, both colloid and metal elution resumed at much higher concentrations than any other treatment. Total metal loads transported in the presence of biosolids were two to twenty-five times higher than those encountered with the reclaimed or the reclaimed spoil combination monoliths (Table 5.3). However, the eluted colloid-bound metal load was not as high, suggesting that organic ligands rather than colloid particles may be responsible for carrying the additional metal load. This is supported by the colloid elution pattern, which was constant but irregular throughout the leaching (Figure 5.3), while metal elution was continuous for Cd, Cu, and Zn, with peak concentrations occurring within the first 0.5 PV. This pattern could have also been influenced by the release of salts leaching from the biosolids, as indicated by eluent EC measurements. Although there was no peak concentration, Cu was eluted throughout the entire leaching cycle, following an erratic colloid elution pattern. Carbonates from the lime-stabilized biosolids probably precipitated any traces of Pb as insoluble $PbCO_3(s)$ within the monoliths, thus reducing the possibility of Pb transport. The entire amount of Pb observed in the eluent was bound to colloid materials. Colloid transport of Cd, Cr, Cu, Ni, and

TABLE 5.3

Total (mg) and Colloid Bound (%) Metal Loads in Virginia and Kentucky Treatments

	Cd		Cr		Cu		Ni		Pb		Zn	
	Total (mg)	Colloid Bound (%)	Total (mg)	Colloid Bound (%)	Total (mg)	Colloid Bound (%)	Total (mg)	Colloid Bound (%)	Total (mg)	Colloid Bound (%)	Total (mg)	Colloid Bound (%)
VR	0.01	88	0.09	92	0.40	87	0.66	14	0.02	85	3.29	11
VRS	< 0.01	nd	nd	nd	0.04	13	0.12	15	nd	nd	6.78	0.40
VRSB	0.02	1	< 0.01	2	0.93	7	3.31	26	< 0.01	100	21.53	4
KR	< 0.01	39	nd	nd	nd	nd	0.05	87	0.06	56	2.75	12
KRS	0.18	nd	0.03	nd	0.96	nd	7.20	nd	0.03	nd	61.52	nd
KRB	0.36	43	< 0.01	17	3.30	8	2.87	2	< 0.01	100	7.04	9

Note: VR = Virginia Reclaimed; VRS = Virginia Reclaimed + Spoil; VRSB = Virginia Reclaimed + Spoil + Biosolids; KR = Kentucky Reclaimed; KRS = Kentucky Reclaimed + Spoil; KRB = Kentucky Reclaimed + Biosolids.

TABLE 5.4
Selective Properties of Eluents and Colloids for Virginia and Kentucky Treatments

	Virginia Eluents/Colloids			Kentucky Eluents/Colloids		
	Reclaimed	Spoil Amended	Biosolid Amended	Reclaimed	Spoil Amended	Biosolid Amended
pH	6.1	4.9	4.3	7.7	4.2	7.9
EC (μS. cm^{-1})	375	506	859	441	4541	651
Kaolinite (%)	14	—	33	40	—	27
2:1 Minerals (%)	55	—	48	48	—	35
Quartz (%)	6	—	13	10	—	37
Gibbsite	25	—	6	0	—	0
Particle Size (nm)	521	460	2,662	1,125	—	4,704
Total Colloid Mass (mg)	1460	76	871	10,906	0	1,209
Colloid Conc (mg.L^{-1})	208	11	79	1,016	0	104

Zn accounted for less than 5% of their accumulated mass, although colloids were continuously released from biosolid-amended monoliths (Table 5.4). Total Cu, Ni, and Zn were higher in eluents of biosolid-amended monoliths than any other Virginia treatment, suggesting that initial leaching may flush high levels of soluble metals through to groundwater.

Metal Elution in Kentucky Monoliths

Reclaimed Monoliths

The Kentucky reclaimed monoliths (Figure 5.4) released trace amounts (<0.06 mg) of Cd, Ni, and Pb in single flushing events, mostly coincidental to those of colloids. Ni was detected only in the first eluent sample. Although Cu was present in the bulk soil extractions and showed a high affinity for reclaimed colloids, no Cu was observed in eluents from reclaimed monoliths. Overall, Zn was the only metal consistently eluted throughout the entire leaching cycle at quantities somewhat smaller than the Virginia reclaimed monoliths and with only 12% being colloid bound (Table 5.3). This may be due to the lower overall metal content of the Kentucky reclaimed mine sites (Table 5.1). The pattern of Zn elution coincided with flushing events at the beginning of each leaching cycle.

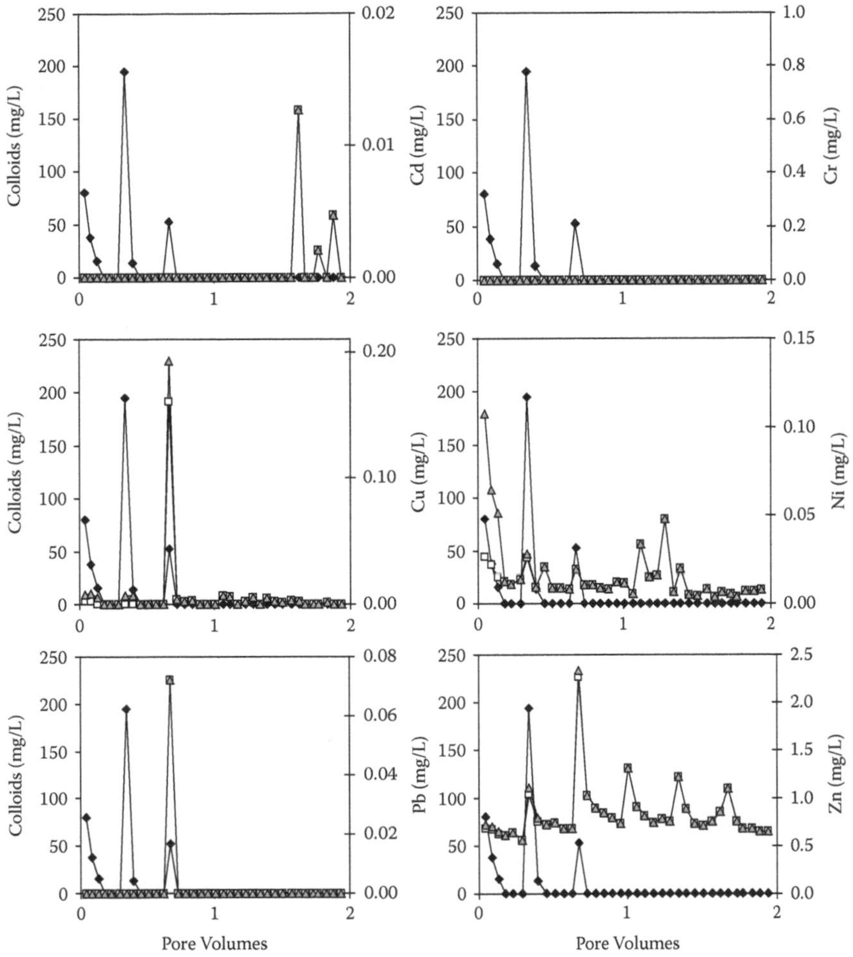

FIGURE 5.2
Inhibition of colloid transport in reclaimed Virginia monoliths.

Spoil Monoliths

When unweathered spoil material was placed below the reclaimed material, colloid transport was completely inhibited (Figure 5.5). All metals showed a gradually or abruptly decreasing concentration with advance leaching. Both Cr and Pb dropped to levels below ICP detection within 1 pore volume (PV), showing the lowest extracted concentrations of all six metals. Cu and Ni descended to steady concentrations of 0.03 and 0.20 mg/L, respectively, after 2 pore volume (PV). The elution of Cd and Zn was irregular, with spikes generally following the resumption of each leaching cycle. Total Zn eluted from the Kentucky reclaimed-spoil combination monoliths was the highest across all treatments (Table 5.3).

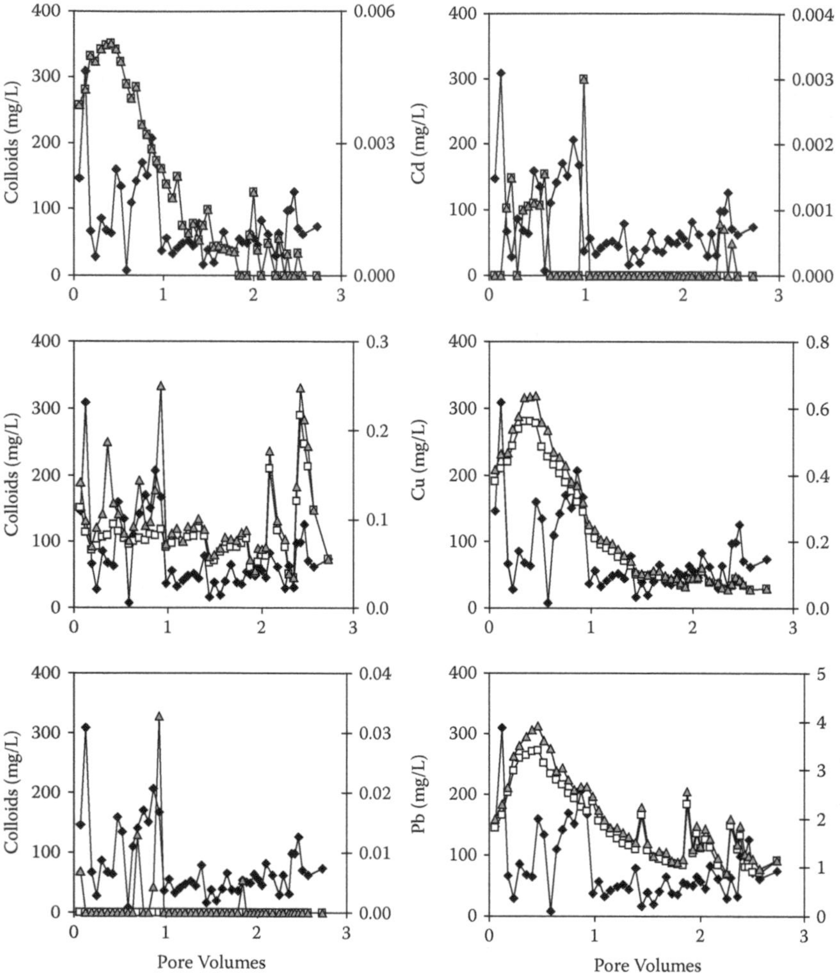

FIGURE 5.3
Colloid elution pattern during leaching for reclaimed spoil combination monoliths.

Biosolid Monoliths

The application of biosolids to the Kentucky reclaimed monoliths resulted in negligible releases of Cd, Cr, and Pb, but notable increases in Cu, Ni, and Zn elution compared to the reclaimed without spoil monoliths (Figure 5.6). Cu, Ni, and Zn elution reached maximum concentrations within the first 1 to 2 PV, descending thereafter to levels similar to the biosolid-amended Virginia monoliths. Colloid-associated transport of Cu, Ni, and Zn was relatively low, but nearly 100% of Pb and between 17% and 43% of Cr and Cd eluted from the biosolid-amended Kentucky monoliths was colloid bound (Table 5.3). Total Cu was higher than all other treatments, while total Ni was similar to

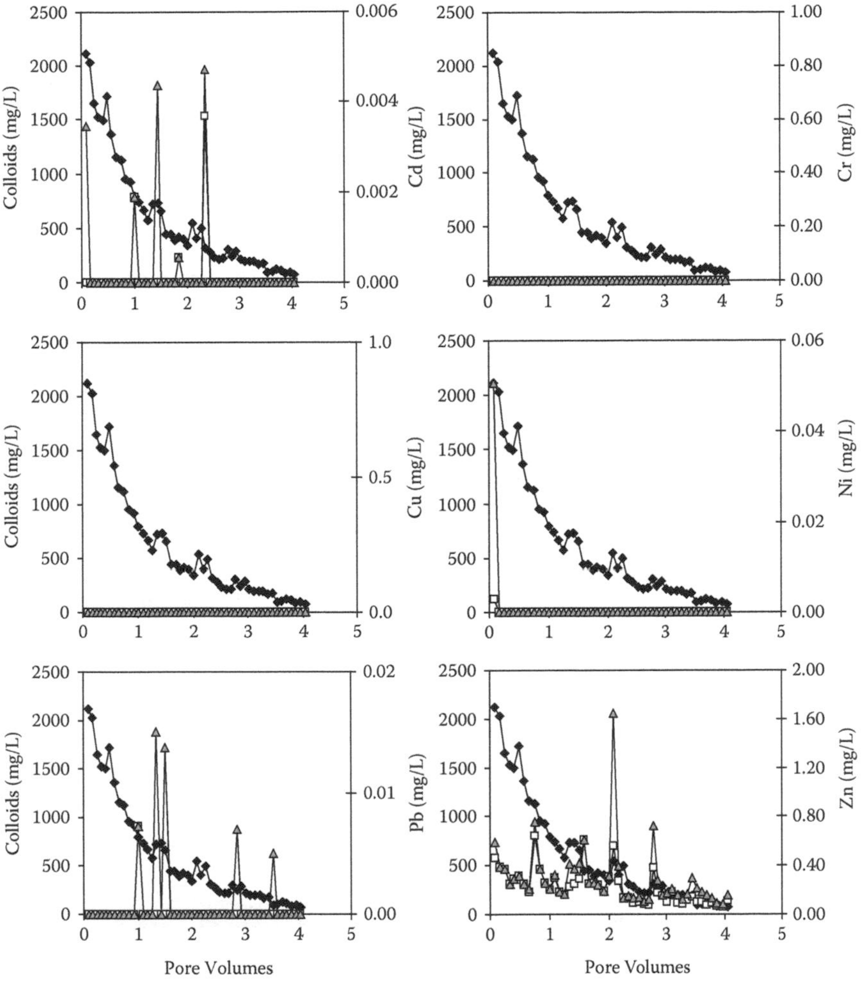

FIGURE 5.4
Elution results for Cd, Cr, Cu, Ni, Pb, and Zn from Kentucky reclaimed monoliths.

that of Virginia biosolid monoliths. Several peaks in Ni and Zn elution coincided with the beginning of leaching cycles.

Metal Associations

Selected colloid samples were treated with ammonium acetate and with 1 M HCl/HNO_3 to determine if the colloid-bound metals were exchangeable. Colloid-bound Cr and Pb were not detected when NH_4^+ was used as an exchangeable cation due to low overall eluent concentrations (<0.01 $mg.L^{-1}$). Colloid-bound Cd, Ni, and Zn concentrations were similar for ammonium and

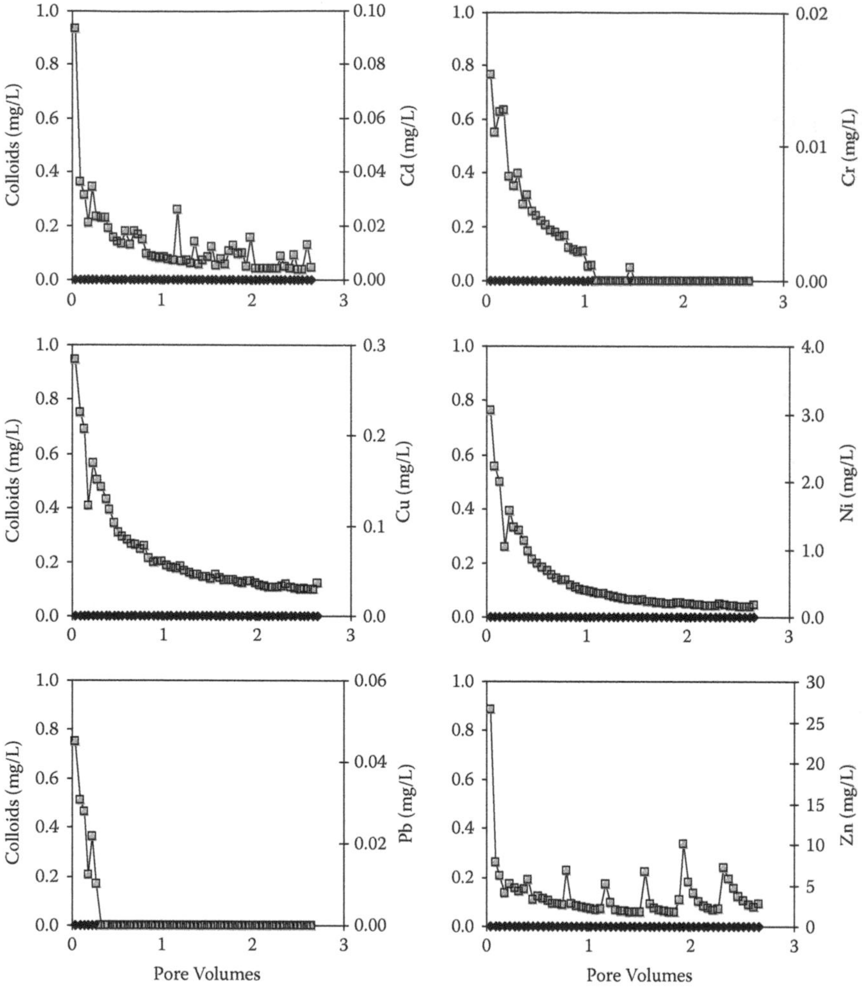

FIGURE 5.5
Inhibition of colloid transport in biosolid-amended Kentucky monoliths.

acid extractions, and only Cu had significantly higher ammonium-extractable than acid-extractable levels. This would indicate that the metals detected in the Virginia and Kentucky monolith eluents were primarily exchangeable, thus having an increased potential to be released by colloids to water resources.

Selected samples from the Kentucky eluents were analyzed by ion chromatography to determine anion content, which was combined with data for all six metals plus Ca, Mg, K, Na, Al, Fe, and DOC. The concentrations of various complexed forms of Cd, Cr, Cu, Ni, Pb, and Zn were calculated by Visual Minteq in the solution phase. For the Kentucky reclaimed monoliths, all of the dissolved Cd, Cu, and Zn were found to be complexed by the DOC

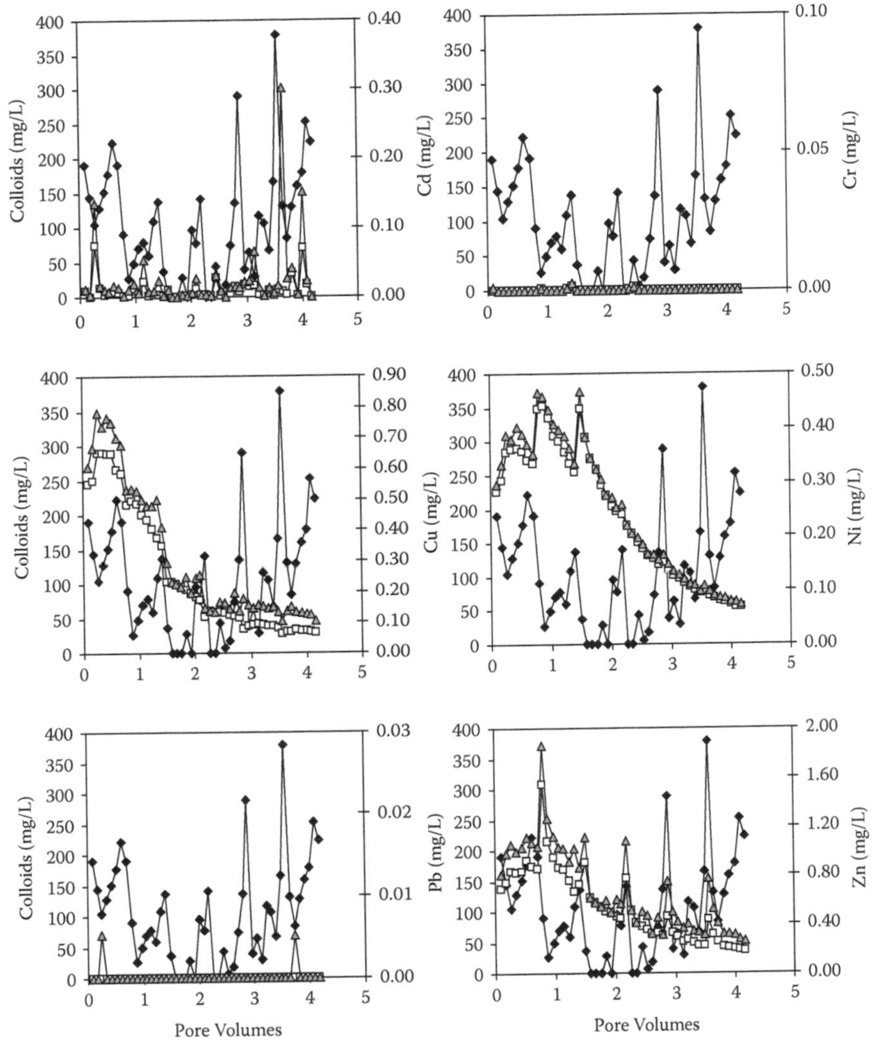

FIGURE 5.6
Colloid elution patterns for reclaimed monoliths without spoil.

(Table 5.5). The selected reclaimed samples did not have Cr, Ni, or Pb above the ICP detection limits to determine speciation. The Kentucky reclaimed monoliths underlain by spoil did have all six metals present, but only Cr was entirely complexed by DOC (Bartlett, 1991). Up to 50% of the Cd, Ni, or Zn was free in solution, with the remaining 40% associated with SO_4^{2-} and less than 10% was complexed by DOC. Both Cu and Pb were either complexed by DOC or associated with SO_4^{2-}, without any free solution species. The Kentucky reclaimed and reclaimed-spoil combination monoliths had

TABLE 5.5

Metal Complexation of Metals by DOC (%) in Kentucky Reclaimed, Reclaimed-Spoil Combination, and Biosolid Amended Monoliths as Indicated by Visual Minteq 2.23 Software

	Reclaimed	Spoil	Biosolids
		(%)	
Cd	100	11	98
Cr	nd	100	nd
Cu	100	77	100
Ni	nd	1	68
Pb	nd	59	nd
Zn	100	0	37
DOC ($mg.L^{-1}$)	44.5	42.2	162.4

Note: nd = none detected, otherwise not determined.

similar average concentrations of DOC, but the extremely high SO_4^{2-}-content in spoil materials may have outcompeted the complexing ability of DOC (Vega, Covelo, and Andrade, 2006). Kentucky reclaimed monoliths amended with biosolids also showed variability between the metals and DOC association. All of the Cu and a majority of the Cd and Ni in solution were complexed by DOC, but only 37% of the soluble Zn was associated with organic carbon. Most of the Zn was in free divalent form, although SO_4^{2-} and DOC contents were comparable to those seen in reclaimed monoliths.

Concluding Remarks

In general, the transport of Cd, Cr, Cu, Ni, Pb, and Zn was increased by their association with colloids in most treatments. The results varied between metals and treatments, but generally the presence of biosolid amendments contributed to higher metal release in soluble and colloid-bound metal fractions. Monoliths with greater colloid release had higher percentages of colloid-bound metals. Recently, reclaimed Kentucky soils following SMCRA guidelines appeared to have lower levels of metal release and contamination potential than the 30-year-old nonregulated Virginia reclaimed soil sites. The presence of spoil layers, particularly in unweathered form, increased the elution of certain metals, especially that of Ni and Zn. Zinc was the dominant metal released in all treatments, while Cu and Ni were present in all but one treatment each. Trace amounts of Cd, Cr, and Pb were observed in most eluents, with Pb being predominantly colloid bound. Nearly all the colloid-bound metals were exchangeable, thus increasing the importance of colloid transport to the overall health of a watershed. Dissolved metals were mostly associated with DOC, unless high sulfate contents were present, particularly in spoil materials.

The results showed conclusive evidence that the addition of biosolid amendments to reclaimed soils could increase transport of metals through both dissolved and colloid phase mobilization. Therefore, reclamation methods must be carefully planned to limit the amount of heavy metals transported to groundwater resources, as metal concentrations in these systems are inherently elevated and can be mobilized further following the disturbance. The application of biosolids should follow strict EPA guidelines governing the concentration of metals within the materials applied. While it may not be possible to restrict the amount of toxic spoil material present within the reclaimed soils, metal composition tests are necessary so that cautionary predictions can be made on the possibility of groundwater contamination.

References

Bartlett, R.J. 1991. Chromium cycling in soils and water: Links, gaps, and methods. *Environ. Health Perspec.* 92, 17–24.

Bertsch, P.M., and Seaman, J.C. 1999. Characterization of complex mineral assemblages: Implications for contaminant transport and environmental remediation. *Proc. Natl. Acad. Sci. USA* 96, 3350–3357.

Brown, S., and Chaney, R. 1997. Subsurface liming and metal movement in soils amended with lime-stabilized biosolids. *J. Environ. Qual.* 26, 724–732.

De Jonge, L.W., Kjaergaard, C., and Moldrup, P. 2004. Colloids and colloid-facilitated transport of contaminants in soils: An introduction. *Vadose Zone J.* 3, 321–325.

Geidel, G., and Caruccio, F.T. 2000. Geochemical factors affecting coal mine drainage quality. In *Reclamation of drastically disturbed lands.* R.I. Barnhisel et al. (Eds.) Agron. Monogr. 41. ASA, CSSA, and SSSA. Madison, WI. pp. 105–130.

Gove, L., Cooke, C.M., Nicholson, F.A., and Beck, A.J. 2001. Movement of water and heavy metals (Zn, Cu, Pb, and Ni) through sand and sandy loam amended with biosolids under steady-state hydrological conditions. *Biores. Technol.* 78, 171–179.

Grolimund, D., Borkovec, M., Barmettler, K., and Sticher, H. 1996. Colloid-facilitated transport of strongly sorbing contaminants in natural porous media: a laboratory column study. *Environ. Sci. Technol.* 30, 3118–3123.

Haering, K.C., Daniels, W.L., and Feagley, S.E. 2000. Reclaiming mined lands with biosolids, manures, and papermill sludges. In *Reclamation of drastically disturbed lands.* R.I. Barnhisel (Ed.) Agron. Monogr. 41. ASA, CSSA, and SSSA. Madison, WI. pp. 615–644.

Haigh, M.J. 1995. Soil quality standards for reclaimed coal-mine disturbed lands: A discussion paper. *Int. J. of Mining, Recl. Environ.* 9, 187–202.

Karathanasis, A.D. 1999. Subsurface migration of copper and zinc mediated by soil colloids. *Soil Sci. Soc. Am. J.* 63, 830–838.

Karathanasis, A.D., and Ming, D.W. 2002. Colloid-mediated transport of metals associated with lime-stabilized biosolids. *Devel. Soil Sci.* 28, 49–62.

Karathanasis, A.D., Johnson, D.M.C., and Matocha, C.J. 2005. Biosolid colloid mediated transport of copper, zinc, and lead in waste amended soils. *J. Environ. Qual.* 34, 1153–1164.

Karathanasis, A.D., and Johnson, D.M.C. 2006. Subsurface transport of Cd, Cr, and Mo mediated by biosolid colloids. *Sci. of the Total Environ.* 35, 157–169.

Konig, N., Baccini, P., and Ulrich, B. 1986. The influence of natural organic substances on the distribution of metals over soil and soil solution. *Z. Pflanzeneraehr. Bodenkd.* 149, 69–82.

Levin, J.M., Herman, J.S., Hornberger, G.M., and Saiers, J.E. 2002. The effect of soil water tension on colloid generation within an unsaturated, intact soil core. *Workshop: Colloids and Colloid Facilitated Transport of Contaminants in Soils and Sediments.* DIAS report. pp. 107–111.

McCarthy, J.F., and Zachara, J.M. 1989. Subsurface transport of contaminants. *Environ. Sci. Technol.* 23, 496–502.

McGrath, S.P., and Lane, P.W. 1989. An explanation for the apparent losses of metals in a long-term field experiment with sewage sludge. *Environ. Pollut.* 60, 235–256.

Pohlman, A.A., and McColl, J.G.. 1986. Kinetics of metal dissolution from forest soils by soluble organic acids. *J. Environ. Qual.* 15, 86–92.

Seta, A.K., and Karathanasis, A.D. 1997. Water dispersible colloids and factors influencing their dispersability from soil aggregates. *Geoderma* 74, 255–266.

Skousen, J.G., Sexstone, A., and Ziemkiewicz, P.F. 2000. Acid mine drainage control and treatment. In R.I. Barnhisel et al. (Eds.) *Reclamation of drastically disturbed lands.* Agron. Monogr. 41. ASA, CSSA, and SSSA. Madison, WI.

Sopper, W.E. 1993. *Municipal sludge use in land reclamation.* Lewis Publishers, Boca Raton, FL.

Stehouwer, R, Day, R.L., and Macneal, K.E. 2006. Nutrient and trace element leaching following mine reclamation with biosolids. *J. Environ. Qual.* 35, 1118–1126.

Streck, T., and Richter, J. 1997. Heavy metal displacement in a sandy soil at the field scale. I. Measurements and parameterization of sorption. *J. Environ. Qual.* 26, 49–56.

Sukkariyah, B.F., Evanylo, G., Zelazny, L.W., and Chaney, R.L. 2005. Recovery and distribution of biosolids-derived trace metals in a clay loam. soil. *J. Environ. Qual.* 34, 1843–1850.

USEPA, 1996. Test Methods for Evaluating Solid Waste, Physical/Chemical Methods, 3rd ed.; USEPA SW-846, U.S. Government Printing Office: Washington, DC.

Vega, F.A., Covelo, E.F., and Andrade, M.L. 2006. Competitive sorption and desorption of heavy metals in mine soils: Influence of mine soil characteristics. *J. Coll. Interface Sci.* 298, 582–592.

6

Trace Element Biogeochemistry in the Rhizosphere

Walter W. Wenzel, Eva Oburger,
Markus Puschenreiter, and Jakob Santner

CONTENTS

Introduction

Trace elements (TEs) commonly occur in the environment at very low concentrations but may exceed their normal concentration range (i.e., background levels) up to several orders of magnitude in soils developed on geogenic anomalies (e.g., ultramafic rocks, ores) or polluted by human activities such as mining and smelting operations. A number of TEs such as Zn, Cu, or Fe are essential nutrients for plants, microorganisms, and other biota including humans; some may be beneficial while others such as Cd have no proven biological function. Depending on their environmental concentration and the sensitivity of the biological receipient, essential as well as nonessential TEs may become toxic and accumulate in the food chain (Marschner, 1995; Adriano, 2001).

A focal point of entry of TEs into the food chain is the so-called plant rhizosphere. Generally, the rhizosphere—located at the root–soil interface—is defined as the zone of soil around roots that is affected by root activities (Hinsinger, 1998; Hinsinger et al., 2005). It is now well established that growing, active roots are able to modify their soil environment in terms of physical, chemical, biochemical, and biological properties and processes. The intensity and the extension of these changes into the bulk soil largely depend on the specific interaction between the plant species and its physiological status; the physical, chemical, and biological properties of the soil; and the considered rhizosphere process or characteristic (Hinsinger et al., 2005; Hinsinger and Courchesne, 2008; Hinsinger et al., 2009). Root activities and related rhizosphere processes can significantly affect the chemical solubility and phytoavailability of both nutrients and pollutants in the rhizosphere, thus constituting an important control of the environmental fate of chemical elements in terrestrial ecosystems. Rhizosphere processes are known to result in element accumulation or depletion, to induce transport processes such as diffusion and mass flow, and result in changes in chemical speciation and solubility. Interactions with rhizosphere microorganisms and their activities may induce further changes in the behavior of chemical elements in the rhizosphere (Hinsinger, 1998; Hinsinger and Courchesne, 2008; Wenzel, 2009).

Although rhizosphere soil at a given time commonly constitutes only a limited part of the total soil mass even in densely rooted topsoils, there is evidence for impacts of rhizosphere processes on the fate of chemical elements in the entire rooted soil zone. In particular, the uptake of water and dissolved elements may decrease leaching of elements from the rootzone and is an important process involved in the cycling of elements in terrestrial ecosystems (Adriano, 2001). On the other hand, root exudates may modify the leachability of elements both in terms of mobilization and immobilization.

In addition to root activities, root architecture and root plasticity in response to environmental conditions (e.g., trace element toxicity) may further change the fate of elements in the rootzone. Soil properties, including the distribution of chemical elements, are typically nonhomogeneous, resulting in hotspots of higher and lower concentrations and/or bioavailability. Plants have developed strategies to explore soil patches with higher nutrient availability or to avoid hotspots of pollution, resulting in related root distribution patterns. Spatial patterns of root activities and rhizosphere development within the rootzone require attention if we want to translate and upscale rhizosphere information obtained on a single root scale to whole root (plant) systems or beyond. However, a review of these aspects is outside the scope of this chapter.

For decades, the bulk of investigations of chemical elements in the rhizosphere had focused on the macronutrients N, P, and K, and the essential micronutrient Fe, whereas comparably little work was done on the impact of root-induced processes on the solubility, speciation, bioavailability, and mobility (including leaching) of other TEs in the rootzones of natural and managed ecosystems. Earlier compilations of the limited information on TEs in the rhizosphere and rootzones have been published in books edited by Gobran, Wenzel, and Lombi (2000), and Huang and Gobran (2005).

This chapter provides an overview of the more recent advancements in the biogeochemistry of TEs in plant rhizospheres and rootzones and their environmental implications.

Biogeochemistry and Bioavailability of Trace Elements as Affected by Rhizosphere Processes

The bioavailability of total TEs in soil is one of the key factors that will rule over micronutrient deficiencies up to toxic effects in plants. TE bioavailability is governed by soil pH, redox conditions, soils sorption capacity and buffer power, dissolved organic matter content and quality, and speciation of all dissolved metals (Antunes et al., 2006). The presence of living plant roots can significantly alter these factors, inducing changes in TE bioavailability. Here we aim to briefly summarize the most important root-induced changes in soil biogeochemistry affecting TE bioavailability in soil.

Root-Induced Changes in Soil Water Content Affecting Trace Element Bioavailability

Water uptake by plant roots induces convective transport of solutes toward the root. Depending on plant uptake rates, accumulation or depletion of solutes in the rhizosphere may occur (Hinsinger, 1998). While transpiration-driven mass flow in soil mainly affects the transport of ions present in the soil solution at large quantities (e.g., Ca, Mg, or metal pollutants), diffusional transport along a concentration gradient driven by rhizosphere depletion represents a major proportion of the actual plant uptake flux of ions that are present only in small concentrations in the soil solution (e.g., P, K, and most TEs if not present at pollutant levels: Fe, Zn, Cu, Mo, Mn, Ni, Co, etc.) (Lorenz, Hamon, and McGrath, 1994). Plant uptake resulting in ion accumulation or depletion in close proximity to the root, however, may also affect the solid–solution partitioning equilibrium of TE and is likely to induce desorption. Zhu and Alva (1993) showed that Ca and Mg significantly inhibited Cu and Zn adsorption, hence an increased Ca^{2+} concentration in rhizosphere solution might effectively out-compete TE sorption to the mineral surface.

Water uptake by plants can decrease the soil's matrix potential to a point where it will restrict diffusional and mass transport toward the roots and can therefore significantly affect TE bioavailability. Plants adapted to arid conditions have been shown to counter-balance drought-induced ion deficiencies by transporting water from deeper, wet horizons through the root system to dry surface horizons (hydraulic lift) (Dawson, 1993; Liste and White, 2008).

Root-Induced Changes in pH

Changes in soil solution pH can significantly affect the surface charge of variable charge minerals such as oxyhydroxides of Fe, Al, and Mn via protonation or deprotonation. For cations, an increase in positive surface charge due to rhizosphere acidification (surface protonation) can potentially result in TE desorption while alkalinization will cause the reverse effect. This mechanism works in the opposite direction for anionic TE species.

Root-induced changes in rhizosphere pH primarily result from different uptake rates of anions and cations where an excess uptake of negative or positive charges is balanced by the release of hydroxyl and/or bicarbonate ions in the first and by the release of protons in the latter case to maintain a neutral charge balance within the cell. The form of nitrogen taken up by the plant (NO_3^- or NH_4^+) appears to play a central role in root-induced pH changes as it is taken up by the root in large quantities. Many authors reported significant rhizosphere acidification by NH_4^+-fed plants and rhizosphere alkalinization by NO_3^--fed plants (Haynes, 1990; Hinsinger et al., 2003). Additionally, the release of CO_2 from root and microbial respiration might also decrease rhizosphere pH, especially in highly compacted soils. Nutrient deficiencies, particularly for P (Wang, Kelly, and Kovar, 2005; Hinsinger and Gilkes, 1996) and for Fe (strategy I plants, all nongraminacous plants) (Marschner and Römheld, 1994; Santi and Schmidt, 2009) often trigger the release of protons into the rhizosphere and are likely to increase the bioavailability of other TEs as well. Rhizosphere alkalinization has often been observed as a plant response to alleviate metal toxicity, particularly in acidic soils. For example, Bravin et al. (2009) showed that rhizosphere alkalinization of durum wheat (*Triticum turgidum durum* L.) significantly decreased Cu bioavailability in a Cu-contaminated acidic vineyard soil. Furthermore, changes in pH may also cause changes in TE speciation, including complexation, in soil solution, that consequently may affect sorption affinity and bioavailability in the rhizosphere.

Root-Induced Changes in Redox Conditions

Changes in redox potential (Eh) might significantly affect the solubility of redox-sensitive trace metals (Fe, Mn, Mo, As, Cu, Se) in soil. Fe and Mn play a central role in redox-controlled changes in trace element bioavailability in soil as they occur as oxyhydroxide minerals and precipitates with a huge sorption potential for other TEs in aerated soils. Under reduced conditions, Fe and Mn, as well as associated ions (adsorbed or occluded), are solubilized.

In well-aerated soils, reducing activity is mainly limited to the plant root surface (strategy I plants) where membrane-bound reductases promote the reduction of Fe^{3+} to Fe^{2+}, shifting the dissolution–precipitation equilibrium of

Fe oxides toward dissolution. In poorly aerated, submerged, or compacted soil, the consumption of O_2 by roots and associated rhizosphere organisms may cause a decrease in redox potential, resulting in solubility changes of redox-sensitive TEs and the concomitant release of associated ions. In paddy field conditions, the oxidation of the rhizosphere through the leakage of O_2 from plant roots adapted to water-logged conditions (e.g., rice, *Oryza sativa*) can significantly affect TE bioavailability. O_2 release by roots causes re-oxidation of the rhizosphere, resulting in the precipitation Fe and Mn oxides at the root surface (iron plaque). Adsorption of TEs to these Fe and Mn precipitates has been shown to significantly decrease their bioavailability, for example, for Se (Zhou and Shi, 2007), As (Hu et al., 2007), and Zn (Shuman and Wang, 1997). Moreover, a recent study showed that an increase in rhizosphere E_h of rice enhanced Cu biavailability compared to the reduced bulk soil due to transformations of sulfur and organic compounds (Lin et al., 2010).

Complexation and Chelation of Trace Elements in Rhizosphere

Living roots actively or passively release large amounts of various organic ligands (e.g., aliphatic organic acids, phenols, phytosiderophores, mucilage) into the rhizosphere that can form strong complexes (effective stability constant $K \sim 10^4$ to 10^7) or chelates (effective stability constant $K \sim 10^8$ to 10^{20}; (Reichmann and Parker, 2005) with a range of metals. Several authors showed that the presence of active roots decreased the free metal concentrations of Cu and Cd by 50% up to 100% (Merckx et al., 1986; Hamon et al., 1995). A decrease in the concentration of free metal species can induce metal solubilization and therefore increase micronutrient (e.g., Mench and Martin, 1991; Dessureault-Rompré et al., 2008; Degryse, Verma, and Smolders, 2008) as well as pollutant bioavailability. The formation of metal–ligand complexes can however also restrict plant uptake of toxic metal concentrations and therefore serve as a detoxifying mechanism (e.g., Pinto, Simoes, and Mota, 2008; Johansson et al., 2008).

The release of carboxylic acids such as citrate, malate, oxalate, and malonate has been found to significantly increase in P- and Fe- (strategy I plants only) deficient plants, and several authors have reported a concomitant increase of other trace metals such as Cu, Mn, and Zn (Dessureault-Rompré et al., 2008; Dakora and Phillips, 2002). Phenolics and mucilage have also been suggested to play an important role in enhancing trace metal bioavailability via complex formation in the rhizosphere (Morel, Mench, and Guckert, 1986; Quartacci et al., 2009). The strongest metal-chelating affinity has been observed for phytosiderophores. Phytosiderophores are nonproteinogenic amino acids that are exuded by graminaceous plants (grasses) by Fe acquisition mechanism (strategy II). Their release is enhanced under Fe but also under Zn and Cu deficiency, and phytosiderophore release has been reported to increase the solubility of many other trace metals such as Cu, Zn, Mn, Ni, and Cd (Treeby, Marschner, and Römheld, 1989; Reichman and Parker, 2005). Despite higher trace metal-complex stabilities, competition with large quantities of alkaline earth metals (Ca, Mg) for organic ligands will significantly increase the concentration of free trace metal ions in the solution, as shown by Hamon et al. (1995). However, as shown for Cu in the following section, increased Ca concentrations may compete for apoplastic binding of trace elements and thus decrease their bioavailability. This was demonstrated in a study on Cd uptake in various wheat cultivars (Wenzel et al., 1996).

Recent Experimental Evidence for Rhizosphere Effects on Trace Element Solubility and Chemical Speciation

Although principal mechanisms of rhizosphere-mediated changes in TE biogeochemistry had been postulated and partially revealed in earlier work (e.g., Gobran, Wenzel, and Lombi, 2000; Fitz and Wenzel, 2002), more direct experimental evidence has only been collected more recently. The main results of selected recent publications are compiled in Table 6.1.

From these works, it appears that the release of protons and/or organic ligands can indeed be considered a key factor in controlling trace element solubility and chemical speciation in the rhizosphere. Trace element solubility was typically found to increase under different experimental conditions (rhizobox, pot, and field experiments and surveys) in the rhizosphere of various plants, including wild herbaceous and grass species, agricultural crops, and forest trees (Table 6.1). These findings were commonly related to exudation of protons and/or organic compounds and associated TE mobilization by rhizosphere acidification, complexation of cationic TEs, or ligand exchange in the case of oxyanions such as As (Table 6.1). An increase in E_h in the rhizosphere was likely to explain enhanced Cu solubility in the rhizosphere of rice in a paddy soil (Lin et al., 2010), whereas decreased E_h was suggested as a possible mechanism contributing to sustained or enhanced As solubility via reductive dissolution of iron oxides (Fitz et al., 2003; Vetterlein et al., 2007).

TE mobilization by exudation of organic ligands appeared to be even more pronounced in some metal hyperaccumulator plant species, compared to metal solubilization by excluders and nonaccumulators grown in the same experimental conditions (Dessureault-Rompré et al., 2010; Wenzel et al., 2003a).

However, there are also reports on decreased TE solubility in plant rhizospheres (Table 6.1). Cd solubility in the rhizospheres of wheat cultivars with differential Cd accumulation potential was consistently lower than in corresponding bulk soils (Greger and Landberg, 2008), but the mechanism of immobilization remains unclear. Cu, Mn, and Zn in *Lupinus albus* rhizosphere decreased when the plant was grown on an acidic soil but this was the case

TABLE 6.1
Overview of Recent Literature on Trace Element Solubility and Speciation as Influenced by Root Activities in the Rhizopshere

	Reference	Plant	Strategy	Growth Substrate	As	Cd	Cu	Cr	Fe	Mn	Ni	Pb	Se	Zn	Main Results
1	Almeida et al. (2008)	*Halimione portulacoides*		Estuarine sediments		x	x					x		x	Substantial Cu and Zn mobilization into solution due to presence of (Cu-) complexing ligands in the rhizosphere sediments; virtually no changes in Cd and Pb solubility
2	Blossfeld et al. (2010)	*Lolium perenne* L., cv Prana		Cd-Pb-Zn contaminated soils, non-contaminated soil		x									Permanent rhizosphere alkalization measured by planar optodes by up to 1.5 pH units without differentiation along the root axis; using total Cd and Kd values, a 2 to 2.1-fold decrease of Cd solubility due to rhizosphere alkalization was calculated
		Thlaspi caerulescens J&C Presl, Viviez population	H (Cd, Zn)			x									Permanent rhizosphere alkalization by up to 1.7 pH units without differentiation along the root axis; 1 to 171-fold decrease in Cd solubility
		Zea mays L., cv INRA MB862				x									Temporary rhizosphere acidification in the elongation zone by up to 0.7 pH units; 1.47 to 5-fold increase of Cd solubility

—*continued*

TABLE 6.1 (Continued)

Overview of Recent Literature on Trace Element Solubility and Speciation as Influenced by Root Activities in the Rhizopshere

	Reference	Plant	Strategy	Growth Substrate	As	Cd	Cu	Cr	Fe	Mn	Ni	Pb	Se	Zn	Main Results
3	Cattani et al. (2006)	*Zea mays* L.		Two soils (uncontaminated forest soil, pH 7.89; Cu-polluted vineyard soil, pH 8.52)			x								Mobilization of Cu in the rhizosphere of the vineyard soil as indicated by soil solution and DGT measurements; tendency of Cu immobilization in the forest soil; three-fold increase of DOC in the rhizosphere in both soils; no correlation of DGT measurements with uptake; DOC is thought to contribute to alleviation of Cu toxicity through formation of Cu-DOC complexes
	Cloutier-Hurteau et al. (2008)	*Populus tremuloides* Michx. Dominated stands	Luvisols (pH 4.97 to 7.10)			x									Cu speciation in water extracts of all samples was dominated by organic forms of Cu; organic Cu increased with increasing pH while the reverse applied for free Cu^{2+}; concentrations of both Cu forms were generally higher in rhizosphere as compared to bulk soils; Cu speciation was to a higher extent related to microbial characteristics microbial biomass C and N, urease and dehydrogenase activities) in the rhizosphere than in bulk soils; relationships between microbial biomass N

							pH and Cu^{2+} indicate that rhizosphere microorganisms modified pH and thus Cu speciation through N assimilation; relations between urease activity, biomass variables, solid and liquid phase organic carbon and Cu in the water extracts suggest that microbial mineralization could partly supply Cu into the soluble phase through root decay
5	Courchesne et al. (2006)	*Populus tremuloides* Michx. Dominated stands	A (Cd, Zn)	Luvisols polluted from Cu smelting	x	x	DOC was generally increased in rhizospheres compared to bulk soils wheras pH was decreased; total dissolved Zn and labile Zn (Znl) measured by differential pulse anodic stripping voltammetry was generally increased in rhizosphere soil as was the free Zn^{2+} activity calulated by the WHAM model; total dissolved Cu was higher in the rhizosphere whereas Cu^{2+} activities did not significantly vary among bulk and rhizosphere soils; acid-extractable Zn was only marginally increased while the corresponding Cu fraction was not changed in the rhizosphere compared to the bulk soils
		Pinus strobus L., *Betula papyrifera* marsh stands		Podzols, Brinisols polluted with Ni and Cu from smelting activities	x	x	
		Abies balsamea, *Acer saccharum* marsh, *Betula papyrifera* marsh stands		Podzols with relatively low level of pollution	x	x	

—continued

TABLE 6.1 (Continued)

Overview of Recent Literature on Trace Element Solubility and Speciation as Influenced by Root Activities in the Rhizopshere

	Reference	Plant	Strategy	Growth Substrate	As	Cd	Cu	Cr	Fe	Mn	Ni	Pb	Se	Zn	Main Results
6	Degryse et al. (2008)	*Spinacia oleracea* L. cv Géant d´Hiver		Resin-buffered nutrient solution and a calcareous, uncontaminated topsoil			x							x	Root exudates of the two dicotyledonous plants were able to mobilize Cu and Zn; plants appeared to respond to Zn deficiency by exuding compounds with higher aromaticity and metal affinity; the obeserved Cu uptake and Zn uptake at low Zn^{2+} activity could not be fully explained by diffusive transport of the free ion through the unstirred layer surrounding the root; it was calculated that complexes with root exudates contributed 0.4% and 20% to the uptake of Cu and Zn, respectively
		Lycopersicon esculentum cv Pyros F1					x							x	

7	Dessoureault-Rompré et al. (2008)	*Lupinus albus* L., cv "Weißblühende Tellerlupine"		Carbonate-free loam soil (pH 6.4)		x	x	x	x	x	During organic acid anion exudation bursts, the investigated trace elements were mobilized in the cluster root rhizospheres; DOC derived from SOM was also mobilized during the exudation bursts; speciation calculations indicate that Fe, Mn and Zn were mainly bound to exuded citrate whereas Cu and Pb formed strong complexes with SOM-derived DOC
8	Dessureault-Rompré et al. (2010)	*Thlaspi perfoliatum*		Agricultural topsoil (pH 5.2) contaminated by septic tank wastes with Cd, Cu, Pb and Zn	x					x	Labile (free metal ions and small, labile complexes) Cd and Zn decreased during the cropping period in the rhizospheres of the respective hyperaccumulator ecotypes; this was associated with higher UV absorptivities as compared to the non-accumulator *T. perfoliatum* and non-rhizosphere soil, indicating that labile and mobile metal-DOM complexes play a key role in the rapid replenishment of available metal pools in the rhizosphere of hyperaccumulating *T. caerulescens* ecotypes
		Thlaspi caerulescens J&C Presl, Prayon population	H (Zn)		x					x	
		Thlaspi caerulescens J&C Presl, Ganges population	H (Cd, Zn)		x					x	

—continued

TABLE 6.1 (Continued)

Overview of Recent Literature on Trace Element Solubility and Speciation as Influenced by Root Activities in the Rhizopshere

	Reference	Plant	Strategy	Growth Substrate	As	Cd	Cu	Cr	Fe	Mn	Ni	Pb	Se	Zn	Main Results
9	Fitz et al. (2003)	*Pteris vittata* L.	H (As)	Calcareous soil with high As concentration (2270 mg kg^{-1}) from geogenic sources (arsenosiderite)	x				x						As fluxes measured by DGT were substantially decreased in the rhizopshere of *P. vittata* after 41 days of cropping; the observed decrease in labile (0.05 M $(NH_4)_2SO_4$-extractable As explained only 8.9% of the amount of arsenic removed by the fern, indicating mobilization from more recalcitrant As fractions; DGT-based modeling using the DIFS model indicated limited re-supply of As in the rhizopshere from such less available pools with rates of release amounting only to one third of that in the bulk soil; virtually sustained As concentrations in soil solution and enhanced Fe solubility in the rhizosphere may be explained by concurrent chemical changes in the rhizosphere, i.e., increased DOC and lower redox, suggesting DOC-triggered dissolution of Fe oxides, Fe complexation and blocking of sorption sites for As (anion competition), which may explain the sustained As concentratioins in soil solution regardless of the reduced resupply kinetics

10	Greger and Landberg (2008)	*Triticum aestivum* (6 cultivars)		x					Cd concentrations in rhizosphere solutions were generally lower than in corresponding bulk soils; the decrease in soluble Cd concentration was more pronounced as Cd accumulation by the cultivars increased, expcept for the *T. aestivum* cultivars; at elevated Cd levels in soil, root CEC was increased; higher root CEC was also observed for the high accumulating cultivars, indicating a potential role of higher root CEC in the release of Cd from soil particles
		Triticum durum (6 cultivars)		x					
		Triticum spelta (4 cultivars)		x					
11	Lin et al. (2010)	*Oryza sativa* L.	Uncontaminated paddy soil (pH 5.63), flooded and non-flooded treatments		x				In the rice rhizosphere, higher E_h along with redox-related transformations of S and SOM contributed to increased soluble and exchangeable Cu which prevailed as Cu(II); in the more reduced bulk soil, part of the Cu was also present as Cu(I)
12	Martínez-Alcalá et al. (2010)	*Lupinus albus* L.	Two similar soils polluted by mining (Cd, Zn, Cu, Pb) but contrasting pH (4.2 and 6.8)	x	x	x	x	x	DOC increased in the rhizospheres of both soils; $CaCl_2$-extractable Mn, Zn and Cu were decreased in the rhizosphere of the acid soil, possibly due to increased retention by Fe(III) oxyhydroxides; in the neutral soil, only extractable Zn was decreased in the rhizosphere

—*continued*

TABLE 6.1 (Continued)

Overview of Recent Literature on Trace Element Solubility and Speciation as Influenced by Root Activities in the Rhizopshere

	Reference	Plant	Strategy	Growth Substrate	As	Cd	Cu	Cr	Fe	Mn	Ni	Pb	Se	Zn	Main Results
13	Nakamaru et al. (2005)	*Glycine max* (L.) Merryl		Andosol (pH 6.4)									x		Decreased Se concentration in the soil solution of rhizospheres could be related to release of protons by soybean roots and associated decrease of Se sorption (i.e., increase of Kd)
14	Puschenreiter et al. (2005)	*Thlaspi goesingense* Hálácsy, Redlschlag population	H (Ni)	Geogenically Ni- and Cr-rich serpentine soil (pH 6.6)							x				Water-soluble Ni was enhanced while labile (exchangeable) Ni was depleted in the rhizosphere of *T. goesingense* in a rhizobox experiment; as ion competition with the large amounts of Ca and Mg present in the soil solutions or complexation with DOM could not explain this finding, it was assumed that oxalate exudation by *T. goesingense* or associated rhizopshere microorganisms may decrease the (apparent) selectivity of the exchange complex for Ni as indicated by a lower Vanselov coefficient

15	Puschenreiter et al. (2003)	*Thlaspi goesingense* Hálácsy, Redlschlag population	H (Ni)	Two calcareous soils, one Cd/Pb/Zn-contaminated by industrial activities, the other one uncontaminated			x	x	Hyperaccumulation of Cd, Zn or Ni was not reflected by equivalent changes of labile metal forms (1 M NH_4NO_3-extractable) in the rhizospheres of the two hyperaccumulators; there were no major changes of pH or DOC; the non-accumulator *T. arvense* showed enhanced DOC concentrations in the rhizosphere, possibly indicating a detoxification mechanisms by metal complexation
		Thlaspi caerulescens J&C Presl, Ganges population	H (Cd,Zn)				x	x	
		Thlaspi arvense L.					x	x	
16	Vetterlein et al. (2007)	*Zea mays* L.		Artificial substrate composed of sand and varying amounts of goethite	x	x			Arsenate (As-V) in soil solution collected with microsuction cups from a rhizobox system was not affected in the rhizosphere whereas As(III) solubility was enhanced in the rhizosphere of some treatments; it remains unclear whether this can be attributed to reductive dissolution (lower redox potential in the rhizosphere) or rather to release of As(II) from roots; the observed mobilization of Fe in some of the treatments may be explained either by release of siderophores or by reductive dissolution

—*continued*

TABLE 6.1 (Continued)

Overview of Recent Literature on Trace Element Solubility and Speciation as Influenced by Root Activities in the Rhizopshere

	Reference	Plant	Strategy	Growth Substrate	As	Cd	Cu	Cr	Fe	Mn	Ni	Pb	Se	Zn	Main Results
17	Vyslouzilová et al. (2006)	*Salix x rubens* (clone C from Kutná Hora)	A (Cd, Zn)	Calcareous, Pb/Zn/Cd-polluted industrial soil		x						x		x	Both metal-accumulating *S. x rubens* clones (derived from a contaminated and a non-contaminated site) were able to mobilize (water extraction, 1 M NH_4NO_3 extraction) Pb and Zn, but not Cd in the rhizosphere; the mobilization of Pb and Zn may be related to the enhanced DOC concentrations in the rhizospheres; the removal of metals by the willows was not reflected by an equivalent decrease of the labile metal fractions, indicating sustained resupply of metals from more recalcitrant fractions; while pH of the calcareous soils was not changed, acid–base titration revealed a decrease of acid neutralization capacity in the rhizosphere which may be related to action of protons in the mobilization of metals from more recalcitrant fractions
		Salix x rubens (clone N from Kurivody)	A (Cd, Zn)			x						x		x	

18	Wenzel et al. (2003)	*Rumex acetosella*		Geogenically Ni- and Cr-rich serpentine soil (pH 6.6)	x	x	x	Water-soluble Ni and DOC in field-collected rhizospheres were enhanced in both excluder and the hyperacccumulator species but most pronounced in the latter; chemical speciation modelling revealed that enhanced Ni solubility was closely related to the higher levels of DOC in rhizosphere soils, suggesting the formation of Ni–organic complexes; Cr solubility and pH were slightly increased in the rhizosphere of the hyperaccumulator; labile (exchangeable) Ni was decreased only in the rhizopshere of *T. goesingense* which can be explained by excessive uptake by the hyperaccumulator; enhanced Ca and Mg solubilities along with increased pH in the rhizosphere appear to be consistent with ligand-promoted dissolution of forsterite-type Mg silicates with potential co-dissolution of Ni
		Silene vulgaris L.			x	x	x	
		Thlaspi goesingense Hálácsy	H (Ni)		x	x	x	

only for Zn on a neutral soil (Martínez-Alcalá, Walker, and Bernal, 2010). The authors suggested that trace element immobilization in the rhizosphere of the acidic soil was related to increased retention by Fe(III) oxyhydroxides. While in this case the explanation remains vague, decreased solubility of oxyanionic Se in soybean rhizosphere could be related to proton exudation and subsequent decrease of Se adsorption, as indicated by an increase in measured K_d values (Nakamaru, Tagami, and Uchida, 2005). Blossfeld et al. (2010) used a novel approach (planar optodes) to produce high-resolution maps of pH in the rhizospheres of three plant species. They found substantial, permanent alkalization in the rhizospheres of ryegrass and pennycress, and temporary acidification in the root elongation zone of corn. Using measured total Cd and K_d values, they calculated substantial decreases of Cd solubility in the alkalized and increases in the acidified rhizospheres (Table 6.1). However, caution is required in translating these results to bioavailability, as will be shown for Cu below.

For Cu, probably the most phytotoxic TE, comprehensive research conducted by Philippe Hinsinger's group has revealed a complex interaction between initial soil pH, pH-dependent alkalization or acidification and apoplastic binding of Cu to root cell walls, associated changes of Cu solubility and speciation, and Cu uptake as the rate-limiting process of bioavailability in the rhizospheres of various plants grown on Cu-polluted soils from abandoned vineyards (Figure 6.1 and Table 6.2). Overall, this work suggests that, across the investigated large range of soil pH, the associated differential Cu solubility and speciation were not reflected by concomitant changes in Cu bioavailability (Figure 6.1). It appears that the investigated plants were using root–rhizosphere processes to maintain Cu bioavailability at a beneficial, nontoxic level. However, other work, including field surveys, on Cu solubility and speciation in polluted and nonpolluted soils (Table 6.1) showed generally increased Cu concentrations in the rhizospheres of corn (Cattani et al., 2006) and various tree species (Cloutier-Hurteau, Sauvé, and Chourchesne, 2008; Courchesne, Kruyts, and Legrand, 2006), indicating the need for further research to elucidate the factors to explain these contrasting findings.

Some researchers tried to also identify changes of TE forms in the solid phase of rhizosphere soils. While there is clear evidence for enhanced association of several TEs, including As, with Fe plaques in the rhizospheres of plants adapted to submerged soil conditions (Zhou and Shi, 2007; Hu et al., 2007; Shuman and Wang, 1997), the widely used fractionation of trace elements by sequential extraction delivers only limited information (Table 6.3). Assignment of chemical extractants to specific chemical forms of TEs is often done without critical evaluation (Wenzel et al., 2001). From the work compiled in Table 6.3 it appears that organic forms of TEs may become bioavailable in plant rhizospheres of metal accumulator (Hammer and Keller, 2002) and nonaccumulator plants (Tao et al., 2004). However, results are difficult to generalize and vary depending on the plant–soil system considered (Table 6.3).

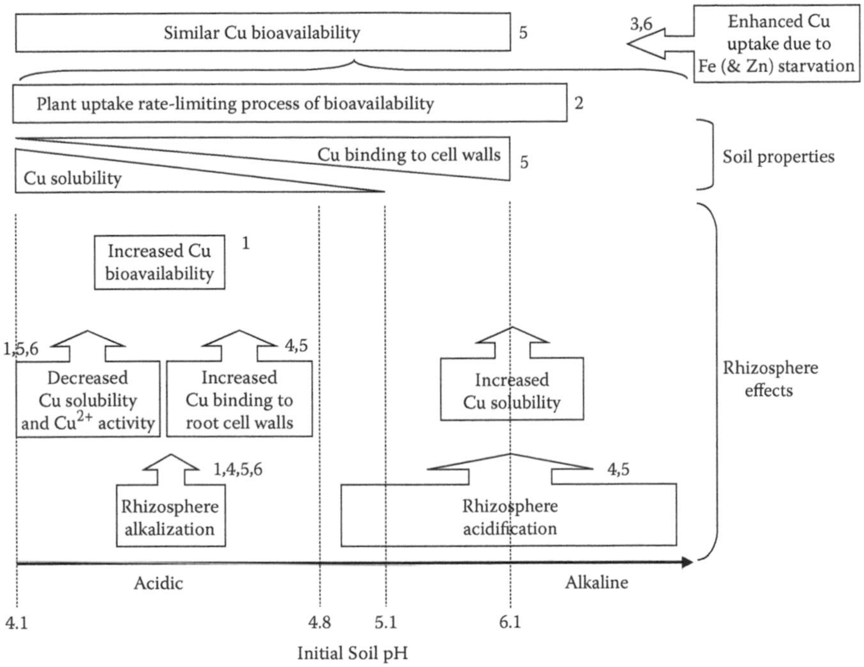

FIGURE 6.1
The copper case illustrating the complexity of trace element biogeochemistry in the rhizosphere (numbers relate to the references listed in Table 6.2).

Alleviation of Trace Element Toxicity in Rhizosphere

Three main mechanisms have been reported to be involved in the alleviation of TE rhizotoxicity: (1) root exudation, (2) ion competition for binding sites on the root surface, and (3) activities of rhizosphere bacteria and mycorrhizal fungi.

The first mechanism was initially described in the context of Al toxicity in acid soils. Delhaize, Ryan, and Randall (1993) found that the exudation of malate and succinate acids from wheat plants leads to increased complexation of Al, which protects the root apex from phytotoxic Al concentrations. The root apex was found to be the primary source of carboxylate exudation. Root exudation may also alleviate the toxicity of several TEs. Dissolved organic matter (DOM) derived from root exudates can influence the speciation of TEs in soil solution, and thus their toxicity. Pinto, Simoes, and Mota (2008) reported that the release of malate and citrate by roots of sorghum and maize led to increased complexation of Cd, thus reducing the deleterious effects of free Cd^{2+} on root growth. A decrease in Mn toxicity to *Lolium perenne* after release of citrate was reported by De la Luz Mora (2009). Differences in Mn tolerance of *Lolium perenne* cultivars were

TABLE 6.2

Literature Related to the Copper Case (Compare Figure 6.1)

	Reference	Plant	Growth Substrate	Main Results
1	Bravin et al. (2009)	*Triticum turgidum durum* L. cv Acalou	Cu-contaminated, acidic (pH-CaCl2 4.1) vineyard soil	Strong rhizosphere alkalization up to 2.8 units due to OH- release caused three-orders-of-magnitude decrease of free Cu^{2+} activity and three-fold decrease of Cu solubility; DGT-assessed labile Cu was halved
2	Bravin et al. (2010)	*Triticum turgidum durum* L. cv Acalou	Seven soils of variable pH-$CaCl_2$ (4.1-7.8) and Cu contamination level	A comparison of measured Michaelis-Menten uptake kinetic for Cu and Cu supply (DGT flux) from soil reveals Cu uptake as the rate limiting process for Cu bioavailability to *T. turgidum* durum (except of moderately contaminated calcareous soils)
3	Chaignon et al. (2002a)	*Triticum aestivum*, cv Aroona and cv Songlen	Cu-polluted, calcareous vineyard soil	Iron starvation resulted in enhanced phytosiderophore release (but no change in soils pH) by both cultivars and related increase in Cu uptake in roots and shoots; Zn starvation was only effective for cv Aroona
4	Chaignon et al. (2002b)	*Lycopersicon esculentum* cv S. Pierre *Brassica napus* cv Goeland	Two Cu-polluted soils (one calcareous, pH 8.7; one acidic, pH 4.3)	Under varying forms of nutrate supply, both plant species induced a systematic acidification in the calcareous soils and alkalization in the acidic soil; *B. napus* was more effective in taking up Cu and alkalizing the rhizosphere in the acidic soil than *L. esculentum*; in the acidic soil, Cu bioavailability increased with increasing pH (alkalization), possibly due to increased Cu binding to root cell walls (apoplasmic Cu)
5	Chaignon et al. (2009)	*Brassica napus* cv Goeland	Cu-polluted, acidic (pH 4.3) vineyard soil	Bioavailability of Cu to *B. napus* was unaffected across the whole range of soil pH adjusted by liming (4.4 - 6.1) although $CaCl_2$-extractable Cu decreased substantially up to pH 5.1; this lack of correlation between chemical Cu availability in soil and plant uptake may be related to the measured rhizosphere alkalization at $pH < 4.8$ and acidification at higher soil pH as well as to the measured increased apoplastic binding of Cu with increasing pH

6	Cornu et al. (2007)	*Lycopersicon esculentum* cv S. Pierre	Two calcareous (pH 8.1 and 8.7) and two acidic (pH 4.3) Cu-contaminated vineyard soils	Cu concentrations in roots were enhanced under Fe deficiency in acid soils while shoot Cu did not vary with the Fe status of the plant; the plant Fe status had generally little effect on pH, redox potential or Cu solubility in the rhizopshere on acidic and calcareous soils; however, independent of Fe status, roots induced rhizopshere alkalization of acidic soils and a decrease of Cu solubility in all soils
7	Michaud et al. (2007)	*Triticum turgidum durum* L.	Forty-two former vineyard soils covering a wide range of pH (4.2 to 7.8) and Cu concentrations (32 to 1030 mg kg^{-1})	Rhizosphere alkalization in the most acidic soils was related to decreased $CaCl_2$-extractable Cu whereas Cu solubility increased in the rhizosphere of all other (calcareous and non-calcareous) soils; Cu toxicity appeared to be greater in calcareous soils due to interaction with Fe nutrition

TABLE 6.3

Overview of Recent Literature on Solid-Phase Chemistry of Trace Elements as Influenced by Root Activities in the Rhizosphere

	Reference	Plant	Strategy	Growth Substrate	As	Cd	Cu	Pb	Se	Zn	Main Results
1	Chen et al. (2010)	*Oryza sativa* L. (2 high and 2 low Se cultivars)		Paddy soil (pH 5.5)					x		Total Se accumulated in the rhizosphere soils; this was associated with increased Se concentrations in the soluble, exchangeable, carbonate and organic-sulfide/elemental Se fractions whereas the residual Se fraction decreased
2	Hammer & Keller (2002)	*Salix viminalis* L. (Swedish clone 78198)	A (Cd, Zn)	Calcareous topsoil contaminated with Cu and Zn by smelter activities		x	x			x	Growth of *T. caerulescens* for 90 days resulted in Cd depletion of $NaNO_3$-, DTPA- and EDTA-extractable pools which was most apparent in the acidic soil; sequential extraction showed that most Cd extracted by the plants from the acidic soil originated from the fraction assigned as organic Cd pool; in the calcareous soil, only a small amount of Cd was removed by *T. caerulescens*, mostly from the fraction assigned as carbonate pool
		Thlaspi caerulescens J&C Presl, Ganges population	HA (Cd, Zn)			x	x			x	
		Thlaspi arvense L.									
3	Silva Gonzaga et al. (2006)	*Pteris vittata* L.	H (As)	Sandy, silicaceous, hyperthermic Paleudult polluted with As from former CCA applications	x						The hyperaccumulator *P. vittata* removed within 8 weeks of growth sustantial amounts of As from various fractions of a 5-step extraction procedure from the rhizosphere soil but did not change the As fractions in the bulk soil. Similar but smaller changes were observed for the non-accumulator *N. exaltata*
		Nephrolepis exaltata L.			x						

					Increased water extractability of As and somewhat higher DOC concentrations was observed in the rhizosphere of both species whereas an increase of pH by 0.4 units only occurred in the rhizosphere of the hyperaccumulator. As the rhizosphere constituted only a very small fraction of the total soil mass, overall phytoextraction of As was marginal for both plants
4	Tao et al. (2003)	*Zea mays* L.	Calcareous soil (pH 8.1) contaminated with metals (Cu) by waste water	x	During the 100 days of cultivation, exchangeable and—with some time lag—carbonate-associated Cu assessed by a sequential extraction procedure initially increased and then decreased in the rhizosphere; oxide-bound Cu initially increased with a concurrent decrease of organic matter-bound Cu but both returned to the initial concentrations after around 40 to 50 days; similarly, DOC initially increased and returned to the initial levels while microbial biomass steadily increased during the whole cultivation period; pH also slightly increased within the initial 50 days with only marginal changes thereafter; redox potential deccreased throughout the cultivation period; the amount of Cu accumulated in the plants exceeded the initial exchangeable Cu pool; from this work it appears that Cu was mobilized from carbonate-associated fractions into the exchangeable fraction, probably aided by the concurrent changes in rhizosphere chemistry

—*continued*

TABLE 6.3 (Continued)

Overview of Recent Literature on Solid-Phase Chemistry of Trace Elements as Influenced by Root Activities in the Rhizosphere

	Reference	Plant	Strategy	Growth Substrate	As	Cd	Cu	Pb	Se	Zn	Main Results
5	Tao et al. (2004)	*Zea mays* L.		Calcareous soil (pH 8.1) contaminated with metals (Cu) by waste water			x				After a 28-day cultivation period, oxide-bound and exchangeable Cu fractions of a 5-step sequential extraction procedure substantially increased at the expense of the organic-matter- and carbonate-associated fractions. This trend was similar for all four plant species. Additional experiments with acid–base titrations of the soil and addition of root exudates of the four plant species collected from hydroponic cultivation in sterile and non-sterile conditions suggest that complexation of Cu rather than pH changes may be responsible for Cu mobilization in the experimental systems and that microbial activity in the rhizosphere may further enhance these effects
		Triticum aestivum L.					x				
		Pisum sativum L.					x				
		Glycine max L. Merr.					x				
6	Wang et al. (2009a)	*Paulownia fortunei* (seem) Hems		Acidic (pH 3.9-5.0) soil contaminated from Pb/Zn smelter activities		x	x	x		x	The proportions of carbonate-, oxide- and organically-bound metals were generally higher in the field-collected rhizosphere soils than in bulk soils whereas the reverse was found for the exchangeable and the residual fractions

7	Wang et al. (2009b)	*Festuca arundinacea*	Copper tailings	x	x	In field-sampled rhizosphere soils from both plant species, the organically-associated and exchangeable Cu fractions increased relative to bulk soils, whereas the reverse trend was observed for the carbonate- and oxide-bound Cu fractions; the changes were accelerated with time of growth; the rhizosphere effect of *T. repens* was more pronounced than that of *F. arundinacea*
		Trifolium repens		x	x	

attributed to different citrate exudation rates. Further examples for the reduction of TE toxicity in the rhizosphere by exudation of carboxylates were reported by, for example, Yang et al. (1997) (Ni); Krotz, Evangelou, and Wagner (1989) (Cd); González and Lynch (1999) (Mn); and Horst et al. (1999) (Mn).

In addition to carboxylates, phytosiderophores may also be involved in Cd detoxification. Hill, Lion, and Ahner (2002) found that Cd increased the rate of 2′-deoxymugineic acid (DMA) exudation of hydroponically grown maize under both Fe-limiting and Fe-sufficient conditions. The complexation of Cd by DMA resulted in decreased Cd accumulation in maize.

The presence of competing ions like Ca or protons may, in addition to DOM, contribute to the inhibition of rhizotoxicity. Voigt, Hendershot, and Sunahara (2006) showed that the biotic ligand model (BLM), including the concentration of Ca and H could better predict the rhizotoxicity of Cu and Cd compared to sole predictions from free ion activities or total concentrations. The important role of Ca and protons in the competition with binding of TEs on the root surface was confirmed in a very recent study by Wu and Hendershot (2010). They showed that Ca, H, and Ni competed for the same binding sites on the root surface of pea. Increasing H^+ (lower pH) and Ca concentrations decreased Ni rhizotoxicity.

The third main mechanism for alleviation of TE toxicity is directly related to the activities of rhizosphere microbes and mycorrhizal fungi. The fungal mycelium may protect plants from toxic TE concentrations by decreasing the uptake in roots (e.g., Joner, Briones, and Leyval, 2000; Sharples et al., 2000; Janoušková and Pavliková, 2010). The underlying mechanisms leading to decreased plant uptake include sequestration of the toxic TEs by the fungus, decreased solubility in the soil due to changes in rhizosphere pH, and dilution in plant tissue due to increased P nutrition and thus plant yield. In addition to the mycorrhizal fungi, rhizosphere bacteria can contribute to the alleviation of TE toxicity, either due to binding of significant amounts to cell walls or complexation of TE by bacterial exudates. Burd, Dixon, and Glick (1998, 2000) reported that the plant growth-promoting bacterium (PGPB) *Kluyvera ascorbata* SUD165 could alleviate Ni toxicity to *Brassica napus* plants by supplementing Fe with the help of siderophores. Someya et al. (2007) found that a *Pseudomonas putida* strain could alleviate the Ni toxicity to *Arabidopsis thaliana*; but while the responsible mechanism is unknown, it was not related to the production of siderophores. The combination of mycorrhizal fungi with specific rhizosphere bacteria has been demonstrated several times to have a beneficial effect on plants due to reduced rhizotoxicity of TEs. This has been shown, for example, for Zn and *Trifolium repens* (Vivas et al., 2006), and for Mn and soybean (Nogueira et al., 2007).

Trace Element Transport and Leaching in Rhizosphere and Rootzone

Transport of TEs toward the root surface either occurs as mass flow or as diffusion of solutes (Barber, 1995). Mass flow is convectional flow of the soil solution and trace elements dissolved therein toward the root. Root water uptake and thus plant transpiration is the driving force of convectional water movement in the rhizosphere. Along with the concentration of solutes, the volume of water taken up by the root determines the mass of solutes that are delivered to the root via mass flow.

Diffusional transport of solutes occurs along a concentration gradient toward the zone of lower solute concentration. By constant removal of solutes from the soil solution, roots induce concentration gradients that drive the diffusive transport of trace elements toward the root surface.

TE uptake is considered to mainly occur as uptake of the free ion, with the exception of the uptake of Fe–phytosiderophore complexes via specific transporters in strategy II plants (Marschner, 1995). However, observations of increased uptake of the micronutrients Cu, Zn, and Mn, and of Cd in nutrient solutions containing organic (Bell, Chaney, and Angle, 1991) as well as inorganic ligands (Smolders and McLaughlin, 1996), showed that metal complexes can enhance TE uptake in plants. It was thus assumed that metal–ligand associations could enter the xylem via the apoplasm at locations where the Casparian bond is either not yet developed (i.e., at the root tip), or where it is disrupted, which was also observed at some occasions (Collins et al., 2002; Vassil et al., 1998). In contrast to these findings, recent investigations in nutrient solutions came to a different conclusion. Degryse et al. (2006a,b) showed for Cd, Cu, and Zn that increased phytoavailability in the presence of labile metal complexes was not explained by the uptake of the complex into the root, but rather by enhanced diffusional transport of the TEs to the root surface. If TE transport toward the root is limited by diffusion, which is likely at the usually low TE concentrations in soil solutions, diffusion to the root as a metal–ligand complex, dissociation of the labile complex near the root, and subsequent uptake of the free ion can enhance the total diffusional flux and thus uptake of TEs in roots compared with systems without complexes. It seems very likely that the formation of labile complexes with root exudates contributes to the increased phytoavailability of TEs, together with enhanced desorption from soil. This is supported by the recent findings of Dessureault-Rompré et al. (2010) who identified labile, mobile complexes of Cd and Zn as main mechanisms of replenishment of these metals in the depleted rhizospheres of two ecotypes of the metal hyperaccumulator pennycress. Very recent results for enhanced diffusional supply of *Brassica napus* roots with phosphate by phosphate sorbed onto Al oxide nanoparticles (Santner et al., 2010) show that this mechanism also applies to colloid-bound elements. This could especially be important for anionic TEs as ligands complexing anions are, unlike anion-sorbing colloids, not common in soils.

Leaching of TEs from the rhizosphere after mobilization by root exudates to deeper soil layers has attracted some attention in recent research. Seuntjens, Nowack, and Schulin (2004) performed a modeling study on Cu uptake into roots and Cu leaching in the presence of EDTA and oxalate. EDTA was found to stabilize Cu at pH 6 due to the formation of Cu–EDTA surface complexes on goethite. At pH 7.5, increased leaching of Cu below the rootzone was found. Oxalate exudation led to reduced leaching of Cu due to sorption of Cu oxalate to goethite if no DOC was present. In contrast, with DOC present, oxalate exudation resulted in increased leaching of Cu due to the displacement of Cu–DOC complexes from goethite by Cu oxalate. Enhanced Cr leaching from a Cr-spiked soil planted with *Festuca arundiacea* was observed in column experiments (Banks, Schwab, and Henderson, 2006). The leaching of Cd from a moderately contaminated soil was reduced when the soil was planted with *Thlaspi caerulescens, Brassica juncea,* or with both species as a co-cropping treatment (Ingwersen et al, 2006). However, Cd concentrations in leachates of soils planted only with *T. caerulescens* were higher than those of the other plant treatments, pointing at a higher Cd mobilization potential of this metal hyperaccumulator compared with nonaccumulator plants. In batch and nonplanted column experiments, Schwab, He, and Banks (2005) investigated the potential of organic acids for leaching Pb from soil. Pb solubility was enhanced in the presence of citrate, but leaching occurred only at very high flow rates. This could be explained by rapid degradation of citrate, resulting in impeded Pb transport down the soil column at lower flow rates. Salicylate decreased Pb concentrations in effluents. In a similar experiment, Schwab, He, and Banks (2008) observed increased Cd and Zn leaching following increasing citrate concentrations. These studies show that root activities may influence the leaching behavior of TEs; but as the available data is limited, more general conclusions on the risk of TEs leaching from the rootzone to deeper soil layers and the groundwater can hardly be drawn. Whereas limited information is available on TE leaching from the rootzone due to mobilization in the rhizosphere, numerous studies of the effect of natural and artificial chelants suggested for use in enhanced phytoextraction have demonstrated the potential of TE mobilization by these compounds (e.g., Madrid, Liiphadzi, and Kirkham, 2003; Nowack, Schulin, and Robinson, 2006; Wenzel et al., 2003b). More research is required to evaluate the contribution of rhizosphere processes to TE leaching in soils. Our own unpublished data from a field lysimeter study indicates enhanced leaching of some TEs from the rootzone of metal-accumulating willows as compared to nonplanted soil, regardless of a substantial reduction in leachate due to the higher transpiration from the planted lysimeters.

Summary

Some trace elements such as the micronutrients Fe and Zn play an essential role in living organisms, while others such as Cd or As have no known biological function. However, any trace element may become toxic to plants and other biota when exceeding certain thresholds. A focal point of entry of trace elements into the food chain and cycling in terrestrial ecosystems is the plant rhizosphere at the soil–root interface.

The biogeochemistry of TE can be substantially influenced by root activities and related rhizosphere processes. Apart from the relatively well-known case of Fe, the fate of most other TEs in plant rhizospheres has been elucidated only more recently. Among the multitude of interacting rhizosphere processes, it appears from our review of recent literature that root (and microbial) exudations of protons or hydroxyl ions and organic ligands are among the major drivers of modifications of TE solubility, speciation, and bioavailability across a large variety of plant-soil systems. In the majority of cases, TEs appear to be mobilized by rhizosphere acidification and/or action of organic ligands. However, the presented case of Cu biogeochemistry in polluted vineyard soils clearly demonstrates the complexity of the root and rhizosphere processes involved, indicating that bioavailability and rhizotoxicity may not be simply related to changes in TE solubility and speciation as suggested by the free ion activity model (FIAM) but are also determined by the sorption capacity of the absorbing root system (BLM) and the initial pH of the soil. These interactions are further confounded by the activities of rhizosphere microorganisms, including rhizobacteria and mycorrhizal fungi that may decrease or enhance the phytoavailability of TEs.

Although advances have been made in the past decade to unravel some of the multiple processes and interactions of TEs in the rhizosphere of terrestrial plants, further research is required to better address the inherent complexities. Here, mathematical modeling along with novel analytical tools is deemed to provide new insights and allow for more generalization of the experimental findings.

A better understanding of the role of root and rhizosphere processes will allow for improved micronutrients management in agriculture and forestry. Apart from plant nutrition, there is growing interest in the alleviation of micronutrient deficiencies in animal and human nutrition, for example, in the case of Se. Using rhizosphere processes may significantly contribute to alleviating malnutrition through enhancing the bioavailability of micronutrient resources stored in soils. Moreover, a better knowledge of the rhizosphere will be crucial in estimating the transfer of pollutant elements from soil to groundwater and the food chain, and to develop more efficient plant-microbial based in situ treatment and remediation technologies to stabilize or clean up contaminated land (Wenzel, 2009).

References

Adriano, D.C. 2001. Trace elements in terrestrial environments. *Biogeochemistry, bioavailability and risk of metals*. New York: Springer.

Almeida, C.M.R., A.P. Mucha, A.A. Bordalo, and M.T.S.D. Vasconcelos. 2008. Influence of a salt marsh plant (*Halimione portulacoides*) on the concentrations and potential mobility of metals in sediments. *Science of the Total Environment* 403: 188–195.

Antunes, P.M.C., E.J. Berkelaar, D. Boyle, B.A. Hale, W. Hendershot, and A. Voigt. 2006. The biotic ligand model for plants and metals: Technical challenges for field application. *Environmental Toxicology and Chemistry* 25: 875–882.

Banks, M.K., A.P. Schwab, and C. Henderson. 2006. Leaching and reduction of chromium in soil as affected by soil organic content and plants. *Chemosphere* 62: 255–264.

Barber, S.A. 1995. *Soil nutrient bioavailability. A mechanistic approach*. New York: John Wiley & Sons, Inc.

Bell, P.F., R.L. Chaney, and J.S. Angle. 1991. Free metal activity and total metal concentrations as indices of micronutrient availability to barley [*Hordeum vulgare* (L.) Klages]. *Plant and Soil* 130: 51–62.

Blossfeld, S., J. Perriguey, T. Sterckeman, J.-L. Morel, and R. Lösch. 2010. Rhizosphere pH dynamics in trace-metal-contaminated soils, monitored with planar pH optodes. *Plant Soil* 330: 173–184.

Bravin, M.N., P. Tentscher, J. Rose, and P. Hinsinger. 2009. Rhizosphere pH gradient controls copper availability in a strongly acidic soil. *Environmental Science and Technology* 43: 5686–5691.

Bravin, M.N., B. Le Merrer, L. Denaix, A. Schneider, and P. Hinsinger. 2010. Copper uptake kinetics in hydroponically-grown durum wheat (*Triticum turgidum durum* L.) as compared with soil's ability to supply copper. *Plant and Soil* 331: 91–104.

Burd, G.I., D.G. Dixon, and B.R. Glick. 1998. A plant growth-promoting bacterium that decreases nickel toxicity in seedlings. *Applied and Environmental Microbiology* 64: 3663–3668.

Burd, G.I., D.G. Dixon, and B.R. Glick. 2000. Plant growth-promoting bacteria that decrease heavy metal toxicity in plants. *Canadian Journal of Microbiology* 46: 237–245.

Cattani, I., G. Fragoulis, R. Boccelli, and E. Capri. 2006. Copper bioavailability in the rhizosphere of maize (*Zea mays* L.) grown in two Italian soils. *Chemosphere* 64: 1972–1979.

Chaignon, V., D. Di Malta, and P. Hinsinger. 2002a. Fe-deficiency increases Cu acquisition by wheat cropped in a Cu-contaminated vineyard soil. *New Phytologist* 154: 121–130.

Chaignon, V., F. Bedin, and P. Hinsinger. 2002b. Copper bioavailability and rhizosphere pH changes as affected by nitrogen supply for tomato and oilseed rape cropped on an acidic and a calcareous soil. *Plant and Soil* 243: 219–228.

Chaignon, V., M. Quesnoit, and P. Hinsinger. 2009. Copper availability and bioavailability are controlled by rhizosphere pH in rape grown in an acidic Cu-contaminated soil. *Environmental Pollution* 157: 3363–3369.

Chen, Q., Shi, W., and Wang X. 2010. Selenium speciation and distribution characteristics in the rhizosphere soil of rice (*Oryza sativa* L.) seedlings. *Communications in Soil Science and Plant Analysis* 41: 1411–1425.

Cloutier-Hurteau, B., B. Sauvé, and F. Courchesne. 2008. Influence of microorganisms on Cu speciation in the rhizosphere of forest soils. *Soil Biology and Biochemistry* 40: 2441–2451.

Collins, R.N., G. Merrington, M.J. McLaughlin, and C. Knudsen. 2002. Uptake of intact zinc-ethylenediaminetetraacetic acid from soil is dependent on plant species and complex concentration. *Environmental Toxicology and Chemistry* 21: 1940–1945.

Cornu, J.Y., S. Staunton, and P. Hinsinger. 2007. Copper concentration in plants and in the rhizosphere as influenced by the iron status of tomato (*Lycopersicon esculentum* L.). *Plant and Soil* 292: 63–77.

Courchesne, F., N. Kruyts, and P. Legrand. 2006. Labile zinc concentration and free copper ion activity in the rhizosphere of forest soils. *Environmental Toxicology and Chemistry* 25: 635–642.

Dakora, F. D., and D.A. Phillips. 2002. Root exudates as mediators of mineral acquisition in low-nutrient environments. *Plant and Soil* 245: 35–47.

Dawson, T.E. 1993. Hydraulic lift and water use by plants: Implications for water balance, performance and plant-plant interactions. *Oecologia* 95: 565–574.

Degryse, F., E. Smolders, and R. Merckx. 2006a. Labile Cd complexes increase Cd availability to plants. *Environmental Science and Technology* 40: 830–836.

Degryse, F., E. Smolders, and D.R. Parker. 2006b. Metal complexes increase uptake of Zn and Cu by plants: Implications for uptake and deficiency studies in chelator-buffered solutions. *Plant and Soil* 289: 171–185.

Degryse, F., V.K. Verma, and E. Smolders. 2008. Mobilization of Cu and Zn by root exudates of dicotyledonous plants in resin-buffered solutions and in soil. *Plant and Soil* 306: 69–84.

De la Luz Mora, M., A. Rosas, A. Ribera, and Z. Rengel. 2009. Differential tolerance to Mn toxicity in perennial ryegrass genotypes: Involvement of antioxidative enzymes and root exudation of carboxylates. *Plant and Soil* 320: 79–89.

Delhaize E., P.R. Ryan, and P.J. Randall. 1993. Aluminum tolerance in wheat (*Triticum aestivum* L.). *Plant Physiology* 103: 695–702.

Dessureault-Rompré, J., B. Nowack, R. Schulin, M.-L. Tercier-Waeber, and J. Luster. 2008. Metal solubility and speciation in the rhizosphere of *Lupinus albus* cluster roots. *Environmental Science and Technology* 42: 7146–7151.

Dessureault-Rompré, J., J. Luster, R. Schulin, M.-L. Tercier-Waeber, and B. Nowack. 2010. Decrease of labile Zn and Cd in the rhizosphere of hyperaccumulating *Thlaspi caerulescens* with time. *Environmental Pollution* 158: 1955–1962.

Fitz, W.J., and W.W. Wenzel. 2002. Arsenic transformations in the soil–rhizosphere–plant system: Fundamentals and potential application to phytoremediation. *Journal of Biotechnology* 99: 259–278.

Fitz, W.J., W.W. Wenzel, H. Zhang, J. Nurmi, K. Štipek, Z. Fischerova, P. Schweiger, G. Köllensperger, Lena Q. Ma, and G. Stingeder. 2003. Rhizosphere characteristics of the arsenic hyperaccumulator *Pteris vittata* L. and monitoring of phytoremoval efficiency. *Environmental Science and Technology* 37: 5008–5014.

Gobran, G.R., W.W. Wenzel, and E. Lombi (Eds.). 2000. *Trace elements in the rhizosphere.* Boca Raton, FL: CRC Press.

González, A., and J.P. Lynch. 1999. Subcellular and tissue Mn concentration in bean leaves under Mn toxicity stress. *Australian Journal of Plant Physiology* 26: 811–822.

Greger, M., and T. Landberg. 2008. Role of rhizosphere mechanisms in Cd uptake by various wheat cultivars. *Plant and Soil* 312: 195–205.

Hammer, D., and C. Keller. 2002. Changes in the rhizosphere of metal-accumulating plants evidenced by chemical extractants. *Journal of Environmental Quality* 31: 1561–1569.

Hamon, R.E., S.E. Lorenz, P.E. Holm, T.H. Christensen, and S.P. McGrath. 1995. Changes in trace metal species and other components of the rhizosphere during growth of radish. *Plant, Cell and Environment* 18: 749–756.

Haynes, R.J. 1990. Active ion uptake and maintenance of cation-anion balance: A critical examination of their role in regulating rhizosphere pH. *Plant and Soil* 126: 247–264.

Hill, K.A., L.W. Lion, and B.A. Ahner. 2002. Reduced Cd accumulation in *Zea mays*: A protective role for phytosiderophores? *Environmental Science and Technology* 36: 5363–5368.

Hinsinger, P. 1998. How do plant roots acquire mineral nutrients? Chemical processes involved in the rhizosphere. *Advances in Agronomy* 64: 225–265.

Hinsinger, P., and R. Gilkes. 1996. Mobilization of phosphate from phosphate rock and alumina-sorbed phosphate by the roots of ryegrass and clover as related to rhizosphere pH. *European Journal of Soil Science* 47: 533–544.

Hinsinger, P., C. Plassard, C. Tang, and B. Jaillard. 2003. Origins of root-mediated pH changes in the rhizosphere and their responses to environmental constraints: A review. *Plant and Soil* 248: 43–59.

Hinsinger P., and F. Courchesne. 2008. *Biogeochemistry of metals and metalloids at the soil–root interface.* Eds. A. Violante, P.M. Huang, and G.M. Gadd. *Biophysic-chemical processes of heavy metals and metalloids in soil environments,* pp. 267–311. Hoboken, NJ: Wiley.

Hinsinger, P., G.R. Gobran, P.J. Gregory, and W.W. Wenzel. 2005. Rhizosphere geometry and heterogeneity arising from root-mediated physical and chemical processes. *New Phytologist* 168: 293–303.

Hinsinger, P., A.G. Bengough, D. Vetterlein, and I.M. Young. 2009. Rhizosphere: biophysics, biogeochemistry and ecological relevance. *Plant and Soil* 321: 117–152.

Horst, W.J., M. Fecht, A. Neumann, A.H. Wissemeier, and P. Maier. 1999. Physiology of manganese toxicity and tolerance in *Vigna unguiculata* (L.) Walp. *Journal of Plant Nutrition and Soil Science* 162: 263–274.

Hu, Z.-Y., Y.-G. Zhu, M. Li, L.-G. Zhang, Z.-H. Cao, and F.A. Smith. 2007. Sulfur (S)-induced enhancement of iron plaque formation in the rhizosphere reduces arsenic accumulation in rice (Oryza sativa L.) seedlings. *Environmental Pollution* 147: 387–393.

Huang, P.M., and G.R. Gobran (Eds.). 2005. *Biogeochemistry of trace elements in the rhizosphere.* Amsterdam: Elsevier.

Ingwersen, J., B. Bücherl, G. Neumann, and T. Streck. 2006. Cadmium leaching from micro-lysimeters planted with the hyperaccumulator *Thlaspi caerulescens*: Experimental findings and modeling. *Journal of Environmental Quality* 35: 2055–2065.

Janoušková, M., and D. Pavlíková. 2010. Cadmium immobilization in the rhizosphere of arbuscular mycorrhizal plants by the fungal extraradical mycelium. *Plant and Soil* 332: 511–520.

Johansson, E.M., P.M.A. Fransson, R.D. Finlay, and P.A.W. van Hees. 2008. Quantitative analysis of root and ectomycorrhizal exudates as a response to Pb, Cd and As stress. *Plant and Soil* 313: 39–54.

Joner, E.J., R. Briones, and C. Leyval. 2000. Metal-binding capacity of arbuscular mycorrhizal mycelium. *Plant and Soil* 226: 227–234.

Krotz, R.M., B.P. Evangelou, and G.J. Wagner. 1989. Relationships between cadmium, zinc, Cd-binding peptide and organic acid in tobacco suspension cells. *Plant Physiology* 91: 780–787.

Lin, H., J. Shi, B. Wu, J. Yang, Y. Chen, Y. Zhao, and T. Hu. 2010. Speciation and biochemical transformations of sulfur and copper in rice rhizosphere and bulk soil - XANES evidence of sulfur and copper associations. *Journal of Soils and Sediments* 10: 907–914.

Liste, H.-H., and J.C. White. 2008. Plant hydraulic lift of soil water – Implications for crop production and land restoration. *Plant and Soil* 313: 1–17.

Lorenz, S.E., R.E. Hamon, and S.P. McGrath. 1994. Differences between soil solutions obtained from rhizosphere and non-rhizosphere soils by water displacement and soil centrifugation. *European Journal of Soil Science* 45: 431–438.

Madrid, F., M.S. Liphadzi, and M.B. Kirkham. 2003. Heavy metal displacement in chelate-irrigated soil during phytoremediation. *Journal of Hydrology* 272: 107–119.

Marschner, H. 1995. *Mineral nutrition of higher plants.* London: Academic Press.

Marschner, H., and V. Römheld. 1994. Strategies of plants for aquisition of iron. *Plant and Soil* 165: 261–274.

Martínez-Alcalá, I., D.J. Walker, and M.P. Bernal. 2010. Chemical and biological properties in the rhizosphere of *Lupinus albus* alter soil heavy metal fractionation. *Ecotoxicology and Environmental Safety* 73: 595–602.

Mench, M., and E. Martin. 1991. Mobilization of cadmium and other metals from two soils by root exudates of *Zea mays* L., *Nicotiana tabacum* L. and *Nicotiana rustica* L. *Plant and Soil* 132: 187–196.

Merckx, R., J.H. van Ginkel, J. Sinnaeve, and A. Cremers. 1986. Plant-induced changes in the rhizosphere of maize and wheat. II. Complexation of cobalt, zinc and manganese in the rhizosphere of maize and wheat. *Plant and Soil* 96: 95–107.

Michaud, A. M., M.N. Bravin, M. Galleguillos, and P. Hinsinger. 2007. Copper uptake and phytotoxicity as assessed *in situ* for durum wheat (*Triticum turgidum durum* L.) cultivated in Cu-contaminated, former vineyard soils. *Plant and Soil* 298: 99–111.

Morel, J.L., M. Mench, and A. Guckert. 1986. Measurement of Pb^{2+}, Cu^{2+} and Cd^{2+} binding with mucilage exudates from maize (*Zea mays* L.) roots. *Biology and Fertilization of Soils* 2: 29–34.

Nakamaru, Y., K. Tagami, and S. Uchida. 2005. Depletion of selenium in soil solution due to its enhanced sorption in the rhizosphere of soybean. *Plant and Soil* 278: 293–301.

Nogueira, M.A., U. Nehls, R. Hampp, K. Poralla, and E.J.B.N. Cardoso. 2007. Mycorrhiza and soil bacteria influence extractable iron and manganese in soil and uptake by soybean. *Plant and Soil* 298: 273–284.

Nowack, B., R. Schulin, and B.H. Robinson. 2006. Critical assessment of chelant-enhanced metal phytoextraction. *Environmental Science and Technology* 40: 5225–5232.

Pinto, A.P., I. Simões, and A.M. Mota. 2008. Cadmium impact on root exudates of sorghum and maize plants: A speciation study. *Journal of Plant Nutrition* 31: 1746–1755.

Puschenreiter, M., A. Schnepf, I. Molina Millán, W.J. Fitz, O. Horak, J. Klepp, T. Schrefl, E. Lombi, and W.W. Wenzel. 2005. Changes of Ni biogeochemistry in the rhizosphere of the hyperaccumulator *Thlaspi goesingense*. *Plant and Soil* 271: 205–218.

Puschenreiter, M., S. Wieczorek, O. Horak, and W.W. Wenzel. 2003. Chemical changes in the rhizosphere of metal hyperaccumulator and exluder *Thlaspi* species. *Journal of Plant Nutrition and Soil Science* 166: 579–584.

Quartacci, M.F., B. Irtelli, C. Gonnelli, R. Gabbrielli, and F. Navari-Izzo. 2009. Naturally-assisted metal phytoextraction by *Brassica carinata*: Role of root exudates. *Environmental Pollution* 157: 2697–2703.

Reichman, S.M., and D.R. Parker. 2005. Metal complexation by phytosiderophores in the rhizosphere. In Eds. P.M. Huang, and G.R. Gobran. *Biogeochemistry of trace elements in the rhizosphere*, pp. 129–154. Toronto: Elsevier.

Santi, S., and W. Schmidt. 2009. Dissecting iron deficiency-induced proton extrusion in *Arabidopsis* roots. *New Phytologist* 183: 1072–1084.

Santner, J., F. Degryse, M. Puschenreiter, W.W. Wenzel, and E. Smolders. 2010. Personal communication.

Schwab, A.P., Y. He, and M.K. Banks. 2005. The influence of organic ligands on the retention of lead in soil. *Chemosphere* 61: 856–866.

Schwab, A.P., D.S. Zhu, and M.K. Banks. 2008. Influence of organic acids on the transport of heavy metals in soil. *Chemosphere* 72: 986–994.

Seuntjens, P., B. Nowack, and R. Schulin. 2004. Root-zone modeling of heavy metal uptake and leaching in the presence of organic ligands. *Plant and Soil* 265: 61–73.

Sharples, J.M., A.A. Meharg, S.M. Chambers, and J.W.G. Cairney. 2000. Symbiotic solution to arsenic contamination. *Nature* 404: 951–952.

Shuman, L.M., and J. Wang. 1997. Effect of rice variety on zinc, cadmium, iron, and manganese content in rhizosphere and non-rhizosphere soil fractions. *Communications in Soil Science and Plant Analysis* 28: 23–36.

Silva Gonzaga, M.I., J.A.G. Santos, and L.Q. Ma. 2006. Arsenic chemistry in the rhizosphere of *Pteris vittata* L. and *Nephrolepis exaltata* L. *Environmental Pollution* 143: 254–260.

Smolders, E., and M.J. McLaughlin. 1996. Chloride increases cadmium uptake in Swiss chard in a resin-buffered nutrient solution. *Soil Science Society of America Journal* 60: 1443–1447.

Someya N., Y. Sato, I. Yamaguchi, H. Hammamoto, Y. Ichiman, K. Akutsu, H. Sawada, and K. Tsuchiya. 2007. Alleviation of nickel toxicity in plants by a rhizobacterium strain is not dependent on its siderophore production. *Communications in Soil Science and Plant Analysis* 38: 1155–1162.

Tao, S., Y.J. Chen, F.L. Xu, J. Cao, and B.G. Li. 2003. Changes of copper speciation in maize rhizosphere soil. *Environmental Pollution* 122: 447–454.

Tao, S., W.X. Liu, Y.J. Chen, F.L. Xu, R.W. Dawson, B.G. Li, J. Cao, X.J. Wang, J.Y. Hu, and J.Y. Fang. 2004. Evaluation of factors influencing root-induced changes of copper fractionation in rhizosphere of a calcareous soil. *Environmental Pollution* 129: 5–12.

Treeby, M., H. Marschner, and V. Römheld. 1989. Mobilization of iron and other micronutrient cations from a calcareous soil by plant-borne, microbial, and synthetic metal chelators. *Plant and Soil* 114: 217–226.

Vassil, A.D., Y. Kapulnik, I. Raskin, and D.E. Salt. 1998. The role of EDTA in lead transport and accumulation by Indian mustard. *Plant Physiology* 117: 447–453.

Vetterlein, D., K. Szegedi, J. Ackermann, J. Mattusch, H.-U. Neue, H. Tanneberg, and R. Jahn. 2007. Competitive mobilization of phosphate and arsenate associated with goethite by root activity. *Journal of Environmental Quality* 36: 1811–1820.

Vivas, A., B. Biró, J.M. Ruíz-Lozan, J.M. Barea, and R. Azcón. 2006. Two bacterial strains isolated from a Zn-polluted soil enhance plant growth and mycorrhizal efficiency under Zn-toxicity. *Chemosphere* 62: 1523–1533.

Voegelin, A., F.-A. Weber, and R. Kretzschmar.2007. Distribution and speciation of arsenic around roots in a contaminated riparian floodplain soil: Micro-XRF element mapping and EXAFS spectroscopy. *Geochimica et Cosmochimica Acta* 71: 5804–5820.

Voigt, A., W.H. Hendershot, and G.I Sunahara. 2006. Rhizotoxicity of cadmium and copper in soil extracts. *Environmental Toxicology and Chemistry* 25: 692–701.

Wang, J., C.B. Zhang, and Z.X. Jin. 2009a. The distribution and phytoavailability of heavy metal fractions in rhizosphere soils of *Paulowniu fortunei* (seem) Hems near a Pb/Zn smelter in Guangdong, PR China. *Geoderma* 148: 299–306.

Wang, Y., L. Zhang, Y. Huang, J. Yao, and H. Yang. 2009b. Transformation of copper fractions in rhizosphere soil of two dominant plants in a deserted land of copper tailings. *Bulletin of Environmental Contamination and Toxicology* 82: 468–472.

Wang, Z.Y., J.M. Kelly, and J.L. Kovar. 2005. Depletion of macro-nutrients from rhizosphere soil solution by juvenile corn, cottonwood, and switchgrass plants. *Plant and Soil* 270: 213–221.

Wenzel, W.W. 2009. Rhizosphere processes and management in plant-assisted bioremediation (phytoremediation) of soils. *Plant and Soil* 321: 385–408.

Wenzel, W.W., W.E.H. Blum, A. Brandstetter, F. Jockwer, A. Köchl, M. Oberforster, H.E. Oberländer, C. Riedler, K. Roth, and I. Vladeva. 1996. Effects of soil properties and cultivar on cadmium accumulation in wheat grain. *Journal of Plant Nutrition and Soil Science* 159: 609–614.

Wenzel, W.W., M. Bunkowski, M. Puschenreiter, and O. Horak. 2003a. Rhizosphere characteristics of indigenously growing nickel hyperaccumulator and tolerant plants on serpentine soil. *Environmental Pollution* 123: 131–138.

Wenzel, W.W., R. Unterbrunner, P. Sommer, and P. Sacco. 2003b. Chelate-assisted phytoextraction using canola (*Brassica napus* L.) in outdoors pot and lysimeter experiments. *Plant and Soil* 249: 83–96.

Wenzel, W.W., N. Kirchbaumer, T. Prohaska, G. Stingeder, E. Lombi, E., and D.C. Adriano. 2001. Arsenic fractionation in soils using an improved sequential extraction procedure. *Analytica Chimica Acta* 436: 309–323.

Wu, Y.H., and W.H. Hendershot. 2010. The effect of calcium and pH on nickel accumulation in and rhizotoxicity to pea (*Pisum sativum* L.) root: Empirical relationships and modeling. *Environmental Pollution* 158: 1850–1856.

Yang, X.E., V.C. Baligar, J.C. Foster, and D.C. Martens. 1997. Accumulation and transport of nickel in relation to organic acids in ryegrass and maize grown with different nickel levels. *Plant and Soil* 196: 271–276.

Zhou, X.-B., and W.-M. Shi. 2007. Effect of root surface iron plaque on Se translocation and uptake by Fe-deficient rice. *Pedosphere* 17: 580–587.

Zhu, B., and A.K. Alva. 1993. Differential adsorption of trace metals by soils as influenced by exchangeable cations and ionic strength. *Soil Science* 155: 61–66.

7

Heavy Metals in Agricultural Watersheds: Nonpoint Source Contamination

Moustafa A. Elrashidi

CONTENTS

Introduction

Managing nonpoint sources of contamination from agricultural land is technically complex. Contamination sources often are located over a large geographic area and are difficult to identify. Identifying hotspots within a watershed enables more efficient use of funds to alleviate potential problems and protect water resources. There are models that can estimate the impact of nonpoint sources of contamination from agricultural watersheds. However, these models are complex and expensive because they require very extensive data input.

The NRCS (Natural Resources Conservation Service) developed a technique (Elrashidi et al., 2003, 2004, 2005a,b, 2007a,b, 2008, 2009) to estimate element loss by runoff for agricultural watersheds. The NRCS technique applies the USDA runoff curve number model (USDA/SCS, 1991) to estimate loss of runoff water from soils by rainfall. The technique assumes that dissolved inorganic chemicals are lost from a specific depth of surface soil that interacts with runoff and leaching water. These chemicals may include any essential plant nutrients (i.e., N, P, Cu, Zn, etc.) and environmentally toxic elements such as Pb, Cd, Ni, and As. Geographical Information Systems (GIS; ESRI, 2006) are used to present data spatially in watershed maps.

The NRCS technique is quick and cost-effective because it utilizes existing climatic, hydrologic, and soil survey information. The Soil Survey Geographic Database (SSURGO; USDA/NRCS, 1999) is used to identify major soils, areas, and locations in the watershed. Land cover databases (NLCD, 1992) and the National Agricultural Statistics Service (NASS, 2003) are used to identify areas of crop, pasture, forest, etc. The National Water & Climate Center (NWCC, 2003) is used to access information on precipitation and other climate data. The United States Geological Survey (USGS, 2007) maintains flow gauging stations in major streams and rivers in the United States. The water flow data along with the drainage area can be applied to calculate the observed surface runoff from the watershed, which can be used to validate values predicted by the runoff and percolation models.

NRCS Technique

Estimation of Runoff Water

Rainfall is the primary source of water that runs off the surfaces of small agricultural watersheds. The main factors affecting the volume of rainfall that runs off are the kind of soil and the type of vegetation in the watershed (USDA/SCS, 1991). The runoff equation can be written as follows:

$$Q = (R - 0.2S)^2 \div (R + 0.8S) \quad (7.1)$$

where Q = runoff (inches), R = rainfall (inches), and S = potential maximum retention (inches) after runoff begins.

The potential maximum retention (S) can range from zero on a smooth and impervious surface to infinity in deep gravel. The S value is converted to a runoff curve number (CN) that depends on both the hydrologic soil group and type of land cover by the following equation:

$$CN = 1000 \div (10 + S) \quad (7.2)$$

According to Equation (7.2), CN is 100 when S is zero and approaches zero as S approaches infinity. Runoff curve numbers (CNs) can be any value from zero to 100, but for practical applications are limited to a range of 40 to 98. Substituting Equation (7.2) into Equation (7.1) gives

$$Q = \{R - [2(100 - CN)/\ CN]\}^2 \div \{R + [8(100 - CN)/CN]\} \qquad (7.3)$$

The hydrologic groups of the identified major soils are used to determine CN values for different land covers in the watershed.

The annual rainfall for the watershed is taken from the USDA/NRCS National Water & Climate Center (NWCC, 2003). In Equation (7.3), the effective rainfall (R) is the portion of annual rainfall that could generate runoff (Gilbert et al., 1987). The hydrologic group for a given soil and related CNs for various types of land cover are published in the USDA/NRCS *National Engineering Field Manual* (USDA/SCS, 1991).

For agricultural land in the watershed, the effective rainfall (R) and the runoff curve numbers are determined first; then the runoff equation is applied to estimate the runoff water (Q) for soil under forest, pasture, and crop. The equation calculated runoff water in inches (depth of water). Values are usually converted to millimeters.

Determining Dissolved Elements in Soils and Water

Soil samples are analyzed on air-dried, <2-mm soil by methods described in the Soil Survey Investigations Report (SSIR) No. 42 (USDA/NRCS, 2004). Alphanumeric codes in parentheses next to each method represent specific standard operating procedures. Particle-size analysis is performed by the sieve and pipette method (3A1). Cation exchange capacity (CEC) is conducted by NH_4OAc buffered at pH 7.0 (5A8b). Total carbon (C) content is determined by dry combustion (6A2f), and the $CaCO_3$ equivalent is estimated by the electronic manometer method (6E1g). Organic C in soil is estimated from both the total- and $CaCO_3$-C. Soil pH is measured in a 1:1 soil:water suspension (8C1f). Bulk density (BD) is estimated from particle size analysis and organic matter content (Rawls, 1983). The liquid limit is determined by the American Society for Testing and Materials method D 4318 (ASTM, 1993).

Dissolved elements (nutrients and heavy metals) are determined in soils. Anion exchange resin (AER) extractable-P was determined by the method described in Elrashidi et al. (2003). Soluble nitrate N is extracted with 1.0 M KCl solution and measured by flow injection, automated ion analyzer LACHAT Instruments (6M2a). Water-extractable elements (Al, As, B, Ba, Fe, Ca, Cd, Co, Cr, Cu, K, Mg, Mn, Mo, Na, Ni, P, Pb, Si, Sr, and Zn) for soils are determined according to the Soil Survey Laboratory procedure (4D2b1) (USDA/NRCS, 2004). In this method (4D2b1), the soil:water system (20 g soil and 100 mL d.w.) is allowed to equilibrate at room temperature for 23 hours before shaking the suspension for 1 hour. The supernatant is passed through a 0.45-μm filter. Elements are determined in the filtrate by inductively coupled plasma-optical emission spectrometry (ICP-OES) (Perkin Elmer 3300 DV). Nitrate-N, nitrite-N, sulfate-S, chloride (Cl), and fluoride (F) concentrations in the filtrate are determined by high pressure ion chromatography (6M1c) (HPIC, Dionex Corp.). The pH in the water extract is measured with the combination electrode and digital pH/ion meter, Model 950, Fisher Scientific (8C1a) as described in USDA/NRCS (2004).

Water samples are filtered through disposable nylon filter media (0.45 μm pore size). The phosphorus concentration is determined in the water samples by the modified phospho-molybdate/ascorbic acid method (Olsen and Sommers, 1982) or ICP-OES (Perkin Elmer 3300 DV). Nitrate-N, nitrite-N, sulfate-S, chloride (Cl), and fluoride (F) concentrations in the filtrate are determined by HPIC (6M1c) (HPIC, Dionex Corp.). Element concentrations in the filtrate (Al, As, B, Ba, Fe, Ca, Cd, Co, Cr, Cu, K, Mg, Mn, Mo, Na, Ni, P, Pb, Si, Sr, and Zn) are determined by ICP-OES (Perkin Elmer 3300 DV) (4I3a). The pH in the water is measured with the combination electrode and digital pH/ion meter, Model 950, Fisher Scientific (8C1a) as described in USDA/NRCS (2004).

Estimating Elements Loss by Runoff

Nutrients such as N, K, P, and other agricultural chemicals are released from a thin layer of surface soil that interacts with rainfall and runoff. In chemical transport models, the thickness of the interaction zone is determined by model calibration with experimental data, with depths ranging between 2.0 and 6.0 mm (Donigian et al., 1977). Frere et al. (1980), however, suggested an interaction zone of 10 mm, assuming that only a fraction of the chemical present in this depth interacts with rainfall water. In previous studies in this laboratory, Elrashidi et al. (2003, 2004, 2005a,b, 2007a,b, 2008, 2009) successfully used a fixed soil thickness of 10 mm to estimate the loss of nutrients and heavy metals by runoff from agricultural land.

In this technique, an interaction zone of 10 mm is used to calculate the amount of element released from surface soils by runoff. Also, it is assumed that during the runoff occurrence, water content in the surface 10-mm soil depth is at the liquid limit, the moisture content at which the soil passes from a plastic to liquid state. Thus, during the runoff occurrence, the total amount of water (where the element in the 10-mm soil depth is dissolved) is the sum of water within the soil body (liquid limit) and that on the surface of soil (runoff water). The volume of water in the 10-mm soil depth is usually very small when compared with runoff water. Only the element in runoff water is removed and lost during the runoff occurrence.

GIS Digital Mapping

Digital maps for water and nutrient losses from agricultural land in the watershed are generated by Geographical Information Systems (GIS) software:

ArcView 9.2 (ESRI, 2006). The input data required to generate the GIS map include spatial data layers (soil series and land cover) and the tabular data from both the runoff and leaching (amount of water and nutrient loss from soils and concentrations in both runoff and leaching waters).

The principal spatial data layer used is the Soil Survey Geographic Database (SSURGO) (USDA/NRCS, 1999). Both the National Land Cover (NLCD, 1992), and National Agricultural Statistics Service (NASS, 2003) spatial layers are used to identify areas of forest, pasture, and crop within the watershed. Other types of land cover, such as urban, water, or marsh, are usually not mapped for the watershed. The proposed technique calculated water and nutrient losses as well as concentrations in runoff and leaching water for soils under different types of land cover (forest, pasture, and crop). Thus, GIS mapping of agricultural land in the watershed includes data layers for soils and land cover as well as water or elements.

Application of NRCS Technique

In this chapter we describe how the technique was applied to study the nonpoint source of heavy metals contamination of surface water in the Wagon Train (WT) Watershed, Lancaster County, Nebraska. Information on the watershed, major soils, methods used for soil and water sampling, as well as the description of streams and procedure used to estimate the inflow for WT reservoir are included.

Wagon Train Watershed

The Wagon Train (WT) Watershed has a 315-acre (128-hectare) reservoir located on the Hickman Branch of Salt Creek (Platte River Basin) in Lancaster County, Nebraska (Figure 7.1). The reservoir was constructed primarily as a flood control structure by the U.S. Army Corps of Engineers in 1962. The total drainage area encompasses 9,984 acres (4,042 hectares) of agricultural land. Most of the area (70%) is cultivated with crops: soybean (*Glycine willd*), corn (*Zea mays* L.), wheat (*Triticum aestivum* L.), sunflower (*Helianthus L.*), and alfalfa (*Medicago sativa L.*). The remainder of the watershed is mostly covered with grassland, while forestland, wetland, and urban development account for small areas.

The watershed topography is moderately sloping, and soils are moderate to poorly drained. The land relief consists of uplands, stream terraces, and bottom lands. There are 33 miles (53 kilometers) of streams in the watershed, and 40 ponds ranging in size from 0.3 to 6.5 acres (0.12 to 2.6 hectares). Overland flow enters the reservoir through intermittent tributaries. From the dam, the water flows into the Hickman Branch of Salt Creek, which flows west and north through Lincoln, and eventually to the Platte River near Ashland, Nebraska.

We used the soil survey information SSURGO (USDA/NRCS, 1999) to determine the major soils in the watershed. Both the National Land Cover Data (NLCD, 1992) and National Agricultural Statistics Service (NASS, 2003) were used to identify different land covers. The watershed has three major soil associations:

1. The Wymore-Pawnee association consists of deep, nearly level to sloping soils, located on ridge tops and side slopes: Wymore (fine, montmorillonitic, mesic Aquic Argiudolls); Pawnee (fine, montmorillonitic, mesic Aquic Argiudolls).
2. The Pawnee-Burchard association consists of deep, gently to steeply sloping, loamy and clayey upland soils that developed in glacial till: Burchard (fine-loamy, mixed, mesic Typic Argiudolls).
3. The Kennebec-Nodaway-Zook association contains deep, nearly level or gently sloping silty soils formed in alluvium on flood plains: Kennebec (fine-silty, mixed, mesic Cumulic Hapludolls); Nodaway (fine-silty, mixed, nonacid, mesic Mollic Udifluvents); Zook (fine, montmorillonitic, mesic Cumilic Hapaquolls).

Nine soil series (Wymore, Pawnee, Nodaway, Sharpsburg, Mayberry, Colo, Judson, Burchard, and Kennebec) account for 96.1% of the agricultural land. Nearly three-quarters of the watershed consists of Wymore and Pawnee soils.

Soil and Water Sampling

Soils were sampled from each of three widely existent phases of Wymore (Wymore-WtB, -WtC2, and -WtD3), and two phases of Pawnee (Pawnee-PaC2 and -PaD2) along with the other seven soil series. This approach produced a total of twelve soil map units to sample. Soil samples were collected from cropland and grassland within each map unit. Recently, updated soil survey activities have split Sharpsburg into three soil series (Tomek, Yutan, and Aksarben). The new classification, however, should not affect results given in this study.

Representative soil samples were collected from each of the twelve soil map units. To distribute sampling locations evenly within the agricultural area, the watershed was divided into six sections. An equal number of samples were taken at random from each section. In total, seventy-two soil samples from cropland and twenty-four from grassland were collected (Figure 7.1).

At the randomly selected sampling sites, three cores were taken from the top 30-cm soil depth and mixed thoroughly in a stainless steel tray.

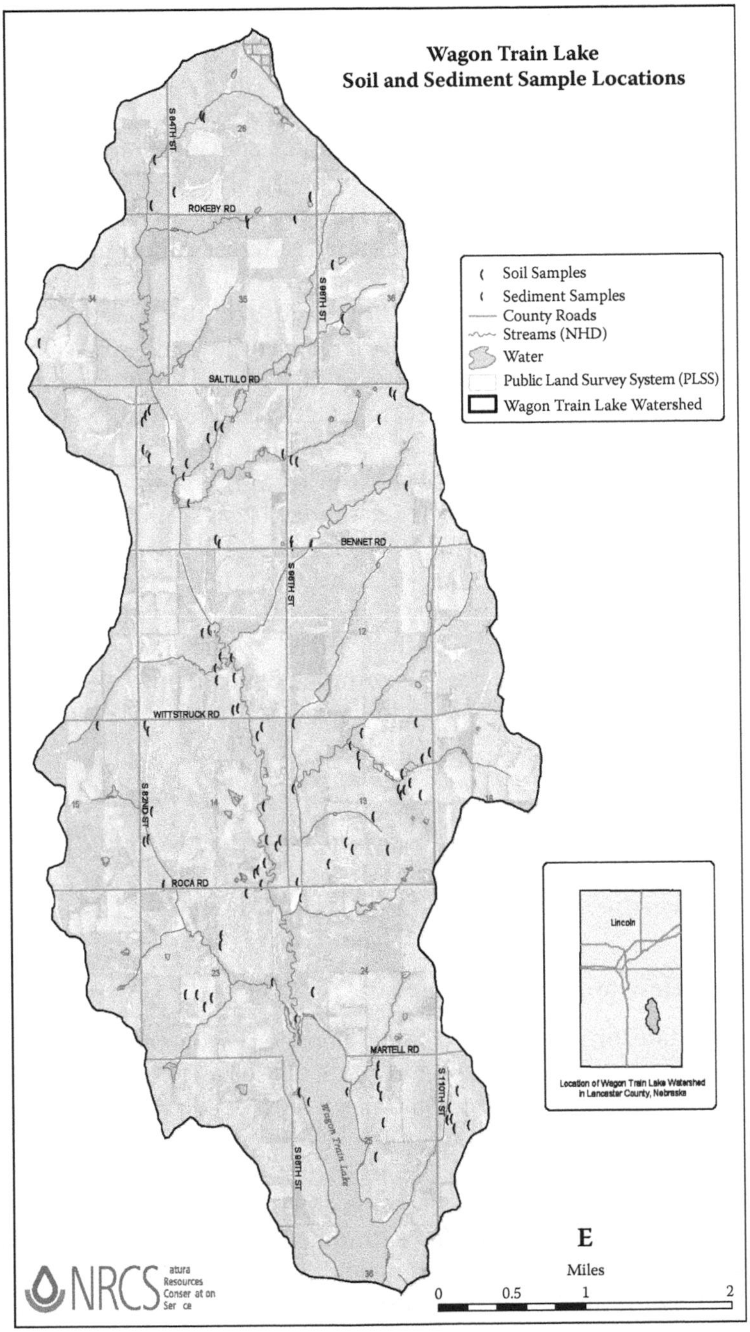

FIGURE 7.1
Soil and water sampling locations in the Wagon Train Watershed, Lancaster County, Nebraska.

Approximately, a 2-kg composite sample was packed in a plastic bag and sealed. Sampling was completed during April 2003 prior to fertilizer application for the summer crop.

Many small streams receive surface water runoff from the agricultural land in the watershed. Eventually, streams located northerly of the reservoir join in a single stream that runs southerly about 0.5 km before entering the reservoir near the north edge. Water samples taken along the main stream were assumed to represent the surface water runoff generated from the entire watershed.

Most of the surface water runoff from the agricultural land in the WT watershed and water inflow for the WT reservoir are expected during the rainy season in the spring, summer, and fall (March through November). During the period from March through November, weekly water samples were collected from the main stream (Figure 7.1). The analysis for major streams proved that samples taken from the main stream are representative of runoff generated from the entire watershed (Elrashidi et al., 2005a,b).

Water samples were collected (grab) in midstream, using 1-L polyethylene bottles that had been rinsed twice with stream water prior to sample collection. The water samples were taken immediately to the laboratory and refrigerated at 4°C. The water analysis was completed within a week. The soil and water sampling locations are shown in Figure 7.1.

Soil and Water Analysis

Soil and water samples were analyzed as described above in the NRCS technique. Classification and selected properties for soils under crop and grass in the WT watershed are given in Table 7.1.

Observed Inflow for WT Reservoir

In 1962, the dam on a tributary of Salt Creek and construction of the WT reservoir were completed. However, the U.S. Geological Survey (USGS, 2001) had monitored the water flow in Salt Creek and streams in the Platte River basin long before the construction of the WT reservoir. The Salt Creek gage at Roca (USGS gage # 06803000, hydrologic unit 10200203, Lancaster County, Nebraska) with a period of record from 1951 to 2000 provided average monthly water flow rate values for a drainage area of 106,880 acres (43,286 hectares) encompassing the WT watershed (USGS, 2001). Recently, the Lower Platte South Natural Resources District (LPSNRD, 2004) used the ratio of the watershed to the Salt Creek drainage area (9.34%) to calculate the average monthly water flow rate values for the WT watershed. In our study, we used these average monthly water flow rate values to calculate the observed inflow for the WT reservoir.

TABLE 7.1

Soil classification and selected properties for 12 major soils under crop and grass cover in Wagon Train Watershed, Lancaster County, Nebraska

Soil (Map Unit)	Classification	Land Use	Clay (%)	OC (%)	CEC (cmol/kg)	pH water	Hydrologic Group
Wymore (WtB)	Fine, montmorillonitic, mesic Aquic Argiudolls	Cropland	37.3	2.14	25.9	5.56	D
		Grassland	32.9	2.44	25.7	5.9	D
Wymore (WtC2)	Fine, montmorillonitic, mesic Aquic Argiudolls	Cropland	37.9	2.23	26.5	5.7	D
		Grassland	35.6	3.46	28.2	5.8	D
Wymore (WtD3)	Fine, montmorillonitic, mesic Aquic Argiudolls	Cropland	41.2	2.16	29.3	5.85	D
		Grassland	34.2	2.78	28.9	6.4	D
Pawnee (PaC2)	Fine, montmorillonitic, mesic Aquic Argiudolls	Cropland	35.2	1.94	24.9	5.64	D
		Grassland	29.3	2.38	21.7	5.55	D
Pawnee (PaD2)	Fine, montmorillonitic, mesic Aquic Argiudolls	Cropland	34.9	1.85	24.5	5.79	D
		Grassland	34.7	2.39	25.5	6.1	D
Nodaway (No, Ns)	Fine-silty, mixed, nonacid, mesic Mollic Udifluvents	Cropland	29.4	2.08	24.4	6.58	B
		Grassland	30.1	2.97	26.4	6.25	B
Sharpsburg (ShC, ShD, ShD2)	Fine, montmorillonitic, mesic Typic Argiudolls	Cropland	39.7	1.94	27.6	5.7	B
		Grassland	37.4	2.05	27	6.15	B
Mayberry (MeC2, MeD2, MhC3)	Fine, montmorillonitic, mesic Aquic Argiudolls	Cropland	31.8	1.96	22.8	5.99	D
		Grassland	26.0	2.08	20.4	6.5	D
Colo (Co, Cp)	Fine-silty, mixed, mesic Cumulic Hapludolls	Cropland	32.1	2.13	25	6.3	C
		Grassland	29.0	2.95	26.1	6.1	C
Judson (JuC)	Fine-silty, mixed, mesic Cumulic Hapludolls	Cropland	32.0	2.26	24.8	6.05	B
		Grassland	30.5	3.06	24	6.0	B
Burchard (BpF, BrD, BrE)	Fine-loamy, mixed, mesic Typic Argiudolls	Cropland	29.8	1.89	21.7	5.96	B
		Grassland	30.1				
Kennebec (Ke)	Fine-silty, mixed, mesic Cumulic Hapludolls	Cropland	27.6	1.94	20.7	5.95	B
		Grassland	24.7				

Case Study

In this study, we investigated eight heavy metals (Al, Fe, Si, Cd, Cu, Ni, Pb, and Zn) in twelve major soils in the Wagon Train Watershed, Lancaster County, Nebraska. We were interested in understanding the role of these agriculture soils as a nonpoint source of metals contamination to surface waters. Heavy storms may generate runoff events that remove dissolved soil chemicals from agricultural land to surface water bodies (nonpoint source of contamination). Most heavy metals have natural input to streams, rivers, and lakes from weathering and dissolution of oxides, carbonate, and silicate minerals in soils. However, anthropogenic activities can introduce greater amounts of heavy metals to soils and natural waters. The anthropogenic inputs that can introduce heavy metals into the environments include the application of commercial fertilizers, liming materials, sewage sludge, manure, animal wastes, soil amendments, pesticides, coal combustion by-products, in addition to auto emissions and fallout from metal-smelting industries. Losses of heavy metals by runoff from agricultural land have received little attention from agronomists and soil scientists. However, from both animal and human health perspectives and an environmental water quality viewpoint, heavy metal concentrations and form as well as total quantity lost from nonpoint sources are important concerns for both agricultural management and subsequent water users.

Chemicals and dissolved elements are released from a thin layer of surface soil that interacts with rainfall and runoff water (Sharpley, 1985). Chemicals transported by runoff water from agricultural land can pose a risk to surface and ground waters (nonpoint sources of contamination). Studying nonpoint sources of contamination from agricultural land is technically complex. Contamination sources often are located over a large geographic area and are difficult to identify. Identifying hotspots within a watershed enables more efficient use of funds to alleviate potential problems and protect water resources. There are models that can be used to estimate the impact of nonpoint sources of contamination from agricultural watersheds. But these models are too complex and expensive because they require very extensive data inputs.

The USDA-NRCS developed an exploratory technique to estimate elements loss by runoff for agricultural watersheds. The technique is quick and cost-effective because it utilizes existing climatic, hydrologic, and soil survey databases. Thus, lengthy and expensive models could be performed on certain areas of high risk. The NRCS technique was applied to estimate losses of phosphorus, nitrogen, and alkaline earth elements by runoff from agricultural land (Elrashidi et al., 2005a,b, 2007). The technique applies the USDA runoff model (USDA/SCS, 1991) to estimate loss of runoff water from soils by rainfall. The technique assumes that water-soluble elements are lost from a specific depth of surface soil that interacts with runoff water. A brief description of the technique is reported in the previous section. The objectives of this study were to apply this technique on the WT watershed (Lancaster County, Nebraska) to estimate (1) losses of dissolved Al, Fe, Si, Cd, Cu, Ni, Pb, and Zn from soils by runoff water, and (2) metals loading into the WT reservoir.

The study was conducted on the WT watershed in Lancaster County, Nebraska. Information on the watershed, major soils, streams, and surface water body, as well as the methods applied for soil and water sampling and analysis, and for estimating heavy metals loss by runoff were given above under the application of the NRCS technique on the WT watershed.

Runoff and Reservoir Inflow

The historic record of monthly rainfall for Lancaster County (NWCC, 2003) was applied in the runoff model (USDA/SCS, 1991) to predict the runoff water. The predicted annual loss of water by runoff (cubic meter per hectare), and water present in the interaction zone (cubic meter per hectare) for twelve major soils under crop and grass in the WT watershed are given in Table 7.2. Generally, the annual loss of water from soil by runoff was higher for cropland than grassland. The predicted average (area weighted) of annual runoff water was 1,122 and 942 $m^3.ha^{-1}.yr^{-1}$ for cropland and grassland, respectively. These results accounted for 15.4% and 12.9% of the annual rainfall for cropland and grassland, respectively. Similar values were reported for thirteen United States soils of humid regions (rainfall > 800 $mm.yr^{-1}$) where the average was 15% for cropland and 12% for grassland (Elrashidi et al., 2003).

However, these values were relatively higher than those reported for Lancaster County, Nebraska, where the watershed is located (Elrashidi et al., 2004). This could be attributed to the slow water infiltration rate (hydrologic group D) for the dominant soils (Wymore, Pawnee, and Mayberry) in the watershed. These three soils occupy approximately 80% of the agricultural land in the watershed.

The results indicated that the Wymore-WtC2 soil map unit, irrespective of the land cover, produced the highest volume of runoff water mainly because of its abundance in the watershed. On the other hand, Kennebec soil, which had very limited area, generated the least volume of runoff water. The total annual loss of runoff water from the twelve major soil map units was 4.15 million m^3. The area of the twelve major soil map units (3885 hectares) incorporated about 96% of the entire watershed. When the entire watershed area (4042 hectares) was considered, the total annual runoff accounted for 4.31 million cubic meters of water. The observed average annual inflow for the WT reservoir for a fifty-year period between 1951 and 2000 is 4.25 million cubic meters (USGS, 2001; Elrashidi et al., 2005a). The predicted annual runoff and the observed annual inflow appeared in good agreement.

TABLE 7.2

Predicted Annual Loss of Water by Runoff ($m^3.ha^{-1}$) and Water Present in Interaction Zone ($m^3.ha^{-1}$) for Twelve Major Soils under Crop and Grass in the Wagon Train Watershed, Lancaster County, Nebraska

	Area		Runoff †		Interaction Zone	
Soil (Map Unit)	Crop	Grass	Crop	Grass	Crop	Grass
	(ha)		($m^3.ha^{-1}.yr^{-1}$)		($m^3.ha^{-1}.yr^{-1}$)	
Wymore (WtB)	391	167	1,167	1,004	59.5	59.5
Wymore (WtC2)	1,270	544	1,167	1,004	59.5	59.5
Wymore (WtD3)	124	53	1,167	1,004	59.5	59.5
Pawnee (PaC2)	240	103	1,167	1,004	60.8	60.8
Pawnee (PaD2)	54	23	1,167	1,004	60.8	60.8
Nodaway (No, Ns)	142	61	901	638	38.4	38.4
Sharpsburg (ShC, ShD, ShD2)	124	53	901	638	57.6	57.6
Mayberry (MeC2, MeD2, MhC3)	110	47	1,167	1,004	51.2	51.2
Colo (Co, Cp)	107	46	1,084	876	64.0	64.0
Judson (JuC)	71	30	901	638	48.4	48.4
Burchard (BpF, BrD, BrE)	24		901	638	54.4	54.4
Kennebec (Ke)	31	13	901	638	44.8	44.8
Weighted Average			1,122	942	57.7	57.7
Total	2,719	1,165	3,050,750	1,097,609	156,976	67,275

Source: From USDA/SCS (1991).

Elements in Soil and Water Phase

The average and standard deviation (%) of water-soluble elements for soils (milligrams or micrograms per kilogram) under crop and grass in the WT watershed are given in Table 7.3. For cropped soils, Al dissolved in water phase ranged between 128 and 352 $mg.kg^{-1}$ with an area-weighted average of 239 $mg.kg^{-1}$ soil. A wider range of Al concentration (14.5 to 349 $mg.kg^{-1}$) was observed for grassland. Meanwhile, the area-weighted average was lower than that of cropland at 136 $mg.kg^{-1}$ soil. For the twelve major soils, irrespective of land cover, the average Al concentration in water phase was 208 $mg.kg^{-1}$ soil.

For cropped soils, Fe dissolved in water phase ranged between 72.8 and 193 $mg.kg^{-1}$, with an area-weighted average of 130 $mg.kg^{-1}$ soil. A relatively wider range of Fe concentration (8.5 to 181 $mg.kg^{-1}$) was observed for grassland. However, the area-weighted average was much lower than that of cropland at 70.9 $mg.kg^{-1}$ soil. For the twelve major soils, irrespective of land cover, the average Fe concentration in water phase was 113 $mg.kg^{-1}$ soil.

For cropped soils, Si dissolved in water phase ranged between 184 and 668 $mg.kg^{-1}$, with an area-weighted average of 266 $mg.kg^{-1}$ soil. A wider range of

Si concentration (25.8 to 568 mg/kg) was observed for grassland. However, the area-weighted average was lower than that of cropland at 173 $mg.kg^{-1}$ soil. For the twelve major soils, irrespective of land cover, the average Si concentration in water phase was 238 $mg.kg^{-1}$ soil. Compared to Al, Fe, and Si, many fewer concentrations were measured for the other elements investigated (Cd, Cu, Ni, Pb, and Zn). For soils under crop, Cd dissolved in water phase ranged between 3.52 and 6.69 $\mu g.kg^{-1}$ with an area-weighted average of 4.96 $\mu g.kg^{-1}$ soil. A similar range of Cd concentrations (2.93 to 8.03 $\mu g.kg^{-1}$) was observed for grassland. However, the area-weighted average was slightly higher than that of cropland at 5.17 $\mu g.kg^{-1}$ soil. For the twelve major soils, irrespective of land cover, the average Cd concentration in water phase was 5.02 $\mu g.kg^{-1}$ soil.

For cropped soils, Cu dissolved in water phase ranged between 380 and 584 $\mu g.kg^{-1}$, with an area-weighted average of 433 $\mu g.kg^{-1}$ soil. A similar range of Cu concentrations (337 to 614 $\mu g.kg^{-1}$) was observed for grassland. However, the area-weighted average was slightly lower than that of cropland at 418 $\mu g.kg^{-1}$ soil. For the twelve major soils, irrespective of land cover, the average Cu concentration in water phase was 429 $\mu g.kg^{-1}$ soil. For soils under crop, Ni dissolved in water phase ranged between 135 and 244 $\mu g.kg^{-1}$, with an area-weighted average of 187 $\mu g.kg^{-1}$ soil. A wider range of Ni concentrations (70.6 to 268 $\mu g.kg^{-1}$) was observed for grassland. However, the area-weighted average was lower than that of cropland at 147 $\mu g.kg^{-1}$ soil. For the twelve major soils, irrespective of land cover, the average Ni concentration in water phase was 175 $\mu g.kg^{-1}$ soil.

For cropped soils, Pb dissolved in water phase ranged between 6.84 and 14.5 $\mu g.kg^{-1}$, with an area-weighted average of 11.5 $\mu g.kg^{-1}$ soil. A slightly wider range of Pb concentrations (5.30 to 18.5 $\mu g.kg^{-1}$) was observed for grassland. However, the area-weighted average was similar to that of cropland at 11.6 $\mu g.kg^{-1}$ soil. For the twelve major soils, irrespective of land cover, the average Pb concentration in water phase was 11.5 $\mu g.kg^{-1}$ soil. For soils under crop, Zn dissolved in water phase ranged between 215 and 496 $\mu g.kg^{-1}$, with an area-weighted average of 346 $\mu g.kg^{-1}$ soil. A wider range of Zn concentrations (8.55 to 600 $\mu g.kg^{-1}$) was observed for grassland. However, the area-weighted average was smaller than that of cropland at 229 $\mu g.kg^{-1}$ soil. For the twelve major soils, irrespective of land cover, the average Zn concentration in water phase was 311 $\mu g.kg^{-1}$ soil.

Elements in Stream Water

During the rainy season (March through November), the concentration of elements was measured in surface water samples collected weekly from the main stream in the WT watershed. In this study, we refer to the average element concentration in stream for the March through November rainy season as an annual average. Elrashidi et al. (2005a,b) found that water samples collected from the main stream in the WT watershed are representative of runoff generated from the entire watershed.

TABLE 7.3

Average (AV) and Standard Deviation Percentage (SD%) of Water-Soluble Elements in Soils (mg or µg.kg^{-1}) under Crop and Grass in Wagon Train Watershed, Lancaster County, Nebraska

Soil (Map Unit)		Al		Fe		Si		Cd		Cu		Ni		Pb		Zn	
		Crop	Grass	Crop	Grass	Crop	Grass	Crop	Grass	Crop	Grass	Crop	Grass	Crop	Grass	Crop	Grass
		(mg.kg^{-1})						(µg.kg^{-1})									
Wymore-WtB	AV	193.63	203.49	104.86	106.90	183.89	218.59	4.83	6.09	388.27	437.67	153.98	182.23	14.51	18.04	272.04	346.46
	SD%	68	65	67	63	67	76	27	25	13	9	40	31	63	82	76	84
Wymore-WTC2	AV	269.52	77.34	147.10	40.98	235.39	97.09	5.03	4.87	423.53	396.89	201.63	119.89	11.54	8.63	383.12	134.43
	SD%	77	56	79	53	70	67	63	13	20	5	61	21	81	54	94	110
Wymore-WTD3	AV	293.71	348.97	172.34	181.20	486.58	706.51	6.69	8.03	507.45	613.71	211.06	267.88	14.01	9.67	495.93	600.23
	SD%	34	57	36	57	66	78	41	17	33	35	33	43	72	52	37	42
Pawnee-PaC2	AV	205.51	155.43	106.79	73.99	223.68	177.60	4.28	5.38	403.09	409.52	181.60	131.05	6.84	16.62	285.12	183.06
	SD%	90	79	93	77	88	88	31	25	6	9	51	34	37	79	88	111
Pawnee-PaD2	AV	351.88	314.99	192.97	152.43	668.12	567.84	5.92	4.53	541.63	511.89	244.48	188.90	8.32	14.20	470.86	404.49
	SD%	37	45	44	40	67	73	56	51	33	44	43	37	43	76	48	56
Nodaway	AV	127.91	188.64	72.78	103.61	236.72	231.13	4.32	6.25	474.85	463.95	135.41	212.19	13.10	18.46	215.05	416.50
	SD%	76	91	74	90	79	89	65	21	35	14	51	46	121	83	109	107
Sharpsburg	AV	221.97	231.36	122.11	116.02	308.05	266.41	5.65	4.67	474.69	398.95	190.84	189.59	13.46	17.16	337.16	380.97
	SD%	64	60	63	55	70	65	32	70	24	22	33	36	101	80	55	52
Mayberry	AV	268.24	109.81	137.22	57.40	262.49	88.12	4.14	3.49	380.06	337.45	196.18	105.43	13.60	9.13	379.12	187.00
	SD%	53	64	55	64	60	39	62	61	32	16	48	28	81	49	70	49
Colo	AV	162.71	176.83	94.05	94.04	377.88	127.85	5.17	5.79	514.95	456.46	157.21	200.42	6.88	8.61	247.07	354.13

	SD%	77	26	79	25	89	21	29	9	29	2	39	7	29	32	74	23
Judson	AV	161.88	76.93	88.53	40.22	215.95	67.93	3.52	3.33	411.45	345.51	152.30	121.21	7.42	5.30	232.87	100.49
	SD%	53	24	53	23	72	5	36	52	28	12	34	9	48	0	65	77
Burchard	AV	304.53	14.48	165.41	8.53	531.79	25.78	5.31	2.98	519.65	340.31	225.26	70.59	10.93	5.30	420.17	8.55
	SD%	55	113	47	112	75	49	42	65	32	20	45	65	40	0	52	35
Kennebec	AV	199.95	101.71	109.87	54.89	487.31	223.17	5.23	2.93	584.42	446.73	175.58	99.70	7.35	10.42	356.90	173.85
	SD%	39	72	39	72	45	89	38	51	23	42	28	38	39	57	41	51
Weighted Average (Crop & Grass)	239.03	136.47	130.46	70.87	265.64	173.47	4.96	5.17	433.21	417.63	187.29	147.32	11.50	11.63	345.58	229.48	
Weighted Average (watershed)	208.30		112.61		238.03		5.02		428.54		175.32		11.54		310.79		

Both Al and Fe are abundant in soils, and their solubility is heavily dependent on the pH (Lindsay, 1979). Therefore, both elements rarely occur in high concentrations in freshwaters at neutral or alkaline pH (Hem, 1989). In this study, the annual average pH in water samples collected from the main stream in the WT watershed was 8.39. Expectedly, the observed Al and Fe concentrations were very low, where Al ranged between 1.80 and 33.1 $\mu g.L^{-1}$, with an annual average of 11.1 $\mu g.L^{-1}$. Crain (2001) reported a wide range of Al concentrations in forty-four Kentucky stream stations, ranging between 0.5 and 49,000 $\mu g.L^{-1}$, with a median of 467 $\mu g.L^{-1}$. The concentration of Fe in the WT watershed water ranged from 6.60 to 16.1 $\mu g.L^{-1}$, and averaging 8.61 $\mu g.L^{-1}$, annually. In a study on four creeks in northeastern Kansas, Schmidt (2004) found similar Fe concentrations ranging from 5 to 30 $\mu g.L^{-1}$, with an average of 6 $\mu g.L^{-1}$. Higher Fe concentrations ranging between 50 and 80 $\mu g.L^{-1}$, with an average of 16 $\mu g.L^{-1}$, were measured for Rattle Snake Creek in south-central Kansas (Christensen, 2001). Also, Apodaca and Bails (1999) published relatively high Fe concentrations in Frazer River, Colorado, ranging from 20 to 450 $\mu g.L^{-1}$, with an average of 180 $\mu g.L^{-1}$.

In contrast to Al and Fe, relatively higher concentrations were measured for Si in stream water samples collected from the WT watershed, ranging from 120 to 1,032 $\mu g.L^{-1}$ with an annual average of 531 $\mu g.L^{-1}$. Fuhrer et al. (1996) conducted a study of the water quality in the Columbia River basin and collected water samples at ten sites. They found that the dissolved Si concentration in waters ranged between 2,337 and 9,817 $\mu g.L^{-1}$, with an average of 4,160 $\mu g.L^{-1}$.

The USEPA (2002) recommended two criteria to determine the quality of freshwater: (1) the Criterion Maximum Concentration (CMC) is an estimate of the highest concentration of a material in surface water to which an aquatic community can be exposed briefly without resulting in an unacceptable effect; and (2) the Criterion Continuous Concentration (CCC) is an estimate of the highest concentration of a material in surface water to which an aquatic community can be exposed indefinitely without resulting in an unacceptable effect. The USEPA (2002) recommended Al concentrations of 750 $\mu g.L^{-1}$ for CMC and 87 $\mu g.L^{-1}$ for CCC. Large amounts of Fe are undesirable in water because it can cause taste problems and forms a red oxyhydroxide precipitate that stains laundry and plumbing fixtures (Hem, 1992). The USEPA (2002) recommended an Fe concentration of 1,000 $\mu g.L^{-1}$ for CCC while no concentration was recommended for CMC. The USEPA (2002) recommended a drinking-water standard of 300 $\mu g.L^{-1}$ for Fe while no regulation was established for either Al or Si. Accordingly, the concentrations of Al, Fe, and Si observed in the stream water samples collected from the WT watershed appear to be too low to result in any ecological or human health concern.

The observed monthly average Cd concentrations in the stream water ranged from 0.20 to 1.13 $\mu g.L^{-1}$, with an annual average of 0.45 $\mu g.L^{-1}$ (Figure 7.2a). Apodaca and Bails (1999) reported similar values for Cd in the

Frazer River watershed (Colorado), where it ranged from less than 0.25 to 3.0 μg.L^{-1}, with an average of 0.25 μg.L^{-1}. Also, He et al. (2004) published similar Cd concentrations for surface runoff generated from vegetable farms and citrus groves (0.0 to 2.8 μg.L^{-1}), and forest land (0.0 to 0.47 μg.L^{-1}) in Florida. The USEPA (2002) recommended Cd concentrations of 2.0 μg.L^{-1} for CMC and 0.25 μg.L^{-1} for the CCC. The U.S. drinking-water standard for Cd is 5.0 μg.L^{-1} (USEPA, 2003). With respect to Cd, these regulations indicate that the stream water in the WT watershed does not pose any risk for human consumption. However, there should be some concern regarding the indefinite exposure of the aquatic community to Cd concentration in the WT watershed stream water because it exceeds the EPA recommended CCC (0.25 μg.L^{-1}).

Relatively higher values were determined for Cu than for Cd concentrations in water samples collected from the WT watershed. The observed monthly average Cu concentration ranged between 50.4 and 85.9 μg.L^{-1}, averaging 66.1 μg.L^{-1} annually (Figure 7.2b). Similar values were reported by Crain (2001) for forty-four stream stations in Kentucky where Cu concentrations ranged between 0.5 and 82 μg.L^{-1}. Lower Cu concentrations (range: <4 to 6 μg.L^{-1}; average: 4 μg.L^{-1}) were published for waters in the Frazer River watershed in Colorado (Apodaca and Bails, 1999), and for waters in the Lower Columbia River basin (range: <1 to 3 μg.L^{-1}; average: 1 μg.L^{-1}) (Fuhrer et al., 1996). Copper is an essential element for plants and animals. The toxicity of Cu to aquatic organisms depends on the alkalinity of the water. Copper is much more toxic to organisms found in waters with low alkalinity than in waters with high alkalinity (USEPA, 1976). The USEPA (2002) recommended Cu concentrations of 13 μg.L^{-1} for CMC and 9.0 μg.L^{-1} for CCC. The recommended Cu concentration for drinking water is 1,300 μg.L^{-1} (USEPA, 2003). The fact that Cu concentrations in the stream water exceed the recommended EPA values for CMC and CCC should raise some concern for the negative effects on aquatic life in the WT watershed.

The observed monthly average Ni concentration in the stream water ranged from 5.25 to 9.27 μg.L^{-1}, with an annual average of 6.31 μg.L^{-1} (Figure 7.2c). He et al. (2004) investigated 1,277 surface runoff samples from eleven sites at vegetable farms and citrus groves in Florida. They found that Ni concentrations ranged between 0.0 and 39.3 μg.L^{-1}. The authors also investigated runoff generated from forest land in South Florida where Ni concentrations ranged from 0.0 to 13.3 μg.L^{-1}, with an average of 1.72 μg.L^{-1}. Nickel is rather toxic for most plants and fungi, and the carcinogenic actions of Ni and its salts have been observed in numerous animal experiments (Berman, 1980). The USEPA (2002) recommended Ni concentrations of 970 μg.L^{-1} for CMC and 52 μg.L^{-1} for CCC. The acceptable Ni level for drinking water in the United States has been established at 610 μg.L^{-1} (USEPA, 2002). These regulations suggest no Ni-related environmental problems are expected in the WT watershed for aquatic organisms and human health.

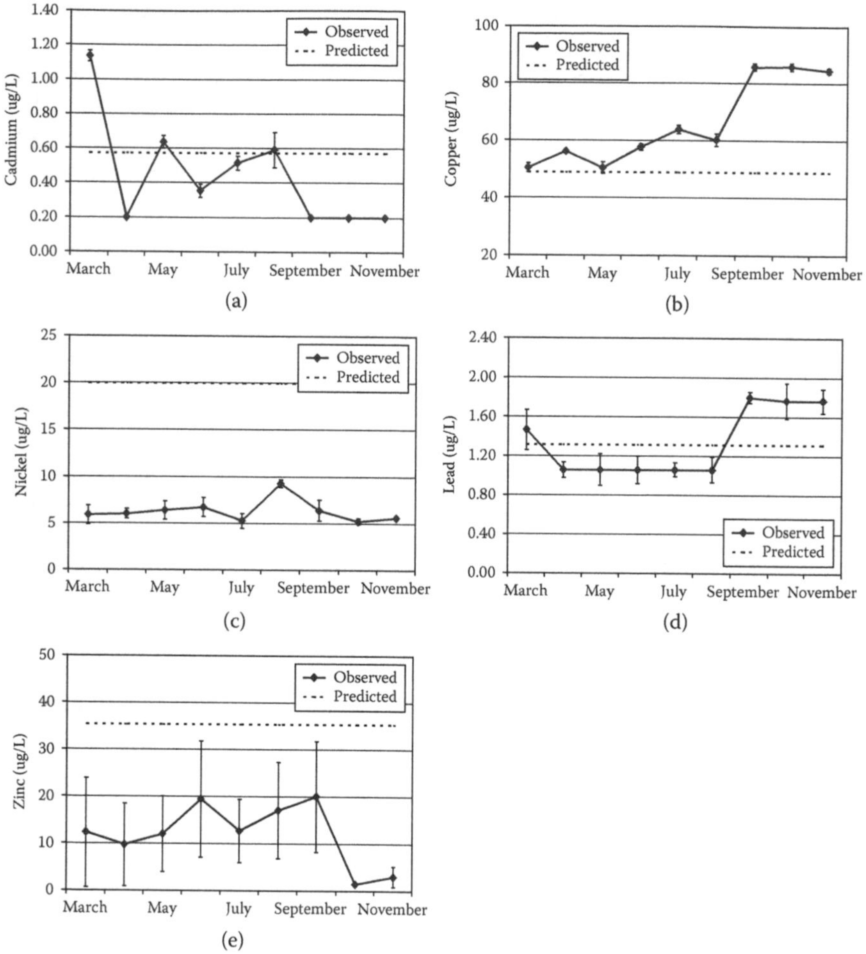

FIGURE 7.2
(a) Predicted Cd in runoff and observed monthly average concentration in stream water (μg.L^{-1}) in Wagon Train Watershed, Lancaster County, Nebraska. (b) Predicted Cu in runoff and observed monthly average concentration in stream water (μg./L^{-1}) in the Wagon Train Watershed, Lancaster County, Nebraska. (c) Predicted Ni in runoff and observed monthly average concentration in stream water (μg.L^{-1}) in the Wagon Train Watershed, Lancaster County, Nebraska. (d) Predicted Pb in runoff and observed monthly average concentration in stream water (μg/L) in the Wagon Train Watershed, Lancaster County, Nebraska. (e) Predicted Zn in runoff and observed monthly average concentration in stream water (μg.L^{-1}) in the Wagon Train Watershed, Lancaster County, Nebraska.

Extremely low concentrations were measured for Pb in water samples taken from the main stream. The observed monthly average Pb concentration ranged between 1.06 and 1.80 μg.L^{-1}, with an annual average of 1.34 μg.L^{-1} (Figure 7.2d). Crain (2001) studied Pb concentrations in water samples collected at forty-four stream stations in Kentucky, and found Pb concentrations to range between 0.3 and 4.5 μg.L^{-1}. Similar Pb concentrations (range: 1 to 7 μg.L^{-1}; average: 5 μg.L^{-1}) were reported for water samples collected from the Frazer River watershed in Colorado (Apodaca and Bails, 1999). Generally, the concentration of Pb in water depends on the solubility of its minerals, which is controlled by pH (Lindsay, 1979). The high pH of 8.39 measured in the water samples might explain these very low Pb concentrations. Lead enters natural water from the atmosphere, runoff, or wastewater discharge. It may enter the environment during mining, refining use, recycling, and coal combustion for fuel (USEPA, 1999). U.S. surface waters, except when subject to contamination, seldom contain Pb in excess of 50 μg.L^{-1} (Adriano, 1986). The USEPA (2002) recommended Pb concentrations of 65 μg.L^{-1} for CMC and 2.5 μg.L^{-1} for CCC. The U.S. drinking-water Maximum Contaminant Level (MCL) for Pb is 15.0 μg.L^{-1} (USEPA, 2003). These regulatory data suggest no Pb-related environmental problems are expected (in the WT watershed) for aquatic organisms and human health.

In comparison to Pb, relatively higher Zn concentrations were measured in the stream water. The observed monthly average Zn concentration ranged from 1.41 to 20.0 μg.L^{-1}, with an annual average of 11.9 μg.L^{-1} (Figure 7.2e). Relatively lower Zn concentrations (range: <1 to 14 μg.L^{-1}; average: 1 μg.L^{-1}) were found by Fuhrer et al. (1996) in their work on water samples collected at ten sites in the Lower Columbia River basin. Apodaca and Bails (1999), in their study on water samples from Frazer River watershed in Colorado, reported Zn concentrations ranging between <8 and 100 μg.L^{-1}, with an average of 8 μg.L^{-1}. On the other hand, Crain (2001) found a wider range and much higher Zn concentrations (range: 0.5 to 1,650 μg.L^{-1}) for water samples collected at forty-four stream stations in Kentucky. Zn is an essential element for plant, animal, and human nutrition, but a high Zn concentration in water can be harmful for aquatic life and poses a risk to human health (Berman, 1980). The USEPA (2003) has recommended Zn concentrations of 120 μg.L^{-1} for both CMC and CCC, and established a limit of 7.4 mg.L^{-1} of Zn for drinking-water supplies. These EPA regulations suggest no Zn-related environmental problems are expected for aquatic organisms in the WT watershed or for using the water for human consumption.

Elements Loss by Runoff from Soils

Predicted losses of eight elements by runoff from twelve major soils (g or kg.ha^{-1}.yr^{-1}) under crop and grass in the WT watershed are given in Table 7.4. As mentioned above, the loss of an element by runoff water (from the 10-mm interactive zone) should include all element forms dissolved in soil solution. For cropped soils in the watershed, the predicted Al loss ranged between 15.7 and 42.8 kg.ha^{-1}.yr^{-1}, with an area-weighted average of 29.1 kg.ha^{-1}.yr^{-1}. A relatively wider range of Al loss (1.71 to 42.2 kg.ha^{-1}.yr^{-1}) was estimated for grassland. However, the area-weighted average was smaller than that of cropland at 16.4 kg.ha^{-1}.yr^{-1}. Crain (2001) reported relatively lower Al losses by

TABLE 7.4

Predicted Loss of 8 elements by Runoff from 12 Major Soils (g or kg/ha/yr) under Crop and Grass in Wagon Train Watershed, Lancaster County, Nebraska

Soil (Map Unit)	Al		Fe		Si		Cd		Cu		Ni		Pb		Zn	
	Crop	Grass	Crop	Grass	Crop	Grass	Crop	Grass	Crop	Grass	Crop	Grass	Crop	Grass	Crop	Grass
	(kg/ha/yr)						(g/ha/yr)									
Wymore-WTB	23.59	24.59	12.77	12.92	22.40	26.41	0.59	0.74	47.29	52.89	18.75	22.02	1.77	2.18	33.13	41.86
Wymore-WTC2	32.82	9.35	17.92	4.95	28.67	11.73	0.61	0.59	51.58	47.96	24.56	14.49	1.40	1.04	46.66	16.24
Wymore-WTD3	35.77	42.17	20.99	21.90	59.26	85.37	0.82	0.97	61.80	74.16	25.70	32.37	1.71	1.17	60.40	72.53
Pawnee-PaC2	25.00	18.76	12.99	8.93	27.21	21.43	0.52	0.65	49.04	49.43	22.09	15.82	0.83	2.01	34.69	22.09
Pawnee-PaD2	42.81	38.02	23.48	18.40	81.28	68.53	0.72	0.55	65.90	61.78	29.74	22.80	1.01	1.71	57.29	48.82
Nodaway	15.70	22.78	8.93	12.51	29.06	27.90	0.53	0.75	58.30	56.01	16.62	25.62	1.61	2.23	26.40	50.29
Sharpsburg	26.70	27.16	14.69	13.62	37.06	31.28	0.68	0.55	57.11	46.84	22.96	22.26	1.62	2.01	40.56	44.73
Mayberry	32.89	13.37	16.83	6.99	32.19	10.73	0.51	0.43	46.60	41.10	24.06	12.84	1.67	1.11	46.49	22.77
Colo	19.67	21.09	11.37	11.22	45.67	15.25	0.63	0.69	62.24	54.45	19.00	23.91	0.83	1.03	29.86	42.24
Judson	19.67	9.15	10.75	4.79	26.23	8.08	0.43	0.40	49.98	41.11	18.50	14.42	0.90	0.63	28.29	11.96
Burchard	36.76	1.71	19.97	1.01	64.19	3.04	0.64	0.35	62.73	40.14	27.19	8.33	1.32	0.63	50.72	1.01
Kennebec	24.38	12.16	13.40	6.56	59.42	26.69	0.64	0.35	71.26	53.43	21.41	11.92	0.90	1.25	43.52	20.79
Weighted Av. (Crop & Grass)	29.10	16.44	15.89	8.54	32.33	20.91	0.60	0.62	52.74	50.35	22.80	17.75	1.40	1.40	42.08	27.63
Weighted Av. (watershed)	25.30		13.68		28.90		0.61		52.02		21.29		1.40		37.74	

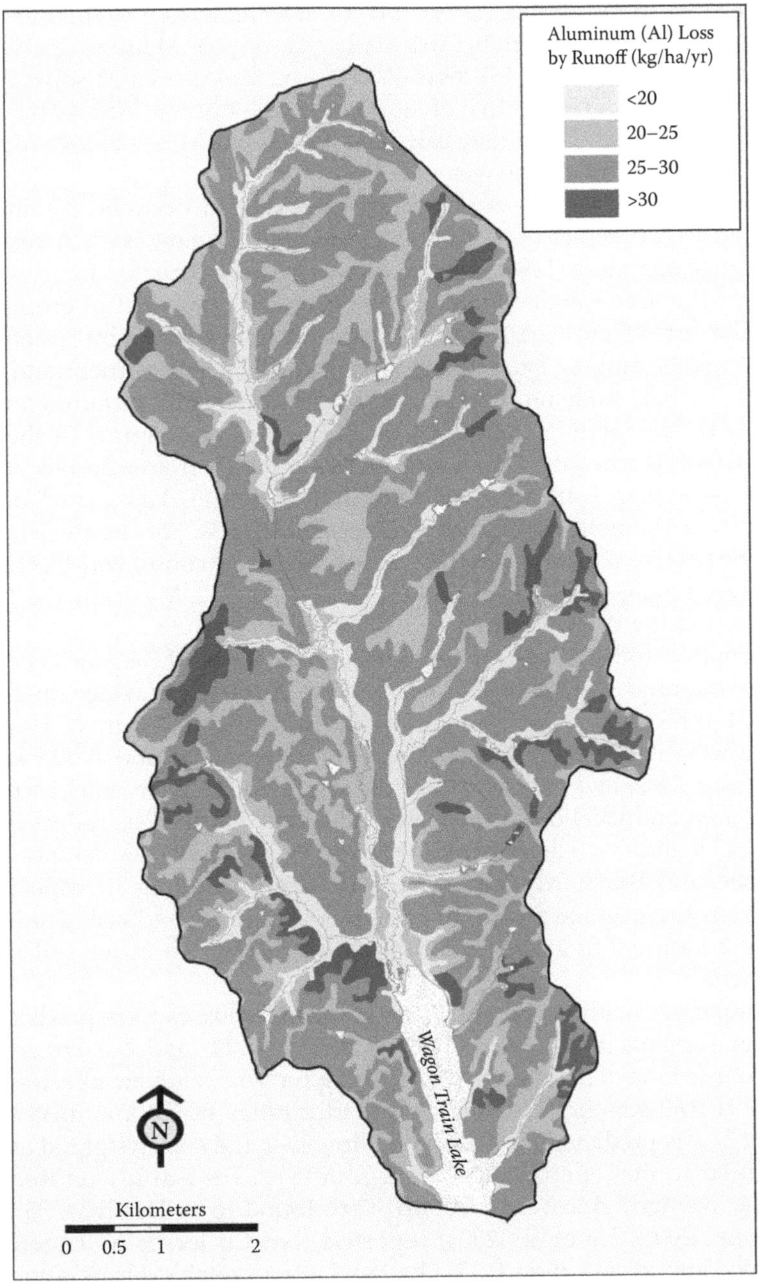

FIGURE 7.3
Aluminum loss by runoff from soils ($kg.ha^{-1}.yr^{-1}$) in the Wagon Train Watershed, Lancaster County, Nebraska.

runoff from Kentucky soils, ranging from 0.04 to 10.9 $kg.ha^{-1}.yr^{-1}$. Figure 7.3 shows Al loss by runoff ($kg.ha^{-1}.yr^{-1}$) from soils as related to their area and location in the watershed map. Dark areas reflect high Al losses (≥30 $kg.ha^{-1}.yr^{-1}$) derived mainly from Wymore-WTD3 and Pawnee-PaD2 soils. For the twelve major soils, irrespective of land cover, the average Al loss by runoff was 25.3 $kg.ha^{-1}.yr^{-1}$. These data calculate an annual Al loss of 98.3 Mg from the twelve major soils in the watershed.

For cropland, Fe loss by runoff from soils ranged between 8.3 and 23.5 $kg.ha^{-1}.yr^{-1}$, with an area-weighted average of 15.9 $kg.ha^{-1}.yr^{-1}$. A relatively wider range of Fe loss (1.01 to 21.9 $kg.ha^{-1}.yr^{-1}$) was calculated for grassland. However, the area-weighted average was smaller than that of cropland at 8.54 $kg.ha^{-1}.yr^{-1}$. Crain (2001) reported similar losses for Fe by runoff from Kentucky soils, ranging from 0.35 to 14 $kg.ha^{-1}.yr^{-1}$. On the other hand, He et al. (2004), in their study on Fe in Florida acidic sandy soils, reported lower Fe losses by runoff ranging from 0.004 to 2.42 $kg.ha^{-1}.yr^{-1}$. Figure 7.4 shows Fe loss by runoff ($kg.ha^{-1}.yr^{-1}$) from soils as related to their area and location in the watershed map. Dark areas reflect high Fe losses (≥17 $kg.ha^{-1}.yr^{-1}$) derived mainly from Wymore-WTD3 and Pawnee-PaD2 soils. For the twelve major soils, irrespective of land cover, the average Fe loss by runoff was 13.7 $kg.ha^{-1}.yr^{-1}$. These results indicate an annual Fe loss of 53.2 Mg from the twelve major soils in the WT watershed.

For cropped soils, Si loss ranged between 22.4 and 81.3 $kg.ha^{-1}.yr^{-1}$, with an area-weighted average of 32.3 $kg.ha^{-1}.yr^{-1}$. A relatively wider range of Si loss (3.04 to 85.4 $kg.ha^{-1}.yr^{-1}$) was predicted for soils under grass. However, the area-weighted average was smaller than that of cropland at 20.9 $kg.ha^{-1}.yr^{-1}$. Figure 7.5 shows Si loss by runoff ($kg.ha^{-1}.yr^{-1}$) from soils as related to their area and location in the watershed map. Dark areas reflect high Si losses (≥45 $kg.ha^{-1}.yr^{-1}$) derived mainly from Wymore-WTD3, Pawnee-PaD2, Kennebec, and Burchard soils. For the twelve major soils, irrespective of land cover, the average Si loss by runoff was 28.9 $kg.ha^{-1}.yr^{-1}$. These data indicate an annual Si loss of 112.3 Mg from the twelve major soils in the watershed.

In comparison with Al, Fe, and Si, much smaller losses were predicted for the other elements investigated (that is, Cd, Cu, Ni, Pb, and Zn). For cropped soils, Cd loss ranged between 0.43 and 0.82 $g.ha^{-1}.yr^{-1}$, with an area-weighted average of 0.60 $g.ha^{-1}.yr^{-1}$. A relatively wider range of Cd loss (0.35 to 0.97 $g.ha^{-1}.yr^{-1}$) was predicted for grassland. However, the area-weighted average was similar to that of cropland at 0.62 $g.ha^{-1}.yr^{-1}$. For a study on Kentucky soils, the average Cd losses by runoff were found to be less than 35 $g.ha^{-1}.yr^{-1}$ (Crain, 2001). He et al. (2004) reported that Cd losses by runoff were extremely low at less than 0.33 $g.ha^{-1}.yr^{-1}$ for vegetable farms and citrus groves in Florida. Figure 7.6 shows Cd losses by runoff ($g.ha^{-1}.yr^{-1}$) from soils as related to their area and location in the watershed map. Dark areas reflect high Cd losses (≥0.7 $g.ha^{-1}.yr^{-1}$) derived mainly from Wymore-WTD3 soil. For the twelve major soils, irrespective of land cover, the average Cd loss by

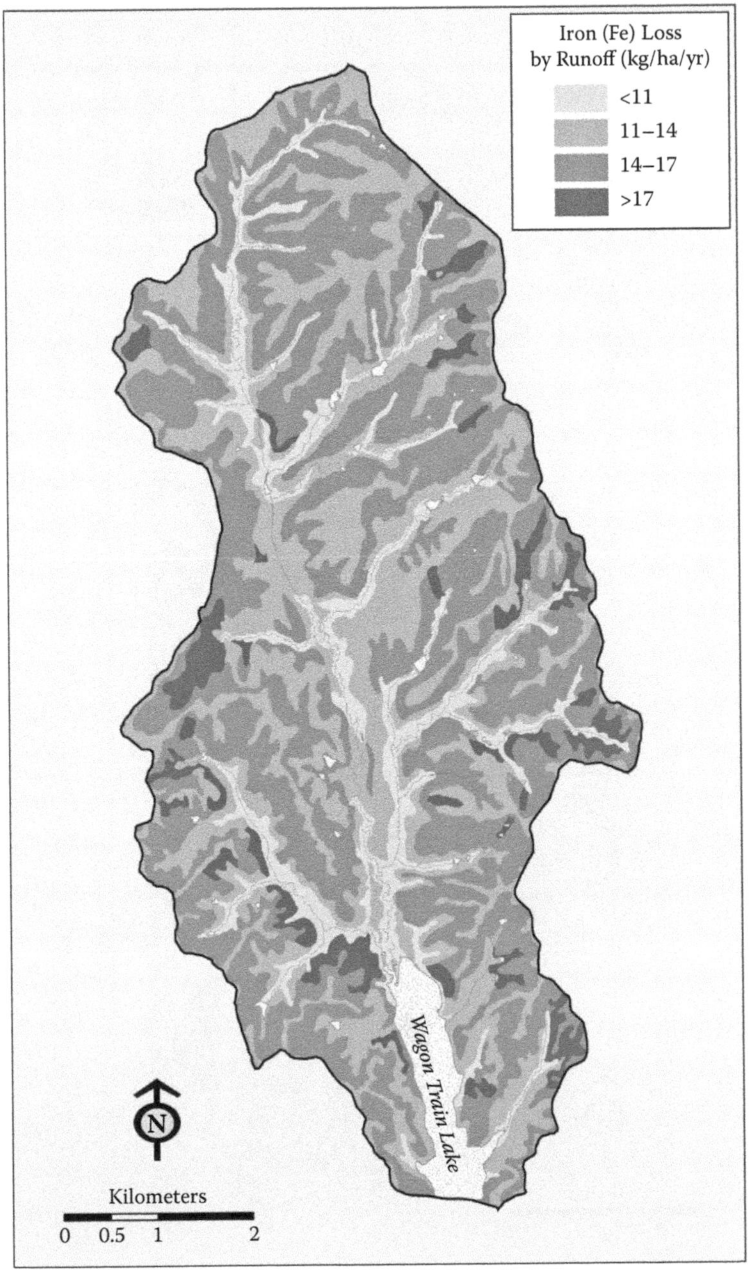

FIGURE 7.4
Iron loss by runoff from soils (kg.ha^{-1}.yr^{-1}) in the Wagon Train Watershed, Lancaster County, Nebraska.

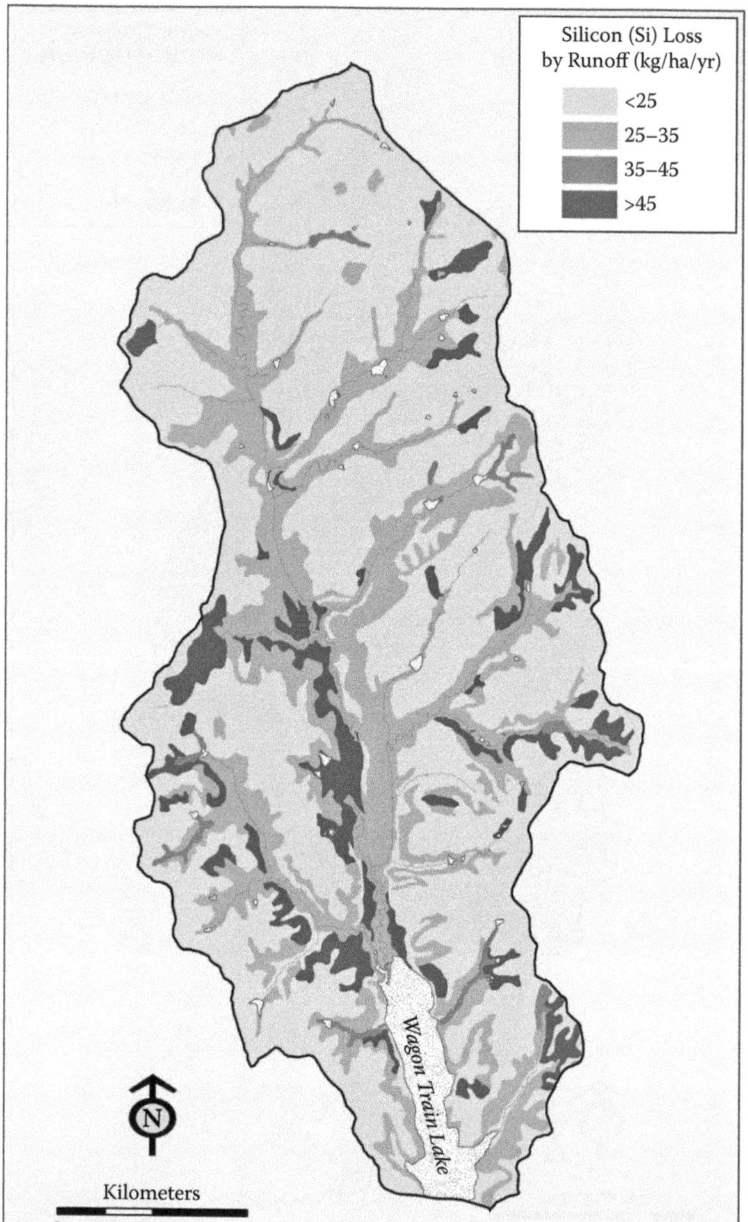

FIGURE 7.5
Silicon loss by runoff from soils (kg.ha^{-1}) in the Wagon Train Watershed, Lancaster County, Nebraska.

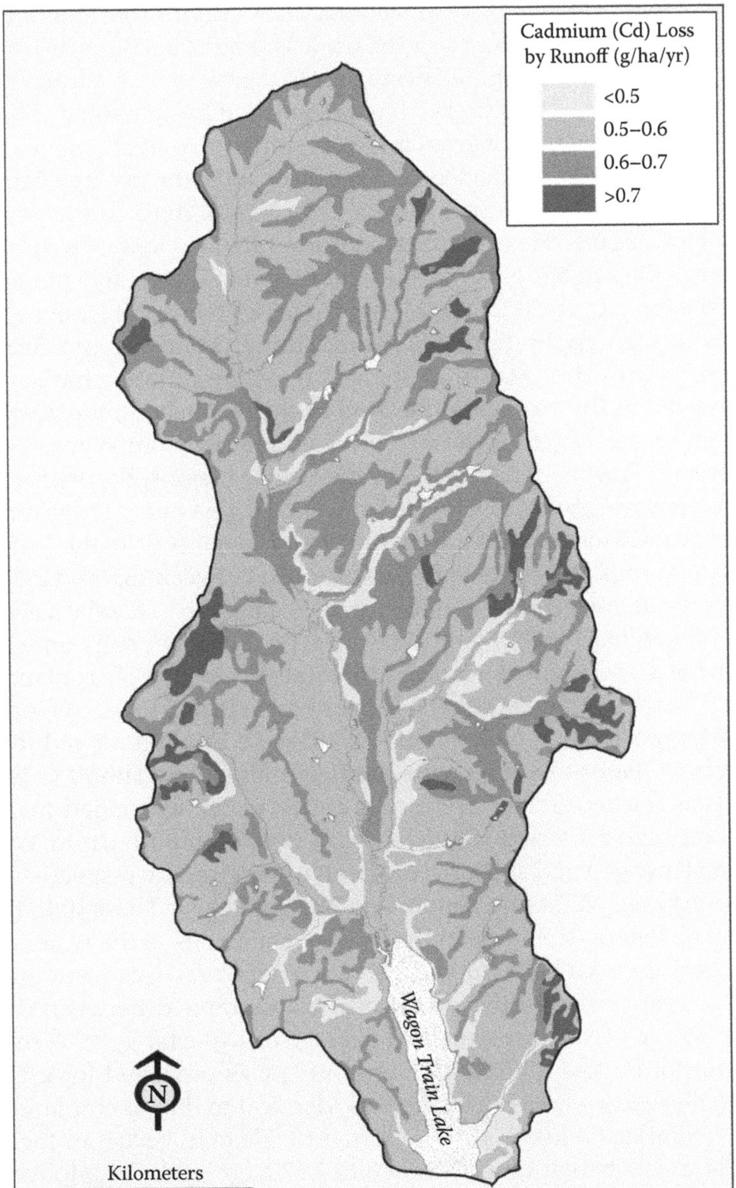

FIGURE 7.6
Cadmium loss by runoff from soils (g·ha^{-1}·yr^{-1}) in the Wagon Train Watershed, Lancaster County, Nebraska.

runoff was 0.61 $g.ha^{-1}.yr^{-1}$. These data indicate an annual Cd loss of 2.37 kg from the twelve major soils in the watershed.

Compared with Cd, relatively greater losses by runoff from soils were estimated for Cu. For soils under crop, the Cu loss ranged between 46.6 and 71.3 $g.ha^{-1}.yr^{-1}$, with an area-weighted average of 52.7 $g.ha^{-1}.yr^{-1}$. A similar range of Cu loss (range: 41.1 to 74.2 $g.ha^{-1}.yr^{-1}$) was predicted for grassland. Therefore, the area-weighted average of grassland was similar to that of cropland at 50.4 $g.ha^{-1}.yr^{-1}$. For a study conducted on Kentucky soils, the average Cu loss by runoff was found to be less than 35 $g.ha^{-1}.yr^{-1}$ (Crain, 2001). In their two-year study on Florida soils, He et al. (2004) reported that Cu losses by runoff had a wide range (3.45 to 657 $g.ha^{-1}.yr^{-1}$) in the first year (2001), and ranged from 5.33 to 336 $g.ha^{-1}.yr^{-1}$ in 2002. The higher Cu losses by runoff from soils were attributed to high applications of Cu-containing pesticides and fungicides (i.e., copper hydroxides). Figure 7.7 shows Cu loss by runoff ($g.ha^{-1}.yr^{-1}$) from soils as related to their area and location in the watershed map. Dark areas reflect high Cu losses (≥64 $g.ha^{-1}.yr^{-1}$) derived mainly from Wymore-WTD3, Kennebec, and Pawnee-PaD2 soils. For the twelve major soils, irrespective of land cover, the average Cu loss by runoff was 52.0 $g.ha^{-1}.yr^{-1}$. These data indicate an annual Cu loss of 202 kg from the twelve major soils in the watershed.

For cropped soils, the predicted Ni loss ranged between 16.6 and 29.7 $g.ha^{-1}.yr^{-1}$, with an area-weighted average of 22.8 $g.ha^{-1}.yr^{-1}$. A relatively wider range of Ni loss (8.33 to 32.4 $g.ha^{-1}.yr^{-1}$) was predicted for soils under grass. However, the area-weighted average was smaller than that of cropland at 17.8 $g.ha.^{-1}.yr^{-1}$. In their study on vegetable farms and citrus groves in Florida, He et al. (2004) reported that Ni losses by runoff were low and ranged from 0 to 8.71 $g.ha^{-1}.yr^{-1}$. Figure 7.8 shows the predicted Ni loss by runoff ($g.ha^{-1}.yr^{-1}$) from soils as related to their area and location in the watershed map. Dark areas reflect high Ni losses (≥22 $g.ha^{-1}.yr^{-1}$) derived mainly from Wymore-WTD3 and Pawnee-PaD2 soils. For the twelve major soils, irrespective of land cover, the average Ni loss by runoff was 21.3 $g.ha^{-1}.yr^{-1}$. These data indicate an annual Ni loss of 82.7 kg from the twelve major soils in the watershed.

In comparison with both Cu and Ni, much smaller losses were predicted for Pb. For cropped soils, the predicted Pb loss ranged between 0.83 and 1.77 $g.ha^{-1}.yr^{-1,}$ with an area-weighted average of 1.40 $g.ha^{-1}.yr^{-1}$. A relatively wider range of Pb loss (0.63 to 2.23 $g.ha^{-1}.yr^{-1}$) was predicted for grassland. However, the area-weighted average was identical to that of cropland at 1.40 $g.ha^{-1}.yr^{-1}$. Similar Pb losses were reported by He et al. (2004) in their work on Florida soils where it ranged from 0 to 2.67 $g.ha^{-1}.yr^{-1}$. For another study conducted on Kentucky soils, the Pb loss by runoff was relatively high and had a wide range between 0 and 35 $g.ha^{-1}.yr^{-1}$ (Crain, 2001). Figure 7.9 shows the predicted Pb loss by runoff ($g.ha^{-1}.yr^{-1}$) from soils as related to their area and location in the watershed map. Dark areas reflect high Pb losses (≥1.6 $g.ha.^{-1}.yr^{-1}$) derived mainly from Wymore-WtB and Nodaway soils. For the twelve major soils, irrespective of land cover, the average Pb loss by runoff

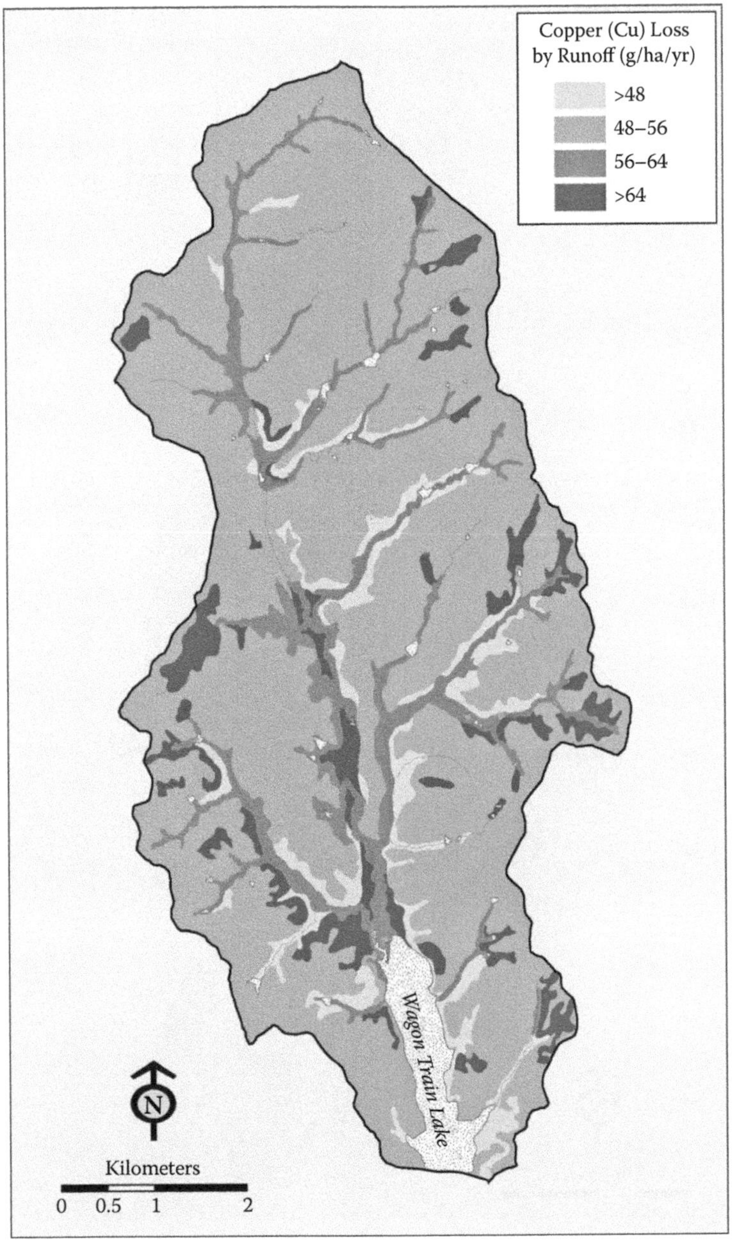

FIGURE 7.7
Copper loss by runoff from soils (g·ha^{-1}) in the Wagon Train Watershed, Lancaster County, Nebraska.

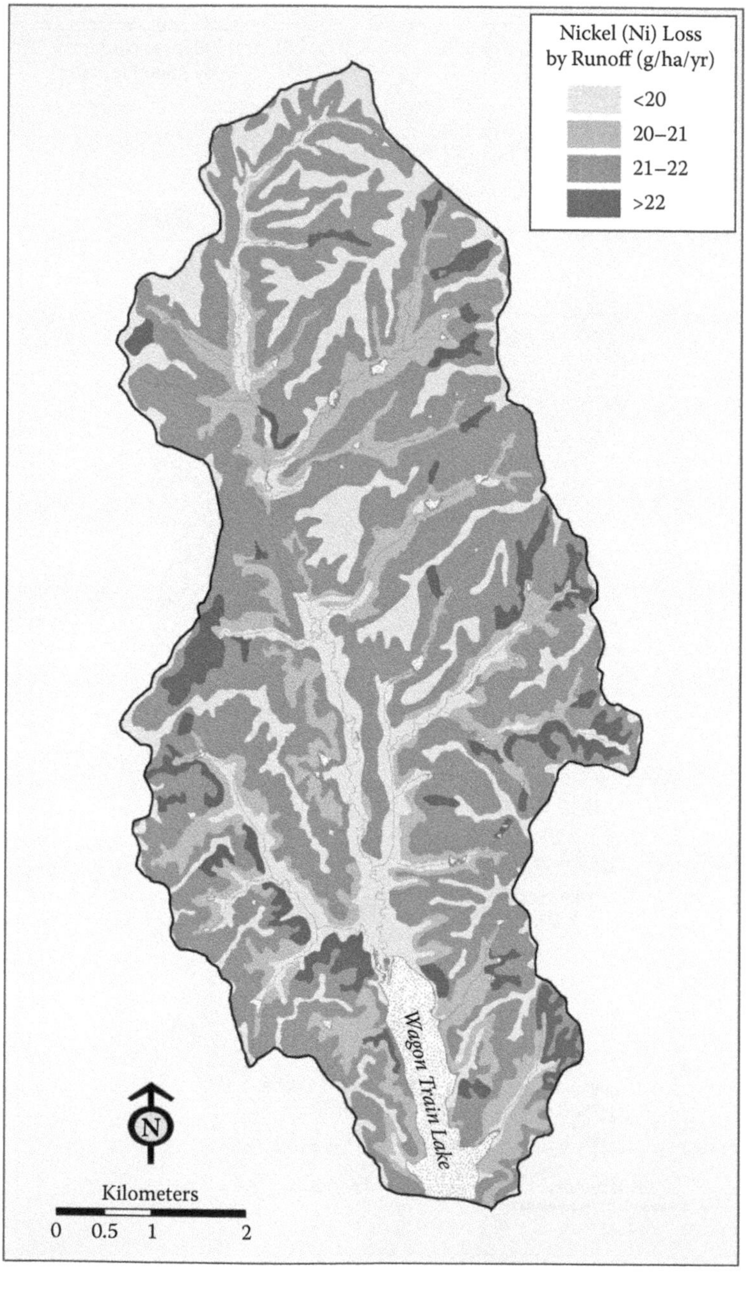

FIGURE 7.8
Nickel loss by runoff from soils ($g \cdot ha^{-1} \cdot yr^{-1}$) in the Wagon Train Watershed, Lancaster County, Nebraska.

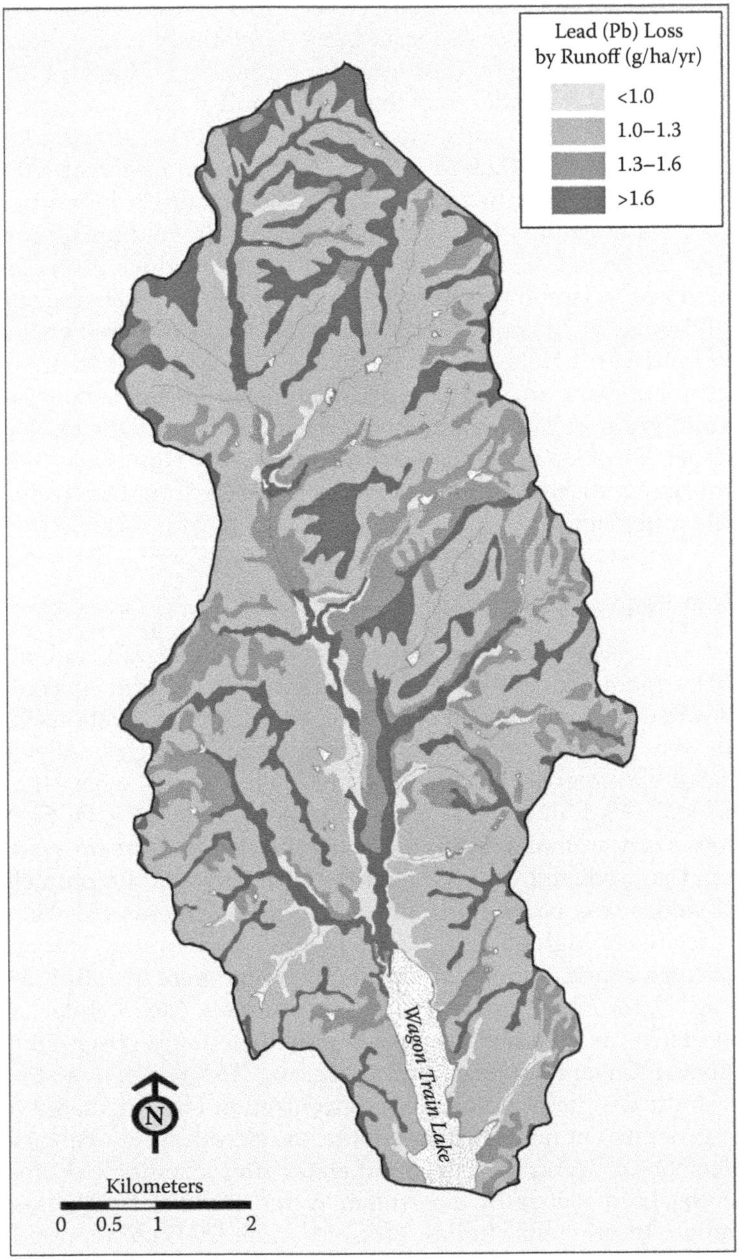

FIGURE 7.9
Lead loss by runoff from soils ($g{\cdot}ha^{-1}{\cdot}yr^{-1}$) in the Wagon Train Watershed, Lancaster County, Nebraska.

was 1.40 $g.ha^{-1}.yr^{-1}$. These data predict an annual Pb loss of 5.44 kg from the twelve major soils in the watershed.

For cropped soils, Zn loss ranged between 26.4 and 60.4 $g.ha^{-1}.yr^{-1}$, with an area-weighted average of 42.1 $g.ha^{-1}.yr^{-1}$. A relatively wider range of Zn loss (1.01 to 72.5 $g.ha^{-1}.yr^{-1}$) was observed for grassland. However, the area-weighted average was smaller than that of cropland at 27.6 $g.ha^{-1}.yr^{-1}$. In their two-year study on Florida soils, He et al. (2004) reported that Zn losses by runoff had a wide range (12.9 to 249 $g.ha^{-1}.yr^{-1}$) in the first year (2001), and ranged from 1.46 to 74.3 $g.ha^{-1}.yr^{-1}$ in 2002. The higher Zn losses by runoff from soils were attributed to high applications of Zn-containing pesticides and fungicides (He et al., 2004). In another study conducted on Kentucky soils, the Zn loss by runoff was relatively high and had a wide range between <35 and 175 $g.ha^{-1}.yr^{-1}$ (Crain, 2001). Figure 7.10 shows the predicted Zn loss by runoff ($g.ha^{-1}.yr^{-1}$) from soils as related to their area and location in the watershed map. Dark areas reflect high Zn losses (≥40 $g.ha^{-1}.yr^{-1}$) derived mainly from Wymore-WTD3 and Pawnee-PaD2 soils. For the twelve major soils, irrespective of land cover, the average Zn loss by runoff was 37.7 $g.ha^{-1}.yr^{-1}$. These data indicate an annual Zn loss of 147 kg from the twelve major soils in the watershed.

Elements in Runoff and Loading

In this study, losses predicted for both water (Table 7.2) and elements (Table 7.4) by runoff from soils were used to estimate element concentrations in runoff water. The predicted average Al, Fe, and Si concentrations in runoff from soils were 23.7, 12.8, and 27.1 $mg.L^{-1}$, respectively. Much smaller average concentrations were predicted for Cd, Cu, Ni, Pb, and Zn, where the values were 0.57, 48.7, 19.9, 1.31, and 35.3 $\mu g.L^{-1}$, respectively. For, Al, Fe, Si, Ni, and Zn, the observed annual average concentrations in the stream water were lower than the predicted values. The difference between the predicted and observed values was particularly great for those elements (Al, Fe, and Si) found in relatively high concentrations in the runoff water. The observed annual average concentrations in the stream water were 11.1, 8.61, 531, 6.31, and 11.9 $\mu g.L^{-1}$, for Al, Fe, Si, Ni, and Zn, respectively. Meanwhile, a reasonable agreement was obtained between the predicted and observed concentrations for Cd, Cu, and Pb (Figure 7.2a, 7.2b, and 7.2d, respectively).

In this study, the predicted element concentration was calculated for runoff water generated at field sites and not in the stream water. Factors affecting an element's concentration in runoff water after leaving field sites might decrease observed values in the stream water and should be taken into consideration. In previous studies, Elrashidi et al. (2005a,b) reported that P and nitrate-N removal by aquatic weeds and algae in streams has decreased observed element concentrations in water. In addition to the removal of elements from water by biological processes, a chemical precipitation of element oxide, hydroxide, and carbonate, could take place at the high pH values

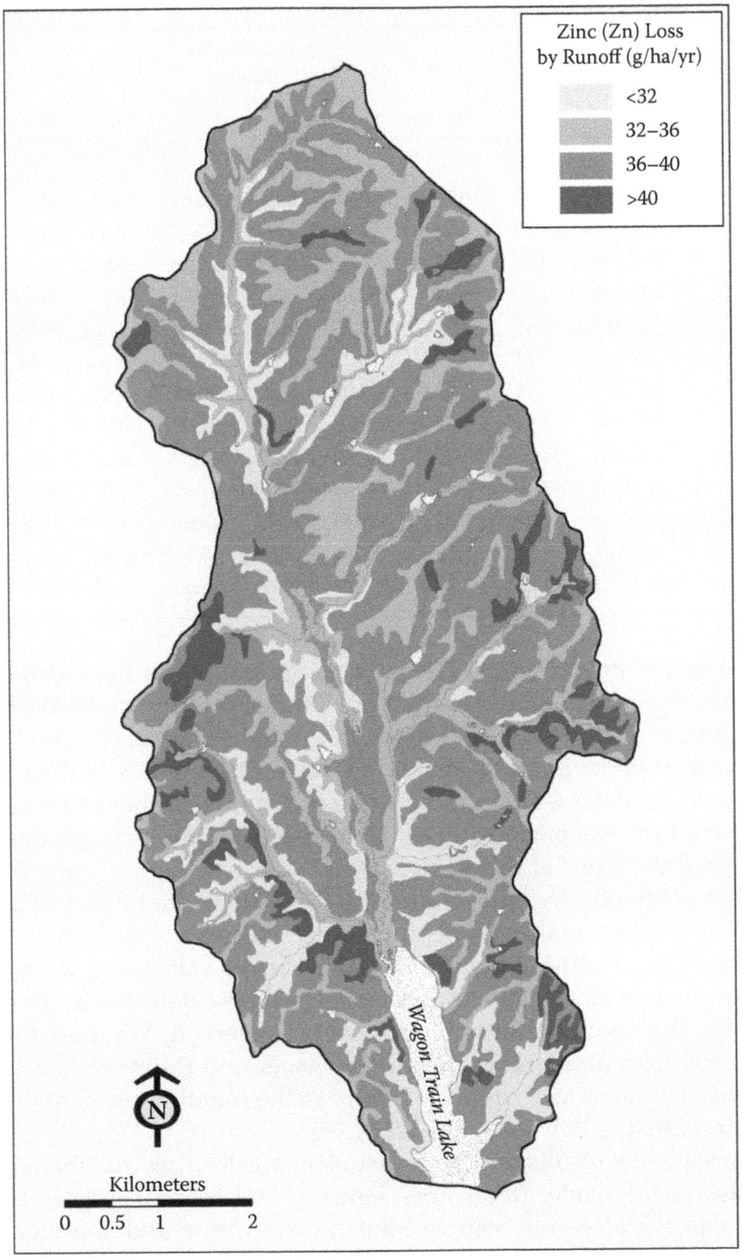

FIGURE 7.10
Zinc loss by runoff from soils (g·ha^{-1}·yr^{-1}) in the Wagon Train Watershed, Lancaster County, Nebraska.

TABLE 7.5

Predicted Average Monthly Elements Loading by Runoff Water (kg) in the Wagon Train Reservoir

Month	Predicted Runoff Water	Al	Fe	Si	Cd	Cu	Ni	Pb	Zn
					Loading By Runoff				
	(m³)				(kg)				
January	91,704	1.02	0.79	48.69	0.04	6.06	0.58	0.12	1.09
February	108,241	1.20	0.93	57.48	0.05	7.15	0.68	0.15	1.29
March	327,729	3.65	2.82	174.02	0.15	21.66	2.07	0.44	3.91
April	446,493	4.97	3.84	237.09	0.20	29.51	2.82	0.60	5.33
May	583,297	6.49	5.02	309.73	0.26	38.56	3.68	0.78	6.96
June	602,841	6.71	5.19	320.11	0.27	39.85	3.80	0.81	7.20
July	461,526	5.14	3.97	245.07	0.21	30.51	2.91	0.62	5.51
August	524,667	5.84	4.52	278.60	0.24	34.68	3.31	0.70	6.26
September	508,130	5.66	4.37	269.82	0.23	33.59	3.21	0.68	6.07
October	323,219	3.60	2.78	171.63	0.15	21.36	2.04	0.43	3.86
November	205,958	2.29	1.77	109.36	0.09	13.61	1.30	0.28	2.46
December	135,301	1.51	1.16	71.84	0.06	8.94	0.85	0.18	1.62
Year	4,314,713	48.02	37.15	2291.11	1.94	285.20	27.23	5.78	51.52

observed in the stream water. The average annual pH in the stream water was 8.39 compared to an average pH of 6.00 measured in soils. We believe a precipitation of oxide and hydroxide of Al and Fe minerals could be the main factor in removing both Al and Fe from water, while biological processes could be playing an important role in removing other elements from water. No effort was made to investigate these assumptions because they were beyond the scope of this study.

One of the objectives of this study was to estimate the impact of agricultural land on water quality (nonpoint source of elements contamination) in the WT reservoir. For the agricultural land in the WT watershed, we assumed that most element loss by runoff from soils and those detected in the stream water were transported eventually to the WT reservoir. We used the average element concentrations in the stream water and the predicted volume of monthly surface water runoff to estimate the monthly elements loading (kilograms) into the WT reservoir (Table 7.5).

As expected, the results indicated that elements loading into the reservoir were least during winter (December, January, and February). Most element loading into the reservoir occurred during the spring and summer (April through September) due to the rainfall pattern. For example, the winter averages of Fe, Cu, and Zn were 0.96, 7.33, and 1.33 $kg.mo^{-1}$, respectively, while there were 4.49, 34.5, and 6.07 $kg{\cdot}mo^{-1}$, respectively, for spring/summer. Silicon at an annual loading of 2.29 $Mg{\cdot}yr^{-1}$ had the greatest quantitative impact on water quality, followed by Cu at 285 kg·yr^{-1}. Meanwhile, Zn, Al, Fe, and Ni appeared to have a moderate impact, where the annual loading ranged between 27 and 52 kg.yr^{-1}. However, when the environmental impact of elements on water quality is evaluated, both the quantitative and qualitative effects of element chemical species should be considered.

References

Adriano, D.C. 1986. *Trace elements in the terrestrial environment*. Springer-Verlag, New York.

American Society for Testing and Materials (ASTM). 1993. Annual book of ASTM standards. Construction. Section 4. Soil and rock; dimension stone; *Geosynthesis*. Vol. 04.08. ASTM, Philadelphia, PA.

Apodaca, L.E., and J.B. Bails. 1999. Frazer river watershed, Colorado-assessment of available water quantity data through water year 1997. U.S. Geological Survey Water-Resources Investigations Report 98-4255, Denver, Colorado, 49 p.

Berman, E. 1980. *Toxic metals and their analyses*. Heyden & Son Ltd., London, United Kingdom.

Christensen, V.G. 2001. Characterization of surface water quality based on real tome monitoring and regression analysis, Quivira National Wildlife Refuge, South Central Kansas. U.S. Geological Survey, Water Resources Investigations Report 01-4248, Lawrence, Kansas, 28 p.

Crain, A.S. 2001. Estimated loads and yields of suspended solids and water quality constituents in Kentucky streams. U.S.Geological Survey, Water Resources Investigations Report 01-4075, Louisville, Kentucky, 125 p.

Donigian, A.S., Jr., D.C. Beyerlein, H.H. Davis, and N.H. Crawford. 1977. Agricultural runoff management (ARM) model version: II. Refinement and testing. EPA 600/3-77-098.m Environ. Res. Lab., U.S. Environmental Protection Agency, Athens, GA.

Elrashidi, M.A., M.D. Mays, and P.E. Jones. 2003. A technique to estimate release characteristics and runoff phosphorus for agricultural land. *Commun. Soil Sci. Plant Anal.* 34:1759-1790.

Elrashidi, M.A., M.D. Mays, S.D. Peaslee, and D.G. Hooper. 2004. A technique to estimate nitrate-nitrogen loss by runoff and leaching for agricultural land, Lancaster County, Nebraska. *Commun. Soil Sci. Plant Anal.* 35:2593-2615.

Elrashidi, M.A., M.D. Mays, J.L. Harder, D. Schroeder, P. Brakhage, S.D. Peaslee, C. Seybold, and C. Schaecher. 2005a. Loss of phosphorus by runoff for agricultural watersheds. *Soil Sci.* 170:543-558.

Elrashidi, M.A., M.D. Mays, A. Fares, C.A. Seybold, J.L. Harder, S.D. Peaslee, and Pam VanNeste. 2005b. Loss of nitrate-N by runoff and leaching for agricultural watersheds. *Soil Sci.* 170:969-984.

Elrashidi, M.A., D. Hammer, M.D. Mays, C.A. Seybold, S.D. Peaslee. 2007a. Loss of alkaline earth elements by runoff from agricultural watersheds. *Soil Sci.* 172:313-332.

Elrashidi, M.A., D. Hammer, M.D. Mays, C. Seybold, and S.D. Peaslee. 2007b. Loss of heavy metals by runoff from agricultural watersheds. *Soil Sci.* 172:876-894.
Elrashidi, M.A., C.A. Seybold, D.A. Wysocki, S.D. Peaslee, R. Ferguson, and L.T. West. 2008. Phosphorus in runoff from two watersheds in Lost River Basin, West Virginia. *Soil Sci.* 173:792-806.
Elrashidi, M.A., L. T. West, C.A. Seybold, D.A. Wysocki, E. Benham, R. Ferguson, and S.D. Peaslee. 2009. Nonpoint source of nitrogen contamination from land management practices in Lost River Basin, West Virginia. *Soil Sci.* 173:792-806.
ESRI. 2006. Environmental Systems Research Institute, ArcGIS Version 9.2 [Online]. Available at http://www.esri.com
Frere, M.R., J.D. Ross, and L.J. Lane. 1980. The nutrient sub-model. Chapter 4. In W.G. Knisel (ed.) CREAMS a field scale model for chemicals, runoff and erosion from agricultural management systems, USDA-SEA-Conserv. Res. Rep. No. 26. 1:65-86.
Fuhrer, G.J., D.Q. Tanner, J.L. Morace, S.W. McKenzie, and K.A. Skach. 1996. Water quality of the lower Columbia river basin: Analysis of current and historical water-quality data through 1994. U.S. Geological Survey Water-Resources Investigations Repot 95-4294. Portland, OR, 138 p.
Gilbert, W.A., D.J. Graczyk, and W.R. Krug. 1987. Average annual runoff in the United States, 1951-1980. Hydrologic investigations. National Atlas HA-710, U.S. Geological Survey, Reston, VA.
He, Z.L., M.K. Zhang, D.V. Calvert, P.J. Stoffella, X.E. Yang, and S. Yu. 2004. Transport of heavy metals in surface runoff from vegetable and citrus fields. *Soil Sc. Soc. Am. J.* 68:1662-1669.
Hem, J.D. 1989. Study and interpretation of the chemical characteristics of natural water (3rd ed.): U.S. Geological Survey, Water Supply Paper 2254, 263 p.
Hem, J.D. 1992. Study and interpretation of the chemical characteristics of natural water (3rd ed.): U.S.Geological Survey Water-Supply Paper 2254, 263 p.
Lindsay, W.L. 1979. *Chemical equilibria in soils.* John Wiley & Sons. New York.
LPSNRD. 2004. Lower Platte South Natural Resources District. A community-based watershed management plan for Wagon Train lake, Lancaster County, Nebraska. Nebraska Department of Environmental Quality, Lincoln, NE.
NASS. 2003. National Agricultural Statistics Service [Online]. Available at http://www.nass.usda.gov/ne. USDA, NASS, Washington, DC.
NLCD. 1992. National Land Cover Data for Nebraska. Version 05-07-00 nominal Thematic Mapper. http://landcover.usgs.gov/natllandcover.html
NWCC. 2003. National Water & Climate Center. http: //www.WCC.NRCS.gov/water/W_CLIM.html
Olsen, S.R., and L.E. Sommers. 1982. Phosphorus. P. 403-430. In A.L. Page et al. (ed.) Methods of soil analysis. Part 2. 2nd ed. ASA and SSSA, Madison, WI.
Rawls, W.J. 1983. Estimating soil bulk density from paricle size analysis and organic matter content. *Soil Sci.* 135:123-125.
Sharpley, A.N. 1985. Depth of surface soil-runoff interaction as affected by rainfall, soil slope, and management. *Soil Sc. Soc. Am. J.* 49:1010-1015.
Schmidt, H.C. 2004. Quality of water on the Prairie Band Potawatomi Reservation, northeastern Kansas, May 2001 through August 2003, U.S. Geological Survey, Scientific Investigations Report 2004-5243, Lawrence, Kansas, 69 p.
USDA/NRCS. 2004. Soil Survey Laboratory Methods Manual. Soil Survey Investigations Report No. 42, Version No. 4 . USDA-NRCS, Washington, D.C.

USDA/NRCS. 1999. Soil Survey Geographic (SSURGO) database for Lancaster County, Nebraska. http://www.ftw.nrcs.usda.gov/ssur_data.html

USDA/SCS. 1991. National Engineering Field Manual. Chapter 2: Estimating Runoff and Peak Discharges. USDA-NRCS, Washington, D.C. P. 1-19.

USEPA 1976. Quality criteria for water, Office of Water. EPA Report , Washington D.C., 256 p.

USEPA 1999. Drinking water and health-What you need to know. Office of Water, EPA 816-K-99-00, Washington D.C.

USEPA 2002. National recommended water quality criteria: 2002. Office of Water, EPA-822-R-02-047, Washington D.C.

UAEPA 2003. National primary drinking water standards. Office of Water, EPA 816-F-03-016, Washington D.C.

USGS. 2001. United States Geological Survey. Water resource data, Nebraska water year 2000. Platte river basin, Lower Platte River Basin, Salt Creek at Roca. pp. 180-181. Water-Data Report NE-00-1. U.S. Department of the Interior, U.S.G.S., Washington, DC.

USGS. 2007. Water Resources. National Water Information System: Web Interface. Available from http://waterdata.usgs.gov/wv/nwis/inventory/. Accessed Oct., 2007.

8

Heavy Metals Transformation in Wetlands

R.D. DeLaune and Dong-Cheol Seo

CONTENTS

Introduction

Heavy metals in wetlands have both natural and anthropogenic sources. Man-made sources include atmospheric deposition, runoff from adjacent uplands, and wastewater discharge from municipalities and industries. Once within a wetland, the ability of a heavy metal to be transported depends on its chemical properties, which influence availability and toxicity. The species of metal (or metal speciation) also determines behavior in wetland environments. Valence, sorption to sediments, complexation with organic matter, precipitation, and interaction with microorganisms are processes governing the availability or toxicity of heavy metals in wetlands.

Some metals are required for plant growth, while others, such as Pb, Cd, and Ni, have no known botanical function. Metals such as Cu, Se, and Zn are essential to plants and animals but can be toxic at high concentrations.

Metals such as Cd, Hg, and Pb are toxic even at relatively low concentrations. Metals such as Hg have a tendency to bioaccumulate at high levels in the food chain. This biomagnification can lead to serious health hazards for animals and humans.

Heavy metals in excessive concentrations can inhibit the activity of microbes that govern important biogeochemical processes in wetlands. Elevated metal concentrations in wetland soils and sediments can interfere with the enzyme functions of microorganisms. Toxicity is related to the form in which the metal is found in wetlands. Toxic metals may precipitate or chelate essential metabolites, act as antimetabolites, or displace essential metals in metalloenzymes. Elevated heavy metal concentrations can influence microbial respiration, organic nitrogen mineralization, denitrification, and methanogenesis. Heavy metals can also influence the degradation of toxic organics, including petroleum hydrocarbons entering the wetland environment (DeLaune et al., 1998).

Metal Forms

The chemical forms of metals in wetland soils can differ in their mobility and bioavailability. Table 8.1 provides a listing of some of the common chemical forms of metals, ranging from most available to least available (Gambrell and Patrick, 1991; Shannon and White, 1991): (1) readily available: dissolved and exchangeable forms; (2) potentially available: metal carbonates, metal oxides, and metal hydroxides, metals adsorbed on or occluded with Fe and Mn oxides, metals strongly adsorbed or chelated with insoluble large molecular humic materials and metals precipitated as sulfide; and (3) unavailable: metals within the crystalline lattice structures of clay and other residual minerals. Table 8.1 shows the general chemical forms or fractions of which metals can be found in wetland soils (Gambrell, 1994).

TABLE 8.1

General Chemical Forms of Metals in Wetland Soils and Sediments

Water soluble:
Soluble free ions (e.g., Mn^{2+})
Soluble inorganic complexes
Soluble organic complexes
Exchangeable (adsorbed on clay and organic complexes)
Precipitated as inorganic compounds
Adsorbed or occluded to precipitated hydrous oxides of Fe and Mn
Complexed with large molecular weight humic material
Primary minerals precipitated as insoluble sulfides (e.g., HgS)

Metals dissolved in pore water are mobile and bioavailable. Adsorbed (exchangeable) metals are also bioavailable due to equilibria with dissolved metals. Both dissolved and exchangeable metals are readily mobilized and bioavailable. On the opposite extreme are metals bound within the crystalline lattice structures of clay and other residual minerals. Metals in this form are essentially permanently immobilized and thus unavailable. Only under long periods of mineral weathering would residual metals become mobile and bioavailable. Between these two extremes are potentially available metals. In metal-contaminated soils, excess metals become primarily associated with these potentially available forms rather than the readily available soluble and exchangeable forms (Feijtel, DeLaune, and Patrick, 1988).

Metal Organic and Clay Interactions

Reactions of metals with organics and clays are important in metal availability and toxicity in wetlands. Organics in wetland soils can form stable complexes with the reduced soluble form of some metals while maintaining the metal in a water-soluble form.

Humic materials have a higher capacity than clays to retain most metals, due to the high density of active functional groups such as -COOH, phenolic-, alcoholic-, enolic-OH, and C=O structures, as well as amino and imino groups. Humic materials are colloidal in nature with very large surface areas and negative charge. The negative charge results from exposed -COOH and -OH groups, where the H^+ is available for exchange with metals. Due to their chemical functional groups, they form stable water-soluble and insoluble complexes with many metal ions.

In contrast, insoluble, large molecular weight humic acids are effective in immobilizing most heavy metals. Humic acids can form insoluble complexes with metal ions, precipitating them under conditions that would otherwise promote migration.

Clays are also important in determining the mobility of the metals in wetlands. Because they consist of extremely small colloidal particles, clays exhibit a very large surface area. With this negatively charged surface area, the clay particle is surrounded by an ionic double layer. The cations in the double layer are subject to interchange with other cations in pore water, giving rise to what is known as the cation exchange capacity (CEC). Cation exchange will occur only if the ion in pore water can be held more strongly than the ion already at the surface, or if a large excess of a cation enters the zone.

Redox and pH Conditions: Effects on Metal Transformation

Metal cycling in wetlands is very dependent on pH. Soil redox conditions (or E_h status) also govern the oxidation and reduction of some trace metals found in wetlands. Trace metals are present in various oxidation states; for example, Cr can exist in several oxidation states—from Cr^0 (the metallic form) to Cr^{6+}. The most stable oxidation states of Cr in the environment are Cr^{3+} and Cr^{6+}. In addition to the elemental metallic form, which is extensively used in alloys, Cr has three important valence forms: Cr^{2+}, Cr^{3+}, and Cr^{6+}. Trivalent Cr^{3+} and hexavalent Cr^{6+} are the most important forms in the environment (Masscheleyn et al., 1992).

Metals bound to oxides, hydroxides, and carbonates are effectively immobilized at near-neutral to somewhat alkaline pH conditions. If, however, the pH becomes moderately to strongly acidic, as can sometimes occur when reduced soils (containing sulfides) become oxidized, these metals may be transformed into readily available forms.

Metals complexed with insoluble, large molecular weight humic compounds are effectively immobilized. There is some evidence that these metals are less effectively immobilized if reduced soils are oxidized. Oxidation of wetland soils can result in a significant release of heavy metal (Gambrell and Patrick, 1988, 1991).

Oxides of Fe, and perhaps Mn and Al, effectively adsorb or occlude most toxic metals. These oxides exist in mineral soils in large quantities. When soils become reducing, the metals bound to Fe and Mn oxides are transformed into readily available forms due to dissolution of the Fe and Mn oxides. During flooding and drainage cycles of wetlands, formation of iron oxyhydroxides is important in retaining metals in surface soils (Gambrell, 1994).

Both the E_h and pH of sediment can regulate the solubility and chemical transformations of trace or heavy metals. Low pH and redox potentials in sediment–water systems tend to favor the formation of soluble species of many metals, whereas in oxidized, nonacidic systems, slightly soluble or insoluble forms tend to predominate. However, pH, and particularly E_h in combination, may regulate other processes such as sulfide formation, which indirectly influence the solubility of metals. In wetlands with high sulfate inputs and reduced sediment environments, the formation and accumulation of sulfides occur. The solubility of divalent metal sulfides in these systems is extremely low. Where large amounts of sulfide are present, sulfide precipitation is thought to be a very effective process for immobilizing trace metals. Thus, a reducing environment, which causes a metal to be present in a soluble ionic form, may also contribute to its being effectively immobilized by sulfide precipitation. Sparingly soluble metal sulfides, which are stable in reduced environments, can oxidize to relatively soluble metal sulfates in aerobic environments or when soil becomes oxidized. Figure 8.1 shows the redox potential at which various metal and metalloid transformations can occur.

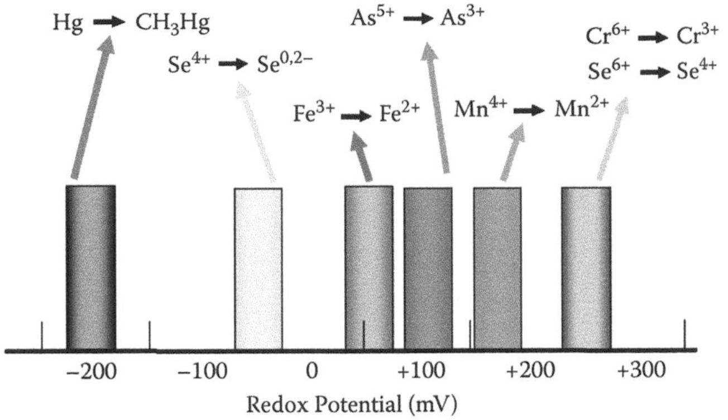

FIGURE 8.1
Soil redox (mV) condition influencing metal and metalloid transformation.

The effects of sediment E_h and pH on As and Se speciation and solubility under controlled redox (500, 200, 0, and −200 mV) and pH (5, natural, and 7.5) conditions have been studied (Masschelyn, DeLaune, Patrick, 1991a). Figure 8.2 shows the species distribution of As at the four redox levels studied in the uncontrolled pH experiment. Both E_h and pH affected the speciation and solubility of As and Se. Under oxidized conditions, As solubility was low, and 87% of the As in solution was present as As^{5+}. Upon reduction, As^{3+} became the major As species in solution, and As solubility increased substantially. No organic arsenicals were detected. The greatest As concentrations in solution were found at −200 mV, the most reduced condition studied. Total As in solution increased approximately 25 times upon reduction of the sediment suspension from 500 mV to −200 mV. Up to 51% of the total As present in the sediment was found to be soluble at −200 mV. Although thermodynamically unstable, there was a considerable amount of As^{5+} present under reduced conditions, indicating that chemical kinetics play an important role in the conversion of As^{5+} to As^{3+}. At −200 mV, As^{5+} comprised 18% of the total soluble As (Masscheleyn, DeLaune, and Patrick, 1991b).

Figure 8.3 summarizes the effect of redox potential and pH on Se speciation and solubility in sediment suspensions. In contrast to As, Se solubility reached a maximum under highly oxidized (500 mV) conditions and decreased significantly upon reduction. Se^{6+} was the predominant dissolved Se species present at 500 mV. At 200 and 0 mV, Se^{4+} became the most stable oxidation state of Se. Under strongly reduced conditions (−200 mV), oxidized Se species were no longer detectable and Se solubility was controlled by the formation of elemental Se and/or metal selenides. Biomethylation of Se was important under oxidized and moderately reduced conditions (500, 200, and 0 mV). More alkaline conditions (pH 7.5) resulted in greater dissolved As and Se concentrations in solution (increased up to ten and six times, respectively) as compared to the more acidic equilibrations (Masscheleyn et al., 1991b).

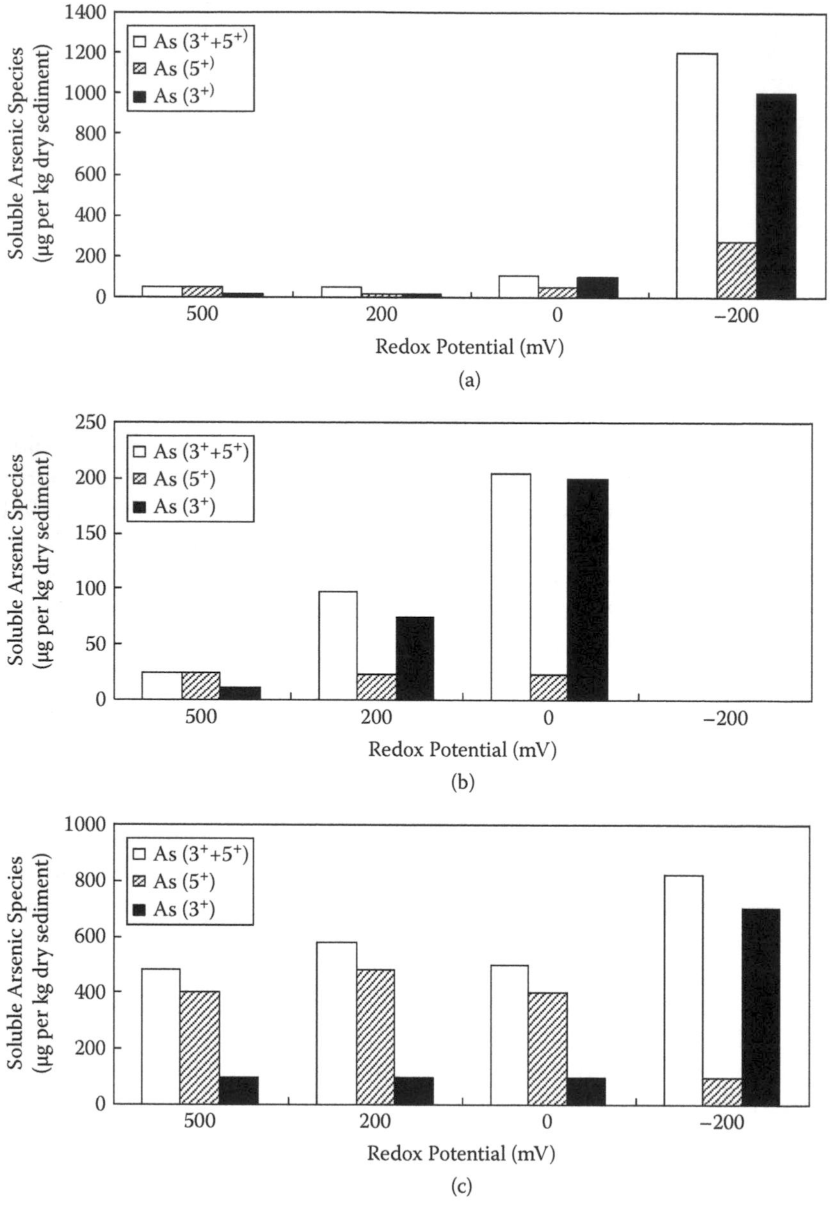

FIGURE 8.2
Distribution of soluble As species under controlled redox conditions. (a) Equilibrations at natural pH (4.0 for 500 mV, 5.3 for 200 mV, 6.1 for 0 mV, and 6.9 for –200 mV). (b) Equilibrations at pH 5.0. (c) Equilibration at pH 7.5. Note changes in scale. (*Source:* Redrawn from Masscheleyn, P.H., R.D. DeLaune, and W.H. Patrick, Jr. *Environ. Sci. Technol.* 25(8): 1414–1419.)

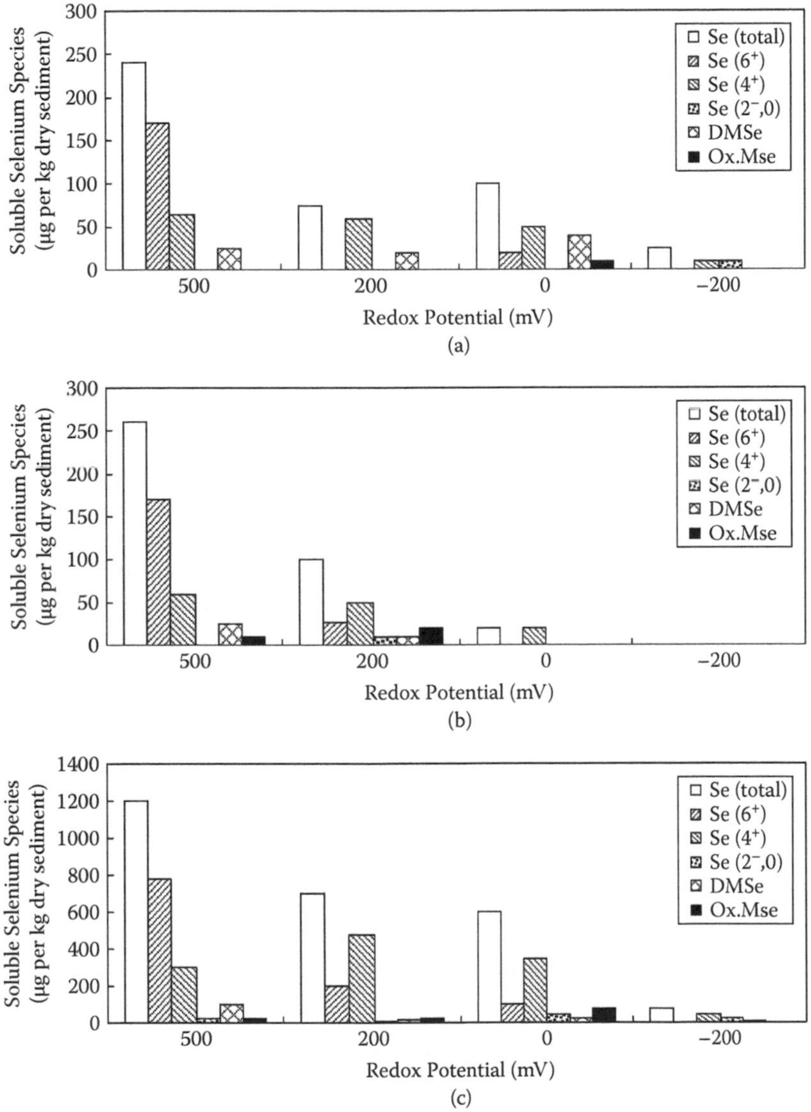

FIGURE 8.3
Distribution of soluble Se species under controlled redox conditions. (a) Equilibrations at natural pH (4.0 for 500 mV, 5.3 for 200 mV, 6.1 for 0 mV, and 6.9 for –200 mV). (b) Equilibrations at pH 5.0. (c) Equilibration at pH 7.5. Note changes in scale. (*Source:* Redrawn from Masscheleyn, P.H., R.D. DeLaune, and W.H. Patrick, Jr. *J. Environ. Qual.* 20: 521–527.)

Se solubility and dominant Se species found at 500, 200, 0, and –200 mV in the uncontrolled pH experiments are shown in Figure 8.2a. In contrast to As, Se solubility was greatest under oxidized (500 mV) conditions and decreased significantly ($P < 0.01$) upon reduction of the sediment suspensions. Furthermore, the presence of methylated Se species suggests that Se chemistry was dominated by chemical transformations among inorganic and organic species.

Se solubility is greater under highly oxidized conditions (500 mV) with approximately 13% of the total Se present in the sediment becoming mobilized (Masscheleyn, DeLaune, and Patrick, 1991a). Both Se^{6+} and Se^{4+} were present, although Se^{6+} is the only thermodynamically stable form. Reduction of the sediment suspensions resulted in a decrease in dissolved Se concentrations. At 200 and 0 mV, Se^{4+} was the most stable oxidation state of Se. At –200 mV, Se^{6+} was not detectable and Se^{4+} concentrations decreased. Only a small amount of $Se^{2-,0}$ was observed under strongly reduced conditions (–200 mV). Methylated Se compounds were present at 500, 200, and 0 mV. Dimethylselenide constituted 5% to 36% of the total soluble Se concentration, depending on the redox status. Oxidized methylated Se compounds were present in only minor amounts.

Experiments were conducted to identify and quantify biogeochemical processes controlling Cr redox chemistry in a seasonally flooded Lower Mississippi Valley forested wetland. Cr speciation, transformations, and solubility were studied in the wetland soil. Wetland soil redox level determined the rate and capacity of the soil to assimilate and retain Cr. Under oxidized and moderately reduced (+500 to +100 mV) soil conditions, Cr behavior was dominated by Cr^{6+} sorption and reduction of Cr^{6+} to Cr^{3+}. Under more reduced soil redox (< +100 mV) levels, Cr chemistry and solubility was controlled by the chemical reduction of Cr^{6+} by soluble ferrous iron. Results obtained suggest that the studied bottomland hardwood wetland or floodplain will serve as a sink for Cr entering the wetland, thereby reducing the contamination of the downstream ecosystem (Masscheleyn et al., 1992).

Chromium Wetland Soil Redox Chemistry: Sorption and Reduction of Cr^{6+}.

In the natural environment, Cr exists mainly in two stable oxidation states: trivalent Cr and hexavalent Cr. In terms of toxicity, mobility, and bioavailability, Cr^{3+} and Cr^{6+} differ remarkably from each other (Bartlett and James, 1988; Calder 1988; Holdway, 1988; Nieboer and Shaw, 1988; Wong and Trevors, 1988; Yassi and Nieboer, 1988). Cr^{6+}, generally in the forms of chromate and dichromate, is toxic and very soluble in water (Xu and Jaffé, 2006).

In a flooded soil or anoxic sediment, microbial mineralization of organic matter proceeds by reduction of electron acceptors like nitrate

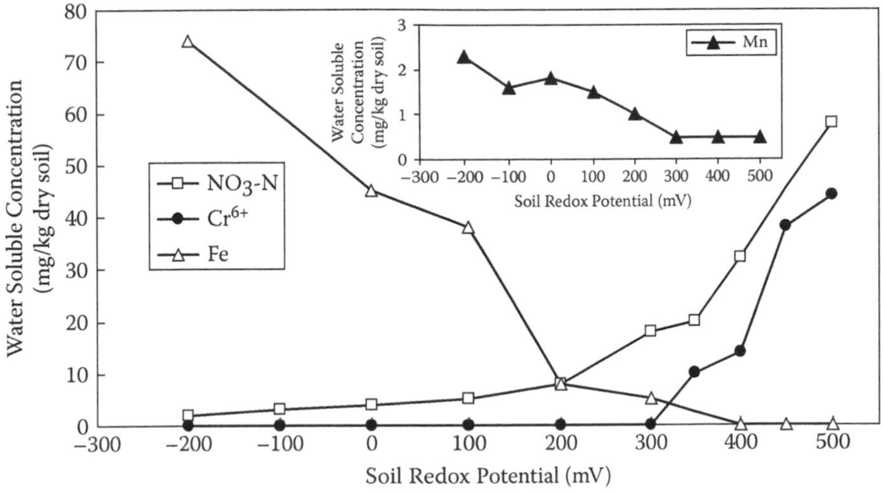

FIGURE 8.4
Critical soil redox potential for nitrate, chromate, manganese, and iron reduction. (*Source:* Redrawn from Masscheleyn, P.H., J.H. Pardue, R.D. DeLaune, and W.H. Patrick, Jr. 1992. *Environ. Sci. Technol.* 26: 1217–1226.).

(denitrification), Mn^{4+}, Fe^{3+}, sulfate, and bicarbonate (methanogenesis). Figure 8.4 illustrates the critical redox potentials for Cr^{6+}, NO_3-N, Mn, and Fe reduction in the bottomland hardwood wetland soil upon flooding (Masscheleyn et al., 1992). No soluble sulfides could be detected. Between 500 and 200 mV, denitrification increased at a rate of 44 mg N (kg of dry soil)$^{-1}$.day^{-1}.mV^{-1}. Nitrate reduction was followed by reduction of Mn^{4+} (E_h <200 mV) and Fe^{3+} (E_h <100 mV). Reduction of spiked Cr^{6+} started at about the same redox levels (E_h <500 mV) as that of nitrate reduction and was completed, at a redox potential of 300 mV, before all the NO_3-N disappeared from solution. In the range of 500 to 300 mV, a linear relationship (r = 0.96) existed between the amount of water-soluble Cr^{6+} and the redox level. For every millivolt decrease in redox potential, the daily amount of Cr^{6+} removed was 56 mg·(kg dry soil)$^{-1}$. In addition to NO_3 and Cr^{6+} reduction, microbial dissimilatory reduction catalyzed Mn^{4+} and Fe^{3+} reduction, leading to increased soluble Fe and Mn concentrations in the soil. Quantitatively, the reduction of Fe^{3+} was the most important chemical change taking place upon the development of anaerobic conditions in the soil. Reduction of the inorganic oxidants (Cr^{6+}, NO_3, Mn, and Fe) increased the pH. The pH of the aerobic (+500 mV) soil suspension was 5.7 and increased to 6.0, 6.8, and 7.3 in the +200-, 0-, and −200-mV soil suspensions, respectively. As discussed above, the reoxidation of reduced Mn and Fe in the aerobic floodwater overlying the submerged soil possibly influences the biogeochemical cycle of Cr by oxidizing Cr^{3+}.

Mercury

Mercury accumulation in fish and other organisms in wetlands occurs as a result of the bioaccumulation of methylmercury in the food chain. Wetlands provide unique biogeochemical conditions controlling methylmercury formation.

The conversion of inorganic Hg to an organic form in wetlands is governed by microbial methylation. Formation of soluble methylmercury in anaerobic sediments by obligate anaerobic bacteria in wetlands serves as a source of bioavailable Hg. Sulfate-reducing bacteria are key participants in the methylation of Hg. The rate of Hg methylation is coupled with the rate of sulfate reduction (King et al., 2001). Methylation enhances the mobility of Hg and greatly increases its toxicity (Ullirich, Tanton, and Abdrashitova, 2001). The organomercurials of environmental interest are monomethylmercury and dimethylmercury. Monomethylmercury present in wetlands readily accumulates in the food chain. Organic matter in sediment has been shown to stimulate bacterial methylation that converts inorganic mercury to methylmercury and increases the methylmercury content of zooplankton up to tenfold (Paterson, Rudd, and Louis, 1998).

The microbial-mediated processes of methylation and demethylation are influenced by factors such as redox potential, pH, sulfate concentration, and microbial activity. A study by DeLaune et al. (2004) showed methylation of added Hg in sediment was greater under reduced conditions as compared to oxidized conditions (Figure 8.5). The majority (over 95%) of mercury

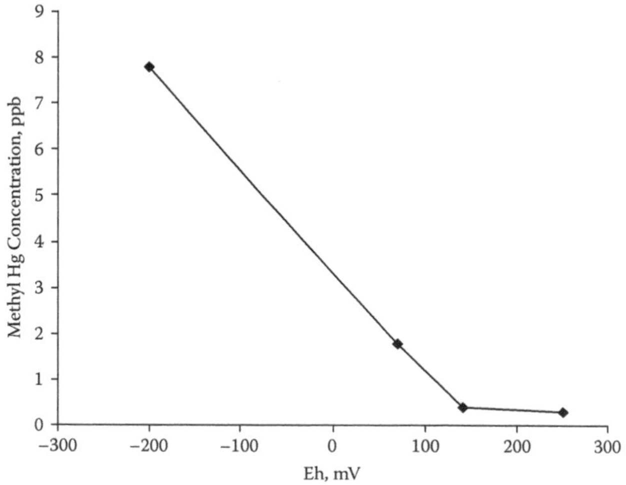

FIGURE 8.5
Effect of sediment redox conditions on changes of methyl Hg concentration in False River sediment. Each point represents an average of two replications. (*Source:* Redrawn from DeLaune, R.D. et al. 2004. *J. Environ. Sci. Health A* 39(8): 1925–1935.)

methylation in an estuarine sediment was attributed to anaerobic sulfate-reducing bacteria such as *Desulfovibrio* (Compeau and Bartha, 1984). High salinity (sulfate) levels inhibited methylation, and methylation increased when sulfate levels were low and concentrations of organic fermentation products were high (Compeau and Bartha, 1984). The extremely low solubility and formation of H_2S help to prevent methylation.

Demethylation is another mercury speciation process in wetlands. There are several pathways for methylmercury to be demethylated. Oxidative demethylation is a dominant process in anaerobic sediment. Demethylation of methylmercury (to elemental mercury and methane) was found to occur at high redox potentials (+110 mV) in estuarine sediments (Compeau and Bartha, 1984). Therefore, demethylation may be the important transformation in the aerobic or moderate aerobic zone in soils and sediments. The dynamics of methylation and demethylation of Hg in sediments is a key factor in the flux of methylmercury from bottom sediments to the overlying water column. Little quantitative information is known about the dynamics of these processes in the wetland ecosystem.

Sedimentation Sinks

Wetlands are effective traps or sinks for metals due to their relative immobility in flooded soils and because the process of soil accretion occurring in wetlands results in burying or placing the metals in an environment where immobilization is even more effective. Vertical accretion processes (organic matter and sediment accumulation) are very effective in trapping metals. Callaway, DeLaune, and Patrick (1998), using ^{137}Cs, dating were able to determine heavy metal chronologies in selected coastal wetlands of Northern Europe. From accretion rates and metal content in the marsh soil, metal sinks and polluting history were quantified. Accretion and sedimentation are also important in isolating metal contaminants in Louisiana coastal wetlands (DeLaune and Gambrell, 1996). It has been reported that marshes as a sink for metals are enhanced by high sedimentation of accretion rates (Hung and Chmura, 2007)

Metal Uptake by Wetland Plants

Wetland plants influence heavy metal uptake transport and mobility. Wetland plants can alter the redox condition, pH, and organic matter content of sediments and thus affect the chemical speciation and mobility of metals.

Metals may be mobilized or immobilized, depending on numerous factors, thus making it difficult to predict actual plant effects on metal mobility for a given set of conditions.

Wetland plants play an important role in heavy metal removal in wetlands. Uptake rates and the tolerance of metals vary among wetland plant species. Metal uptake into plant tissue promotes immobilization in plant tissues. Wetland plants are used for phytoremediation and in constructed wetlands for wastewater treatment (Kadlec and Knight, 1996).

Wetlands are sinks for metals via accumulation by wetland plant adsorption and sedimentation processes. Wetland plants have been utilized for the removal of heavy metals (Williams, Bubb, and Lester, 1994; Weis and Weis, 2004). The value of metal-accumulating plants to wetland remediation has been realized (Black, 1995). The accumulation of heavy metals in many species of wetland plants has also been demonstrated (Dunbabin and Bowmer, 1992; Delgado, Bigeriego, and Guardiola, 1993; Fett et al., 1994; Salt et al., 1995; Zaranyika and Ndapwadza, 1995; Zayed, Gowthaman, and Terry, 1998; Zhu et al., 1999).

Phytostabilization occurs when wetland plants are used to immobilize metals and store them below ground in roots and/or soil, in contrast to "phytoextraction" in which hyperaccumulators may be used to remove metals from the soil and concentrate them in above-ground tissues (Weis and Weis, 2004).

Some wetland plants are hyperaccumulators. For example, *Ceratophyllum demersum* accumulates arsenic with a 20,000-fold concentration factor (Reay, 1972). These plants must be harvested to prevent recycling of accumulated metals when the plants decompose. However, wetland plants are generally not hyperaccumulators. In any case, the mechanical aspects of harvesting plants would be destructive to wetlands comprised of rooted plants (Weis and Weis, 2004).

While most studies have focused on single plant species and few studies have directly compared the species, there are, nevertheless, differences. These differences in heavy metal distribution and release between the species have implications for metal dynamics in wetlands as phytoremediation systems (Weis and Weis, 2004). How heavy metals are distributed within plants and the extent of uptake can have important effects on the residence times of metals in plants and in wetlands and the potential release of heavy metals.

Wetland plants can accumulate heavy metals in their tissues. Some previous studies reported that some wetland plants have the ability to take up greater than 0.5% dry weight (DW) of a given element and bioconcentrate the element in their tissues to 1,000-fold the initial element supply concentration (Yang and Ye, 2009). Translocation within a plant may further alter the metal concentration in specific plant parts. Heavy metals may be translocated via the apoplast, in the phloem, and acropetally in the xylem (Greger, 1999), and such translocation may differ greatly between plant and metal species. One study of the submerged species *Potamogeton pectinatus* reported high translocation of Cd in both directions (Greger, 1999), while another found quite a low, solely acropetal translocation of Cd (Wolterbeek and Van der Meer, 2002).

Heavy Metal Localization, Translocation, and Distribution in Wetland Plants

Vesk, Nockolds, and Allaway (1999) reported metal localization within and around the roots of water hyacinth (*Eichhornia crassipes*) growing in a wetland receiving urban runoff by energy dispersive x-ray microanalysis of sections from freeze-substituted roots. They reported that Cu, Pb, and Zn were not localized at the root surface. On the other hand, Fe was present at high levels at the root surface, and this may have been a root plaque as described for wetland plants with roots anchored in flooded soils. Iron levels decreased centripetally across the root and were higher in cell walls than within cells (Vesk, Nockolds, and Allaway, 1999). MacFarlane and Burchett (2000) reported that in the roots of seedlings of the grey mangrove (*Avicennia marina*), Cu, Pb, and Zn concentrated predominantly in cell walls.

Weis and Weis (2004) reported that the root epidermis served as a barrier to transport of Pb to above-ground tissues, but not to the other metals. The endodermal casparian strip provided a barrier to the movement of all three metals into the stele. Heavy metals in the leaves were highest in the order of xylem > mesophyll > hypodermal tissue. In the cell walls, heavy metal concentrations were also higher than in intracellular locations (Weis and Weis, 2004).

Heavy Metal Translocation and Distribution

Wetland plants can accumulate heavy metals in their leaves and stems when they translocate metals from root tissue to aerial tissue. Many wetland plants accumulated higher concentrations of heavy metals in roots than in shoots (Taylor and Crowder, 1983; Ye et al., 1997a,b; Cheng et al., 2002; Stoltz and Greger, 2002; Deng, Ye, and Wong, 2004, 2006). The pattern of metal distribution in wetland plants is suitable to restrict metals transported from roots to shoots. However, the degree of upward translocation depends on plant species, the particular metal, and a number of environmental conditions such as pH, redox potential (E_h), temperature, salinity, soil particle size, organic matter content, nutrients, and the presence of other ions (Salomons and Förstner, 1984; Chawla, Singh, and Viswanathan, 1991; Greger, 1999; Fitzgerald et al., 2003; Fritioff, Kautsky, and Greger, 2005).

Fitzgerald et al. (2003) found that Cu accumulated primarily in the roots of monocots and dicots, while Pb accumulated mainly in the roots of monocots but also in the shoots of dicots. Salt marsh dicots tended to have similar heavy metal concentrations in roots and shoots, whereas in monocots, the concentrations in the roots were two to three times higher than in the shoots (Otte et al., 1991). Pb uptake into the roots and shoots of rice (*Oryza sativa*) decreased with an increase in E_h and pH, while Cd uptake increased with an increase in E_h and a decrease in pH (Reddy and Patrick, 1977). Some heavy metals (e.g., Zn and Cu) seem to be transported easily in the plant, while other metals (e.g., Pb) appear to be rather immobile (Pichtel, Kuroiwa, and Sawyer, 2000).

Zn accumulation in leaves correlated with sediment concentrations even when they were high, indicating the greatest translocation to aerial portions of the plant. However, Pb levels in leaves remained quite low at all levels of sediment Pb. Lower pH resulted in increased Zn accumulation, and higher levels of Pb and Zn in the sediments resulted in greater Zn accumulation in roots and shoots (Weis and Weis, 2004).

Windham, Weis, and Weis (2001b) reported that in a greenhouse study of *Phragmites australis* and *Spartina alterniflora* grown in sediments to which Pb had been added, concentrations of Pb were greater in *S. alterniflora* leaves and rhizomes than in those of *P. australis* (Figure 8.6). Upper leaves had lower Pb concentrations than lower leaves for both species of plants. Pools of Pb in above-ground biomass were greater in *S. alterniflora* than in *P. australis*, while pools of Pb below ground were not statistically different between the two plant species. Rai (2009) reported heavy metal uptake (Cd, Cr, Cu, Fe, Pb, Mn, Hg, Ni, and Zn) in wetland plants at different sampling sites (Figure 8.7), showing that selected wetland plants accumulated higher concentrations of

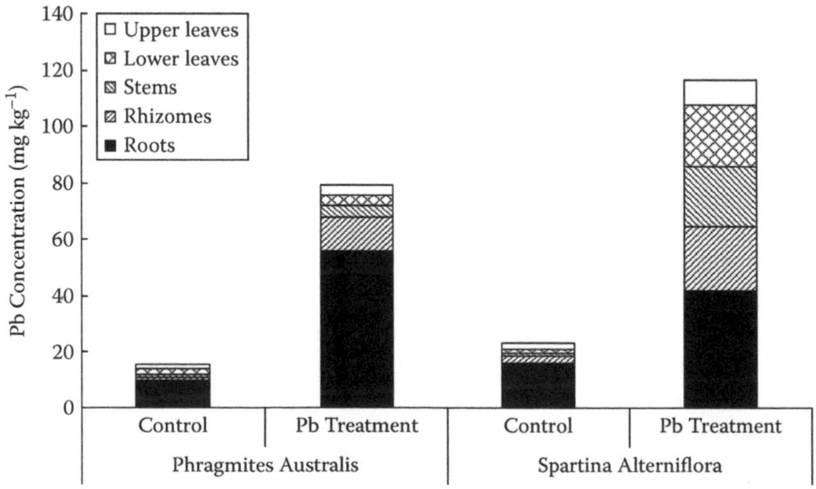

FIGURE 8.6
Concentration of lead in individual tissues of each species grown in each Pb treatment. (*Source:* Redrawn from Windham, L., J.S. Weis, and P. Weis. 2001. *Mar. Pollut. Bull.* 42: 811–816.)

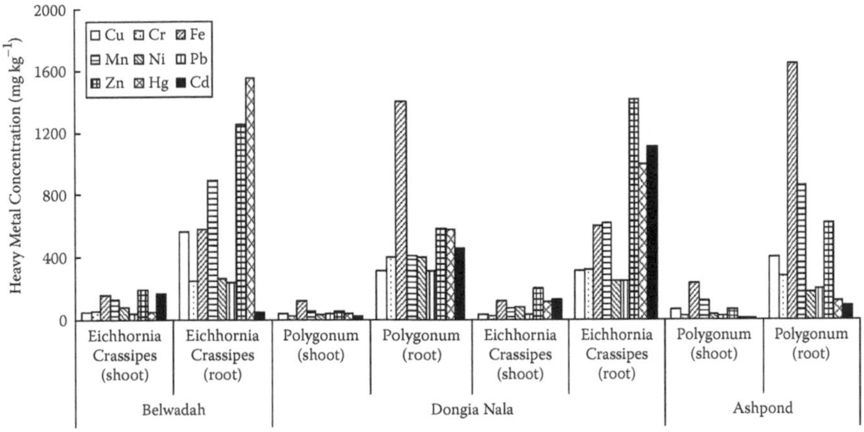

FIGURE 8.7
Heavy metal uptake by water plants at various sampling sites. (*Source:* Redrawn from Rai, P.K. 2009. *Environ. Monit. Assess.* 158: 433–557.)

heavy metals in roots than in shoots. Liu et al. (2007) reported the uptake and distribution of Cd by nineteen wetland plant species in small-scale plot constructed wetlands, showing that the heavy metal concentrations in underground parts were generally higher than those in above-ground parts. But Cd levels also varied considerably with plant species and the kind of metals. On average, the ratios of underground to aboveground heavy metal concentrations were about three times for Cd and Pb, and about two times for Zn. For some species such as *Polygonum hydropiper* and *Monochoria vaginalis,* the metal concentrations in underground parts were much higher than that in aboveground parts. But for other species such as *Cyperus difformis* and *Isachne globosa,* the metal concentrations in underground parts were lower than those in above-ground parts.

On average, more than 60% of the metals absorbed by wetland plants were transferred to above-ground parts, and the portion was higher for Zn than for Cd and Pb on the distribution of heavy metal quantity accumulation. The distribution percentages (%) of the heavy metals to above-ground parts varied greatly with wetland plant species, and they ranged from 32.6% to 90.6% for Cd, from 26.2% to 90.1% for Pb, and from 36.8% to 90.7% for Zn.

Heavy Metal Release by Wetland Plants

Heavy metal release by plants can increase the bioavailability of heavy metals within estuaries, especially in urban and industrialized areas (Berk and

Colwell, 1981). Leaves of seedlings of the mangrove *Avicennia marina* excreted significant quantities of Zn or Cu after exposure to these metals (Waisel, 1977; MacFarlane and Burchett, 2000). Kraus, Weis, and Crow (1986) and Kraus (1988) reported that the metals were associated with salt crystals excreted on the adaxial leaf surface. *Spartina alterniflora* has the theoretical potential to export 145 g Cd, 260 g Pb, 104 g Cr, 260 g Cu, and 988 g Ni·ha^{-1}· yr^{-1} through salt excretion (Kraus, 1988). The relationship of metal release with salt excretion suggests greater metal release at higher salinities, when there is more salt excretion (Weis and Weis, 2004).

Burke, Weis, and Weis (2000) reported leaf excretion by *Spartina alterniflora* and *Phragmites australis* growing together in the same sediment in a contaminated area of the Hackensack Meadowlands in northern New Jersey. Leaves of *S. alterniflora* were found to release two to four times more Pb, Cu, Cr, and Zn than leaves of *P. australis* at the peak of the growing season. Leaf concentrations of Cu and Zn were comparable in the leaves of the two species, while *S. alterniflora* had higher leaf concentrations of Pb and Cr. Thus, *S. alterniflora* can release larger quantities of heavy metals into the wetland environment than *P. australis*.

The release of Hg from leaf tissue was compared between two dominant salt marsh macrophytes, *Spartina alterniflora* and *Phragmites australis* (Windham, Weis, and Weis, 2001a). Leaves of *S. alterniflora* consistently released two to three times more Hg than leaves of *P. australis*. Leaves of *S. alterniflora* also contained greater concentrations of Hg during these months (Figure 8.8). In contrast to *P. australis* leaves, rates of Na release were high for *S. alterniflora* and were correlated with the rate of Hg release. Leaf Hg concentration was highly correlated with Hg release for both species, but the slope was significantly greater for *S. alterniflora*. For both species, the highest Hg content was found in lower leaves in May, followed by upper leaves in May. Hg accumulation in leaf tissue and release from both species appear to be greatest in the spring, although differences between the species persist throughout the peak months of the growing season (Windham, Weis, and Weis, 2001a).

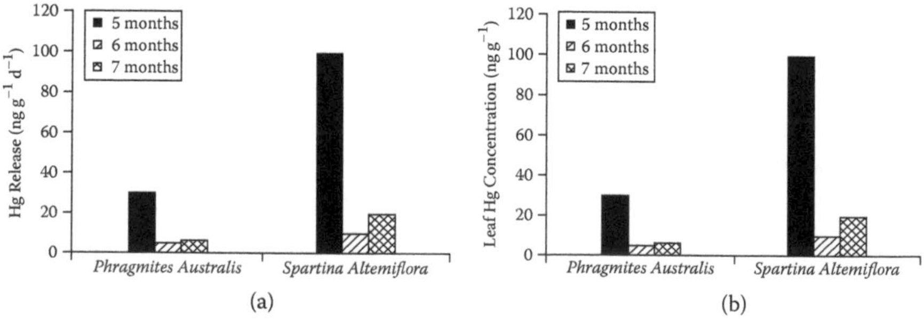

FIGURE 8.8
(a) Hg release from leaf tissue of *S. alterniflora* and *P. australis* and (b) mean concentrations of Hg in leaves across the growing season from May to July. (*Source:* Modified from Windham, L., J.S. Weis, and P. Weis. 2001a *Estuaries* 24: 787–795.)

Toxicological Effects

Heavy metals accumulated in wetland plant tissues can cause toxic effects, especially when translocated to above-ground tissues. Effects can be measured at the biochemical and cellular level, but most studies of effects have focused on growth as the response to the toxicant. Some studies focused on plant growth, showing their responses to heavy metals. When exposed to higher metal concentrations, plants showed a significant reduction in root elongation, height of seedling, leaf number, leaf area, and shoot and root biomass (Snowden and Wheeler, 1993; Ye et al., 1997a,b; Mendelssohn, McKee, and Kong, 2001; MacFarlane and Burchett, 2002).

Grey mangroves seedlings, *Avicennia marina*, when exposed to 100 $\mu g \cdot g^{-1}$ sediment Cu showed a significant reduction in seedling height, leaf number, and leaf area (MacFarlane and Burchett, 2002). On the other hand, very little Pb was translocated to above-ground tissues and minimal negative effects were observed. Zn was translocated to leaves and at 1000 $\mu g \cdot g^{-1}$ sediment Zn, all seedlings died. Significant reductions in height, leaf number, area, biomass, and root growth were seen at 500 $\mu g \cdot g^{-1}$ sediment Zn.

In order to compare a population from a mine site with a reference population to investigate whether tolerance had been acquired by the contaminated population, Ye et al. (1997a,b) used growth as a measure of response of *Phragmites australis* seedlings to Zn, Cd, and Pb. Both populations were found to be very resistant to toxic effects, and no enhanced tolerance of the polluted population was noted. This finding indicated that if this species were used in constructed wetlands, it would not be necessary to select a particular resistant population.

Tolerance to heavy metals in wetland plants can be achieved by sequestering them in tissues or cellular compartments that are insensitive to metals. Restriction of upward movement into shoots and the translocation of excessive metals into old leaves shortly before their shedding can also be considered tolerance mechanisms, as can the increase in the metal-binding capacity of the cell wall (Verkleij and Schat, 1990).

Grill, Winnaker, Zenk (1985) reported that a common biochemical biomarker in plants is the presence of phytochelatins—peptides (polycysteins) that are synthesized rapidly in plant tissues after metal exposure and serve to chelate them. When all free metal ions are chelated, synthesis is terminated. The lower amount of free metals in the cells allows metal-sensitive enzymes to function and the plant to survive. The capacity for chelation is finite, however, and as metal concentrations continue to increase, toxic effects become manifest (Sneller et al., 1999). Other metal-chelating substances may be present in plant exudates, which act to decrease metal uptake and, consequently, toxicity (Verkleij and Schat, 1990).

Padinha, Santos, and Brown (2000) measured thiolic protein concentration, adenylate energy charge (AEC), and photosynthetic efficiency in *Spartina maritima* growing at sites with differing degrees of metal contamination. They found that thiolic protein, which binds metals, was higher in plants grown at metal-contaminated sites than at "clean" sites. Metals bound to thiolic proteins were subsequently excreted back to the environment, maintaining lower concentrations in the leaves. Photosynthesis efficiency and AEC ratio lowered in *Typha domingensis* and *Spartina alterniflora* in response to increasing levels of Cd, while chlorophyll fluorescence was relatively insensitive (Mendelssohn, McKee, and Kong, 2001).

Plant–Sediment Relationships

Wetland sediments are generally considered a sink for metals. A number of studies have analyzed the correlations between metal uptake by wetland plants and metal concentrations in sediment. Most studies reported poor correlations between them (Campbell et al., 1985; Jackson and Kalff, 1993; Cardwell, Hawker, and Greenway, 2002).

The bioavailability of the metals is low compared to terrestrial systems with oxidized soils. Different forms of metals have different availabilities: water-soluble metals and exchangeable metals are the most available, metals precipitated as inorganic compounds, metals complexed with large molecular weight humic materials and metals adsorbed to hydrous oxides are potentially available, and metals precipitated as insoluble sulfide and metals bound within the crystalline lattice of minerals are essentially unavailable (Gambrell, 1994).

Williams (2002) showed that the depth to which plant roots can penetrate is limited and this restricts the uptake of contaminants and rhizosphere actions to shallower levels because of these reducing conditions. In estuaries, many of the metal contaminants are bound to sulfides in anoxic sediments. However, plants can oxidize the sediments in the rootzone through the movement of oxygen downward through aerenchyma tissue (Moorhead and Reddy, 1988), and this oxidation can remobilize the metal contaminants, thus increasing the otherwise low availability of metals in wetland sediments.

Avicennia species of mangroves were found to oxidize the rhizosphere, thus reducing sulfides and enhancing metal concentrations in the exchangeable form (De Lacerda et al., 1993). Heavy metal remobilization may also result from acidification of the rhizosphere by plant exudates (Doyle and Otte, 1997). *Typha latifolia* oxidized the rhizosphere but did not increase pore water metal concentrations (Wright and Otte, 1999).

Changes in sediment E_h and pH conditions can cause changes in metal speciation and solubility, which can result in a flux from sediments to pore water and then to overlying water and/or increase uptake into plants. It is possible that salinity changes may also alter metal speciation and uptake, but Drifmeyer and Redd (1981) found no correlation between salinity and metal content of *Spartina*. Pb uptake into roots and shoots of rice plants (*Oryza sativa*) decreased with increasing redox potential and pH, while Cd uptake increased with increasing redox potential and a decreasing pH (Reddy and Patrick, 1977). Gambrell (1994) found that pH and redox potential affect Pb, Cd, and Hg accumulation in marsh plants. Greater uptake and availability of Cd were seen in a number of wetland species under dry (more oxidized) than under flooded (reduced) conditions (Gambrell, 1994). Jackson and Kalff (1993) also reported that low pH is favorable for heavy metal accumulation in rooted wetland plants. Soil E_h plays an important role in determining the bioavailability of heavy metals to wetland plants. Jackson (1998) found that the greatest metal mobility likely exists within a zone ($-150 < E_h < +200$ mV) between strongly reduced and strongly oxidized sediments. A greater uptake and availability of Cd was observed in a number of wetland plants under dry conditions (higher E_h) than under flooded (lower Eh) conditions (Gambrell, 1994). However, Ye et al. (1998b) found that *Phragmites australis* takes up more Pb, Zn, and Cd under flooded conditions than under dry conditions.

Different forms of the same metal can have different rates of uptake and different effects; for example, marsh sediments tended to reduce the very toxic Cr^{6+} to the less toxic Cr^{3+} form very rapidly (Pardue and Patrick, 1995).

Aquatic plants also synthesize similar lipid-soluble arsenic compounds (Tamaki and Frankenberger, 1992). Submerged plants (*Ceratophyllum demersum* and *Elatine triandra*) in Canada accumulated As, but mostly in organic form (methylarsinic acid and dimethylarsinic acid). Only a small proportion in the wetland plants was inorganic As, while most of the As in the surface water was present as arsenate (Zheng et al., 2003).

A striking feature of the roots of some wetland plants is the presence of metal-rich rhizoconcretions or plaque on the roots (Mendelssohn and Postek, 1982; Vale et al., 1990). These structures are composed primarily of Fe hydroxides and other metals such as Mn that are mobilized and precipitated on the root surface. The metals are mobilized from the reduced anoxic estuarine sediments and concentrated in the oxidized microenvironment around the roots. Their concentrations can reach five to ten times the concentrations seen in the surrounding sediments (Sundby et al., 1998). There have been conflicting reports as to whether the presence of the plaque reduces or increases the uptake of metal by the plants. The presence of such concretions appeared to reduce the amount of Zn taken up by *Aster tripolium* (Otte et al., 1989) and the amount of Mn taken up by *Phragmites australis* (Figure 8.9) (Batty, Baker, and Wheeler, 2000). The mechanism may have been the plaque acting as a physical barrier, although the barrier was

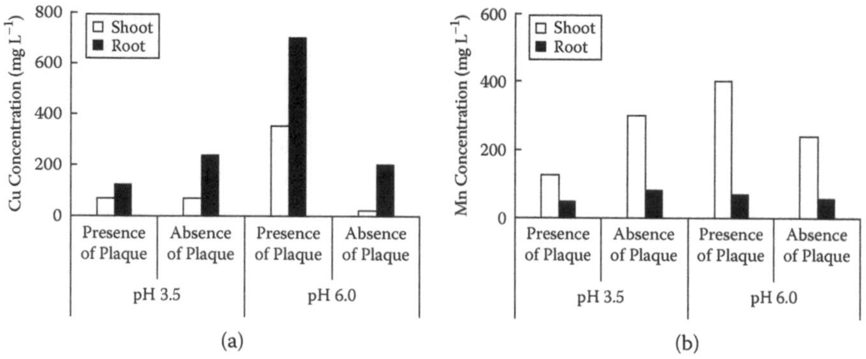

FIGURE 8.9
Heavy metal concentrations in *P. australis* seedlings exposed to (a) 0.5 $mg.L^{-1}$ Cu at pH 3.5 and pH 6.0, and (b) 0.5 $mg.L^{-1}$ Mn at pH 3.5 and pH 6.0. (*Source:* Redrawn from Batty, L.C., A.J.M. Baker, B.D. Wheeler, and C.D. Curtis. 2000. *Ann. Bot.* 86: 647–653.)

not effective at low pH conditions. At higher pH conditions, the presence of plaque enhanced Cu uptake into roots. However, in *Typha latifolia* (cattail), the presence of Fe plaque did not reduce uptake of toxic metals (Ye et al., 1998a). Fe plaque increased Zn uptake by rice (*Oryza sativa*) and movement into shoots (Zhang, Zhang, and Mao, 1998). In contrast, Al was not adsorbed onto the Fe or the Mn plaque but formed a separate phosphate deposit that resembled the Fe and Mn plaques (Batty et al., 2002). They attributed the precipitate to the leakage of oxygen from the roots, and suggested that the P might be immobilized at the root surface. The discrepancies in effects of plaque on metal uptake need to be resolved by further study. Different metals, environmental conditions, or physiologies may account for these differences.

Plants can alter the distribution of metals in wetland sediments by oxidizing the soil in the immediate vicinity of the rhizosphere. The concentrations of several metals were higher in vegetated soils than in unvegetated ones, and were particularly high in soils in the immediate vicinity of the plant roots (Doyle and Otte, 1997). In a study of Cd and Zn concentrations in sediment cores from areas under *Spartina alterniflora* compared with adjacent mud flats without plant cover, the metal concentrations at 5- to 15-cm depth (corresponding to maximum root biomass) were higher by a factor of four under the vegetation (De Lacerda, Freixo, and Coelho, 1997). Plant activity (metal mobilization by oxidation of the rootzone and movement into the rhizosphere) was considered responsible for the increase. Similar results were found at sites with *S. maritima* by Caçador, Vale, and Catarino (1996). They concluded that roots influence metal concentrations in the marsh sediments as a result of producing complex organic compounds and oxidizing the rhizosphere.

Summary

Metals in wetland soils and sediments can exist in solid, exchangeable, and soluble phases. The chemical form of metal in wetland soil and sediment governs bioavailability. The amount of organic matter clay content along with sediment pH and redox conditions also influence the mobility and uptake of metals by aquatic organisms and wetland plants. Wetland plants can also influence soil pH and redox conditions, which affects the speciation, mobility, and availability of metals.

References

Bartlett, R.J., and B.R. James. 1988. Mobility and bioavailability of chromium in soils. In Nriagu, J.O., and E. Nieboer (Eds.) *Chromium in the natural and human environments*. John Wiley & Sons, New York. pp. 267–304.

Batty, L.C., A.J.M. Baker, and B.D. Wheeler. 2002. Aluminium and phosphate uptake by *Phragmites australis*: The role of Fe, Mn, and Al root plaques. *Ann. Bot.* 89: 443–449.

Batty, L.C., A.J.M. Baker, B.D. Wheeler, and C.D. Curtis. 2000. The effect of pH and plaque on the uptake of Cu and Mn in *Phragmites australis* (Cav.) Trin ex. Steudel. *Ann. Bot.* 86: 647–653.

Berk, S.G., and R.R. Colwell. 1981. Transfer of mercury through a marine microbial food web. *J. Exp. Mar. Biol. Ecol.* 52: 157–172.

Black, H. 1995. Absorbing possibilities: Phytoremediation. *Environ. Health Perspec.* 103: 1106–1108.

Burke, D.J., J.S. Weis, and P. Weis. 2000. Release of metals by the leaves of the salt marsh grasses *Spartina alterniflora* and *Phragmites australis*. *Estuar. Coast. Shelf Sci.* 51: 153–159.

Caçador, I., C. Vale, and F. Catarino. 1996. The influence of plants on concentration and fractionation of Zn, Pb, and Cu in salt marsh sediments (Tagus estuary, Portugal). *J. Aquat. Ecosyst. Health* 5: 193–198.

Calder, L.M. 1988. Chromium contamination of groundwater. In Nriagu, J.O., and E. Nieboer (Eds.) *Chromium in the natural and human environments*. John Wiley & Sons, New York. pp. 399–442.

Callaway, J.C., R.D. DeLaune, and W.H. Patrick, Jr. 1998. Heavy metal chronologies in selected coastal wetlands from Northern Europe. *Mar. Pollut. Bull.* 36(1): 82–96.

Campbell, P.G.C., A. Tessier, M. Bisson, and R. Bougie. 1985. Accumulation of copper and zinc in the yellow water lily, *Nuphar variegatum*: relationships to metal partitioning in the adjacent lake sediments. *Can. J. Fish. Aquat. Sci.* 42: 23–32.

Cardwell, A.J., D.W. Hawker, and M. Greenway. 2002. Metal accumulation in aquatic macrophytes from southeast Queensland, Australia. *Chemosphere* 48: 653–663.

Chawla, G., J. Singh, and P.N. Viswanathan. 1991. Effect of pH and temperature on the uptake of cadmium by *Lemna minor* L. B. *Environ. Contam. Tox.* 47: 84–90.

Cheng, S., W. Grosse, F. Karrenbrock, and M. Thoennessen. 2002. Efficiency of constructed wetlands in decontamination of water polluted by heavy metals. *Ecol. Eng.* 18: 317–325.

Compeau, G.C., and R. Bartha. 1984. Methylation and demethylation of mercury under controlled redox, pH and salinity conditions. *Appl. Environ. Microbiol.* 48: 1203–1207.

De Lacerda, L.D., J.L. Freixo, and S.M. Coelho. 1997. The effect of *Spartina alterniflora* Loisel on trace metals accumulation in intertidal sediments. *Mangroves and Salt Marshes* 1: 201–209.

De Lacerda, L.D., C. Carvalho, K. Tanizaki, A. Ovalle, and C. Rezende. 1993. The biogeochemistry and trace metals distribution of mangrove rhizospheres. *Biotropica.* 25: 252–257.

DeLaune, R.D., A. Jugsujinda, I. Devai and W.H. Patrick, Jr. 2004. Relationship of sediment redox conditions to methylmercury in surface sediment of Louisiana Lakes. *J. Environ. Sci. Health A* 39(8): 1925–1935.

DeLaune, R.D., and R.P. Gambrell. 1996. Role of sedimentation in isolating metal contaminants in wetland environments. *J. Environ. Sci. Health A* 31(9): 2349–2362.

DeLaune, R.D., C. Mulbah, I. Devai, and C.W. Lindau. 1998. Effect of chromium and lead on degradation of south Louisiana crude oil in sediment. *J. Environ. Sci. Health A* 33(4): 527–546.

Delgado, M., M. Bigeriego, and E. Guardiola. 1993. Uptake of zinc, chromium and cadmium by water hyacinths. *Water Res.* 27: 269–272.

Deng, H., Z.H. Ye, and M.H. Wong. 2004. Accumulation of lead, zinc, copper and cadmium by twelve wetland plant species thriving in metal contaminated sites in China. *Environ. Pollut.* 132: 29–40.

Deng, H., Z.H. Ye, and M.H. Wong. 2006. Lead and zinc accumulation and tolerance in populations of six wetland plants. *Environ. Pollut.* 141: 69–80.

Doyle, M.O., and M.L. Otte. 1997. Organism-induced accumulation of iron, zinc and arsenic in wetland soils. *Environ. Pollut.* 96: 1–11.

Drifmeyer, J.E., and B. Redd. 1981. Geographic variability in trace element levels in *Spartina alterniflora. Estuar. Coast. Shelf Sci.* 13: 709–716.

Dunbabin, J.S., and K.H. Bowmer. 1992. Potential use of constructed wetlands for treatment of industrial wastewaters containing metals. *Sci. Total Environ.* 111: 151–168.

Feijtel, T.C., R.D. DeLaune, and W.H. Patrick, Jr. 1988. Biogeochemical control on metals distribution and accumulation in Louisiana sediment. *J. Environ. Qual.* 17: 88–89.

Fett, J.P., J. Cambraia, M.A. Oliva, and C.P. Jordao. 1994. Absorption and distribution of Cd in water hyacinth plants. *J. Plant Nutr.* 17: 1219–1230.

Fitzgerald, E.J., J.M. Caffrey, S.T. Nesaratnam, and P. McLoughlin. 2003. Copper and lead concentrations in salt marsh plants on the Suir Estuary, Ireland. *Environ. Pollut.* 123: 67–74.

Fritioff, A., L. Kautsky, and M. Greger. 2005. Influence of temperature and salinity on heavy metal uptake by submersed plants. *Environ. Pollut.* 133: 265–274.

Gambrell, R.P. 1994. Trace and toxic metals in wetlands — A review. *J. Environ. Qual.* 23: 883–891.

Gambrell, R.P., and W.H. Patrick, Jr. 1991. *Handbook — Remediation of contaminated sediments.* EPA/625/6–91/028.

Gambrell, R.P., and W.H. Patrick, Jr. 1988. The influence of redox potential on the environmental chemistry of contaminants in soils and sediments. In Hook, D.D. (Ed.) *The ecology and management of wetlands*. Vol. 1. Timber Press, Portland, OR. pp. 319–333.

Greger, M. 1999. Metal availability and bioconcentration in plants. In Prasad, M.N.V., and J. Hagemeyer (Eds.) *Heavy metal stress in plants: from molecule to ecosystems*. Springer-Verlag, Berlin.

Grill, E., E.L. Winnacker, and M. Zenk. 1985. Phytochelatins: The principal heavy-metal complexing peptides of higher plants. *Science* 230: 674–676.

Holdway, D.A. 1988. The toxicity of chromium to fish. pp. 369–398. In Nriagu, J.O., and E. Nieboer (Eds.) *Chromium in the natural and human environments*. John Wiley & Sons, New York.

Hung, G.A., and G.L. Chmura. 2007. Metal accumulation in surface salt marsh sediments of the Bay of Fundy, Canada. *Estuaries and Coasts* 30(4): 725–734.

Jackson, L.J. 1998. Paradigms of metal accumulation in rooted aquatic vascular plants. *Sci. Total Environ.* 219: 223–231.

Jackson, L.J., and J. Kalff. 1993. Patterns in metal content of submerged aquatic macrophytes: The role of plant growth form. *Freshwater Biol.* 29: 351–359.

Kadlec, R.H., and R.I. Knight. 1996. *Treatment wetlands*. CRC Press, Boca Raton, FL.

King, J.K., J.E. Kotska, M.E. Frischer, F.M. Saunders, and R.A. Jahnke. 2001. A quantitative relationship that demonstrates mercury methylation rates in marine sediments is based on the community composition and activity of sulfate reducing bacteria. *Environ. Sci. Technol.* 35(12): 2491–2496.

Kraus, M.L. 1988. Accumulation and excretion of five heavy metals by the salt marsh grass *Spartina alterniflora*. *Bull-N. J. Acad. Sci.* 33: 39–43.

Kraus, M.L., P. Weis, and J. Crow. 1986. The excretion of heavy metals by the salt marsh cord grass, *Spartina alterniflora* and *Spartina*'s role in mercury cycling. *Mar. Environ. Res.* 20: 307–316.

Liu, J., Y. Dong, H. Xu, D. Wang, and J. Xu. 2007. Accumulation of Cd, Pb and Zn by 19 wetland plant species in constructed wetland. *J. Hazard. Mater.* 147: 947–953.

MacFarlane, G.R., and M.D. Burchett. 2000. Cellular distribution of copper, lead and zinc in the grey mangrove, *Avicennia marina* (Forsk.) Vierh. *Aquat. Bot.* 68: 45–59.

MacFarlane, G.R., and M.D. Burchett. 2002. Toxicity, growth and accumulation relationships of copper, lead and zinc in the grey mangrove, *Avicennia marina* (Forsk.) Vierh. *Mar. Environ. Res.* 54: 65–84.

Masscheleyn, P.H., J.H. Pardue, R.D. DeLaune, and W.H. Patrick, Jr. 1992. Chromium redox chemistry in a lower Mississippi Valley bottomland. *Environ. Sci. Technol.* 26: 1217–1226.

Masscheleyn, P.H., R.D. DeLaune, and W.H. Patrick, Jr. 1991a. Arsenic and selenium chemistry as affected by sediment redox potential and pH. *J. Environ. Qual.* 20: 521–527.

Masscheleyn, P.H., R.D. DeLaune, and W.H. Patrick, Jr. 1991b. Effect of redox potential and pH on arsenic speciation and solubility in a contaminated soil. *Environ. Sci. Technol.* 25(8): 1414–1419.

Mendelssohn, I.A., and M.T. Postek. 1982. Elemental analysis of deposits on the roots of *Spartina alterniflora* Loisel. *Am. J. Bot.* 69: 904–912.

Mendelssohn, I.A., K.L. McKee, and T. Kong. 2001. A comparison of physiological indicators of sublethal cadmium stress in wetland plants. *Environ. Exp. Bot.* 46: 263–275.

Moorhead, K.K., and K.R. Reddy. 1988. Oxygen transport through selected aquatic macrophytes. *J. Environ. Qual.* 17: 138–142.

Nieboer, E., and S.L. Shaw. 1988. Mutagenic and other genotoxic effects of chromium compounds. pp. 399–442. In Nriagu, J.O., and E. Nieboer (Eds.) *Chromium in the natural and human environments.* John Wiley & Sons, New York.

Otte, M.L., J. Rozema, L. Koster, M. Haarsma, and R. Broekman. 1989. Iron plaque on roots of *Aster tripolium* L.: Interaction with zinc uptake. *New Phytol.* 111: 309–317.

Otte, M.L., S.J. Bestebroer, J.M. van der Linden, J. Rozema, and R.A. Broekman. 1991. A survey of zinc, copper and cadmium concentrations in salt marsh plants along the Dutch coast. *Environ. Pollut.* 72: 175–189.

Padinha, C., R. Santos, and M.T. Brown. 2000. Evaluating environmental contamination in Ria Formosa (Portugal) using stress indexes of *Spartina maritima. Mar. Environ. Res.* 49: 67–78.

Pardue, J.H., and W.H. Patrick, Jr. 1995. Changes in metal speciation following alteration of sediment redox status. In Allen, H. (Ed.) *Metal-contaminated aquatic sediments.* Science Publishers, Ann Arbor, MI. pp. 169–185.

Paterson, M., J. Rudd, and V. Louis. 1998. Increases in total mercury in zooplankton following flooding of a reservoir. *Environ. Sci. Technol.* 32: 3868–3874.

Pichtel, J., K. Kuroiwa, and H.T. Sawyer. 2000. Distribution of Pb, Cd and Ba in soils and plants of two contaminated sites. *Environ. Pollut.* 110: 171–178.

Rai, P.K. 2009. Heavy metals in water, sediments and wetland plants in an aquatic ecosystem of tropical industrial region, India. *Environ. Monit. Assess.* 158: 433–557.

Reay, P.F. 1972. The accumulation of arsenic from arsenic-rich natural waters by aquatic plants. *J. Appl. Ecol.* 9: 557–565.

Reddy, C.N., and W.H. Patrick, Jr. 1977. Effect of redox potential and pH on the uptake of Cd and Pb by rice plants. *J. Environ. Qual.* 6: 259–262.

Salomons, W., and U. Förstner. 1984. *Metals in the hydrocycle.* Springer Verlag, Berlin.

Salt, D.E., M. Blaylock, N.P.B.A. Kumar, V. Dushenkov, I. Chet, and I. Raskin. 1995. Phytoremediation: A novel strategy for the removal of toxic metals from the environment using plants. *Biotechnology* 13: 468–474.

Shannon, R.D., and J.R. White. 1991. The selectivity of a sequential extraction procedure for the determination of iron oxyhydroxides and iron sulfides in lake sediments. *Biogeochemistry* 14: 193–208.

Sneller, F.E.C., E.C.M. Noordover, W.M. Ten Bookum, H. Schat, J.J.M. Bedaux, and J.A.C. Verkleij. 1999. Quantitative relationship between phytochelatin accumulation and growth inhibition during prolonged exposure to cadmium in *Silene vulgaris. Ecotoxicology* 8: 167–175.

Snowden, R.E.D., and B.D. Wheeler. 1993. Iron toxicity to fen plant species. *J. Ecol.* 81: 35–46.

Stoltz, E., and M. Greger. 2002. Accumulation properties of As, Cd, Cu, Pb and Zn by four wetland plant species growing on submerged mine tailings. *Environ. Exp. Bot.* 47: 271–280.

Sundby, B., C. Vale, I. Cacador, and F. Catarino. 1998. Metal-rich concretions on the roots of salt marsh plants: mechanism and rate of formation. *Limnol. Oceanogr.* 43: 245–252.

Tamaki, S., and W.T. Frankenberger, Jr. 1992. Environmental biochemistry of arsenic. *Rev. Environ. Contam. Toxicol.* 124: 79–110.

Taylor, G.J., and A.A. Crowder. 1983. Uptake and accumulation of heavy metals by *Typha latifolia* in wetlands of the Sudbury, Ontario region. *Can. J. Bot.* 61: 63–73.

Ullirich, S.M., T.W. Tanton, and S.A. Abdrashitova. 2001. Mercury in the aquatic environment. A review of factors affecting methylation. *Crit. Rev. Environ. Sci. Technol.* 31: 241–293.

Vale, C., F. Catarino, C. Cortesao, and M. Cacador. 1990. Presence of metal-rich rhizoconcretions on the roots of *Spartina maritima* from the salt marshes of the Tagus estuary, Portugal. *Sci. Total Environ.* 97/98: 617–626.

Verkleij, J.A., and H. Schat. 1990. Mechanisms of metal tolerance in higher plants. In Shaw, A.J. (Ed.) *Heavy metal tolerance in plants: evolutionary aspects*. CRC Press, Boca Raton, FL. pp. 179–194.

Vesk, P.A., C.E. Nockolds, and W.G. Allaway. 1999. Metal localization in water hyacinth roots from an urban wetland. *Plant Cell Environ.* 22: 149–158.

Waisel, Y. 1977. Salt excretion in *Avicennia marina*. In Davy, A.J. (Ed.) *Proc. First European Ecology Symposium, Ecological Processes in Coastal Environments*. Norwich, UK: University of East Anglia. pp. 1–21.

Weis, J.S., and P. Weis. 2004. Metal uptake, transport and release by wetland plants: implications for phytoremediation and restoration. *Environ. Int.* 30: 685–700.

Williams, J.B. 2002. Phytoremediation in wetland ecosystems: Progress, problems and potential. *Crit. Rev. Plant Sci.* 21: 607–635.

Williams, T.P., J.M. Bubb, and J.N. Lester. 1994. Metal accumulation within salt marsh environment: A review. *Mar. Pollut. Bull.* 28(5): 277–290.

Windham, L., J.S. Weis, and P. Weis. 2001a. Patterns and processes of mercury (Hg) release from leaves of two dominant salt marsh macrophytes, *Phragmites australis* and *Spartina alterniflora*. *Estuaries* 24: 787–795.

Windham, L., J.S. Weis, and P. Weis. 2001b. Lead uptake, distribution, and effects in two dominant salt marsh macrophytes, *Spartina alterniflora* (cordgrass) and *Phragmites australis* (common reed). *Mar. Pollut. Bull.* 42: 811–816.

Wolterbeek, H.T., and A.J.G.M. van der Meer. 2002. Transport rate of arsenic, cadmium, copper and zinc in *Potamogeton pectinatus* L.: Radiotracer experiments with ^{76}As, $^{109,115}Cd$, ^{64}Cu and $^{65,69m}Zn$. *Sci. Total Environ.* 287: 213–230.

Wong, P.T.S., and J.T. Trevors. 1988. Chromium toxicity to algae and bacteria. In Nriagu, J.O., and E. Nieboer. (Ed.) *Chromium in the natural and human environments*. John Wiley & Sons, New York. pp. 305–315.

Wright, D.J., and M.L. Otte. 1999. Wetland plant effects on the biogeochemistry of metals beyond the rhizosphere. *Biol. Environ.: Proc. Royal Irish Acad.* 99B: 3–10.

Xu, S., and P.R. Jaffé. 2006. Effects of plants on the removal of hexavalent chromium in wetland sediments. *J. Environ. Qual.* 35: 334–341.

Yang, J., and Z. Ye. 2009. Metal accumulation and tolerance in wetland plants. *Front. Biol. China* 4(3): 282–288.

Yassi, A., and E. Nieboer. 1988. Carcinogenicity of chromium compounds. In Nriagu, J.O., and E. Nieboer. (Eds.) *Chromium in the natural and human environments*. John Wiley & Sons, New York. pp. 399–442.

Ye, Z., A.J.M. Baker, M.H. Wong, and A.J. Willis. 1997a. Zinc, lead and cadmium tolerance, uptake and accumulation by the common reed, *Phragmites australis* (Cav.) Trin. ex Steudel. *Ann. Bot.* 80:363–370.

Ye, Z., A.J.M. Baker, M.H. Wong, and A.J. Willis. 1997b. Zinc, lead and cadmium tolerance, uptake and accumulation by *Typha latifolia*. *New Phytol.* 136: 469–480.

Ye, Z., M.H. Wong, A.J.M. Baker, and A.J. Willis. 1998b. Comparison of biomass and metal uptake between two populations of *Phragmites australis* grown in flooded and dry conditions. *Ann. Bot.* 82: 83–87.

Ye, Z.H., A.J.M. Baker, M.H. Wong, and A.J. Willis. 1998a. Zinc, lead and cadmium accumulation and tolerance in *Typha latifolia* as affected by iron plaque on the root surface. *Aquat. Bot.* 61: 55–67.
Zaranyika, M.F., and T. Ndapwadza. 1995. Uptake of Ni, Zn, Fe, Co, Cr, Pb, Cu and Cd by water hyacinth (*Eichhornia crassipes*) in Mukuvisi and Manyame Rivers, Zimbabwe. *J. Environ. Sci. Health A* 30: 157–169.
Zayed, A., S. Gowthaman, and N. Terry. 1998. Phytoaccumulation of trace elements by wetland plants. Duckweed. *J. Environ. Qual.* 27: 715–721.
Zhang, X., F. Zhang, and D. Mao. 1998. Effect of iron plaque outside roots on nutrient uptake by rice (*Oryza sativa* L.). Zinc uptake by Fe-deficient rice. *Plant Soil* 202: 33–39.
Zheng, J., H. Hintelmann, D. Dimock, and M.S. Dzurko. 2003. Speciation of arsenic in water, sediment, and plants of the Moira watershed, Canada, using HPLC coupled to high resolution ICP-MS. *Anal. Bioanal. Chem.* 377: 14–24.
Zhu, Y.L., A.M. Zayed, J.H. Qian, M. Souza, and N. Terry. 1999. Phytoaccumulation of trace elements by wetland plants. II. Water hyacinth. *J. Environ. Qual.* 28: 339–344.

9

Factors Controlling the Dynamics of Trace Metals in Frequently Flooded Soils

Jörg Rinklebe and Gijs Du Laing

CONTENTS

Introduction

Frequently flooded soils are semi-terrestrial and/or semi-aquatic soils with aquic or epiaquic moisture regimes. They are influenced by various aquatic sources, including groundwater, static water, return seepage, and flood water. In general, large variations in flooding frequency as well as in flood water characteristics have been observed. As a result, sedimentation rates of substances also reveal a large variation according to the site-specific conditions.

Frequently flooded soils such as floodplain soils or rice fields are often contaminated by arsenic and/or heavy metals (e.g., Rinklebe, Franke, and Neue, 2007; Roberts et al., 2007). These contaminations can be of natural or anthropogenic origin, or a combination of both. Trace metals and other pollutants can be carried along with the water, suspended particles, and sediments. They can originate from various sources in the catchment area such as industrial or mining activities, discharge of communal waste, diffuse agricultural sources, or natural geogenic processes (e.g., Rinklebe, Franke, and Neue, 2007; Du Laing et al., 2009). Atmospheric deposition can also contribute to the contamination by trace elements.

Generally, three zones can be distinguished in frequently flooded soils: an oxic layer, an anoxic layer, and an oxic–anoxic interface (e.g., Salomons et al., 1987; Neue, 1991). In the oxic surface water or soil layers, stable phases of trace metals are often sorbed species. Important sorbing phases are clay minerals, organic matter, Fe and Mn hydroxides, and carbonates. Soluble organic components and salinity are important factors affecting the desorption from these phases. In the potential flux of metals from frequently flooded soils to surface waters, the oxic–anoxic interface and the anoxic layer beneath the oxic layer are of special importance. In those layers, a special dynamic of redox processes occurs, which can result in the inclusion of metals in precipitates or the dissolution of metal-containing precipitates (e.g., Du Laing et al., 2009).

The kinetics of these processes are of crucial importance for tidal floodplain soils as the location of the oxic–anoxic interface is subject to change due to fluctuating water table levels (Du Laing et al., 2007; Rinklebe et al., 2005). The local variability of water table levels depends on the flood frequency and duration, which are closely related to the topography of the area (Rinklebe, Franke, and Neue, 2007).

Metal fate in frequently flooded soils is governed by numerous processes such as sorption–desorption, precipitation–dissolution, and complexation–decomplexation (Gambrell, 1994; Du Laing et al., 2009). Moreover, the fate of metals in frequently flooded soils is significantly determined by their chemical forms of occurrence. According to Gambrell (1994), those forms include (1) water-soluble metals, as free ions, inorganic or organic complexes; (2) exchangeable metals; (3) metals precipitated as inorganic compounds, including insoluble sulfides; (4) metals complexed with high molecular weight humic materials; (5) metals adsorbed or occluded to precipitated hydrous oxides; and (6) metals bound within the crystalline lattice structures of primary minerals.

In this chapter, several factors are discussed that have an important impact on the dynamics and mobilities of metals in frequently flooded soils. These include redox processes and valence state of metals, oxidation-reduction behavior of Fe and Mn, sulfur cycling, carbonates and pH, adsorption and desorption, salinity, and organic matter. Of course, these factors do not cause isolated impacts on the dynamics and mobilities of metals; instead, the dynamics and mobilities of metals in frequently flooded soils are determined by a combination of factors. Thereby, overlaps and interferences are frequently observed. Additionally, several factors often interact. Thus, the fate of metals in frequently flooded soils can depend on their combination or time-dependent mode of functioning.

Redox Processes

Redox processes are central to the mobility of many metals and metalloids (e.g., Van den Berg et al., 2000; Du Laing et al., 2009). Van Griethuysen et al. (2005) pointed out that the dynamics of many metals in frequently flooded soils are driven by seasonal redox cycles of S, Fe, and Mn. At the oxic–anoxic interface and in the anoxic layers of flooded soils, redox-driven processes often include the precipitation and dissolution of metals. The kinetics of those processes are of great importance for these soils as the location of the oxic–anoxic interface is subject to change due to fluctuating water table levels (Rinklebe et al., 2005; Rinklebe, Franke, and Neue, 2007; Du Laing et al., 2009).

In frequently flooded soils, subsequent oxidation and reduction reactions regulate various biogeochemical processes. Water saturation during extended periods of time usually results in changes in the chemical properties of soils, as well as in microbial populations and processes (e.g., Langer and Rinklebe, 2009). Soils tend to undergo a series of sequential redox reactions when the redox status of the soil changes from aerobic to anaerobic conditions during flooding (e.g., Neue, 1991; Yu et al., 2007). Major reactions include denitrification, Mn(IV) reduction, Fe(III) reduction, sulfate (SO_4^{2-}) reduction, and methanogenesis (Figure 9.1). These processes are mainly catalyzed by microorganisms (Paul and Clark, 1996) and have previously been extensively studied and reviewed (e.g., Patrick and Jugsujinda, 1992; Yu et al., 2007; Du Laing et al., 2009). The intensity of soil reduction can be characterized by the oxidation–reduction potential (redox potential, E_h), which is a measure of the electron availability and allows the prediction of the stability and availability of various metals in

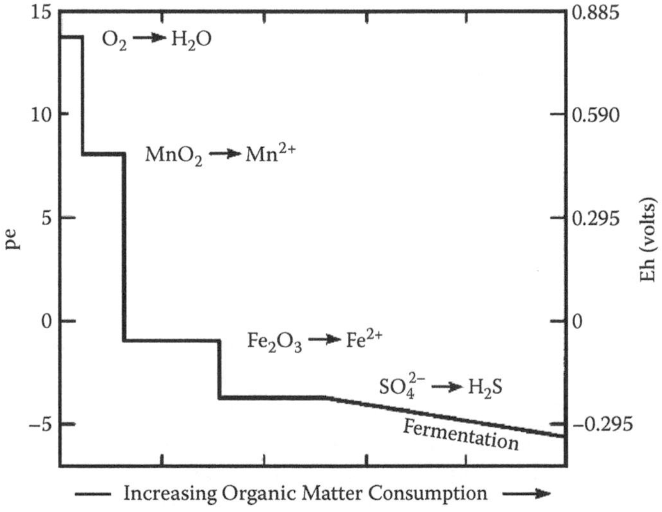

FIGURE 9.1
A hypothetical plot showing change in terminal electron-accepting processes for a soil containing a finite supply of bioavailable organic matter and soil oxidation capacity. (*Source:* Reproduced from Chadwick, O.A., and Charover, J. 2001. *Geoderma* 100: 321–353. With permission from Elsevier.)

frequently flooded soils. It evolves from high to low when the redox status of the soil changes from aerobic to anaerobic conditions during flooding.

Possible redox reactions are dictated by the state of the soil system and the corresponding Gibbs energies of reaction (Chadwick and Chorover, 2001). Some of these reactions are thermodynamically possible. However, those that predominate at any given point in time are determined by redox kinetics, the latter governed to a large degree by soil microbial catalysis (Fenchel, King, and Blackburn, 1998). A sequence of terminal electro-accepting processes is observed typically along redox gradients, both spatially and temporally in soil systems (Patrick and Henderson, 1981; Patrick and Jugsujinda, 1992), which corresponds closely to progressive decreases in the Gibbs energy of the full redox reaction (Chadwick and Chorover, 2001). Figure 9.1 shows this sequence of reactions for a hypothetical soil containing a finite supply of bioavailable organic matter and soil oxidation capacity (Chadwick and Chorover, 2001).

Oxygen (O_2) can be depleted rapidly by microbial and root respiration in soils subjected to limited influx of air or oxygenated water because it is sparingly soluble in water (0.25 mM at 25°C). At this point, dissolved nitrate and available Mn(IV) solids are utilized as alternative electron acceptors during the oxidation of organic material (e.g., Chadwick and Chorover, 2001). If these reactants are exhausted, further reduction results in the successive use of Fe(III) solids (ferric reduction), SO_4^{2-} (sulfate reduction), and eventually organic matter itself (fermentation) or CO_2 (methanogenesis) (Figure 9.1).

Change in Valence State

Metals can directly react to changes in redox conditions. Reduced conditions result in the reduction of Cr (VI) to Cr (III) and the immobilization of chromates (Reddy and DeLaune, 2008). Conversely, Cr(VI) will be mobilized at high E_h values. Marsh sediments tend to rapidly reduce the very toxic Cr(VI) to the less toxic Cr(III) (Pardue and Patrick, 1995). A reduction of Cu(II) to Cu (I) under slightly alkaline and anaerobic conditions was found when a suitable electron donor was available (Simpson, Rosner, and Ellis, 2000). Several electron donors (e.g., Fe(II), sulfur compounds), and bacteria acting as catalysts, may be involved in the Cu(II) to Cu(I) reduction process, subsequently leading to Cu_2S precipitation (Du Laing et al., 2009).

The effect of redox conditions and pH on the release of rare earth elements (REEs) was studied by Cao et al. (2001). Figure 9.2a shows the equilibrium release of lanthanum (La), cerium (Ce), gadolinium (Gd), and yttrium (Y), as well as Fe and Mn at pH 7.5 and two different redox potentials (400 and −100 mV). La, Ce, Gd, Y, Fe, and Mn were sparingly soluble under those conditions, probably due to the more neutral pH. Nevertheless, increasing release of La

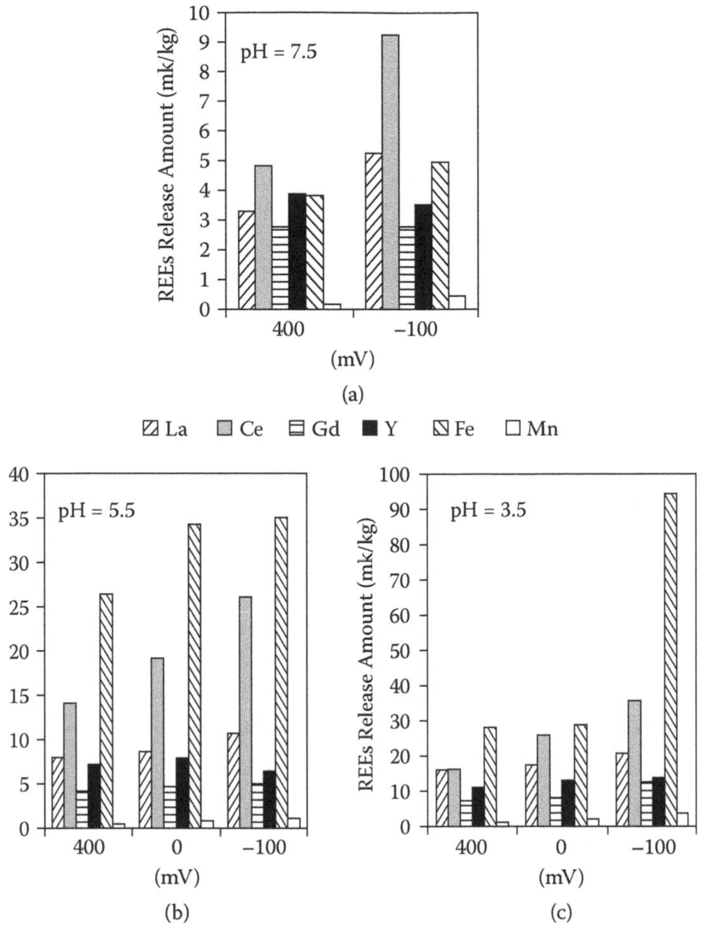

FIGURE 9.2
The effect of E_h and pH on the release of rare earth elements REEs (lanthanum (La), cerium (Ce), gadolinium (Gd), and yttrium (Y)), Fe and Mn from a soil (please notice the different scales). (Reproduced from Cao, X., Chen Y., Wang X., Deng X. Effects of redox potential and pH value on the release of rare earth elements from soil. *Chemosphere* 44, 2001, 655–661, with permission from Elsevier.)

and Ce as well as Fe and Mn under reducing conditions was observed (Cao et al., 2001). At pH 5.5, the solubilities were higher than those under more neutral conditions (Figure 9.2b) and the release of the four selected REEs (La, Ce, Gd, and Y) increased drastically at pH 3.5 (Figure 9.2c; notice the different scales). For instance, the concentrations of La and Ce were increased four times (Cao et al., 2001). This can be attributed to the fact that low pH may favor the conversion of metals from precipitate forms into soluble forms (Harter, 1983). Soluble Ce concentrations significantly increased when the pH was kept at 3.5 and the redox potential was changed from 400 to –100 mV. However, this trend was not found for La, Gd, and Y (Cao et al., 2001). Cao et al. (2001) consider a change of Ce speciation from dissolved Ce^{3+} to stable Ce^{4+} when the redox potential increases while the valences of La, Gd, and Y are unchangeable, as a possible explanation. Although they concluded that some REEs (particularly Ce) were released under reducing conditions, it should be mentioned that they had difficulty in decreasing the redox potential below –100 mV. This problem could be partly overcome using an automated, hermetically sealed biogeochemical microcosm set up as described by Yu et al. (2007) and Frohne et al. (in press). Other researchers have used sodium thioglycollate as a reducing agent to further decrease redox potentials (Smith and Martell, 1976).

Although the mobility and availability of some elements in soils are affected by changes in their valence states as a result of redox changes, this should not be usual for many metals, including the mobile elements Cd and Zn. However, the mobility of these elements can be indirectly affected by redox potential, for instance by (1) redox-induced changes to the metal-binding capacity of humic materials, (2) insoluble metal sulfide formation (or sulfide oxidation), and (3) changes in Fe and Mn oxyhydroxides known to be effective in immobilizing some metals under oxidizing conditions (Du Laing et al., 2008a, 2009). Cd is more mobile and plant-available in oxidized soils and sediments compared to reduced conditions (Gambrell et al., 1991a,b; Vandecasteele, Du Laing, and Tack, 2007; Frohne et al., in press).

Iron and Manganese

As a result of temporal inundations and the establishment of low redox potentials, Fe and Mn (hydr)oxides in the solid phase are reduced to Mn^{2+} and Fe^{2+}, which occur as soluble metals and organic complexes in the liquid phase (Reddy and DeLaune, 2008). Reduction of Fe and Mn oxides can be presented as

$$MnO_2 + 2\,e^- + 4\,H^+ \leftrightarrow Mn^{2+} + 2\,H_2O$$

$$Fe_2O_3 + 2\,e^- + 6\,H^+ \leftrightarrow 2\,Fe^{2+} + 3\,H_2O$$

Some bacteria can catalyze these reductions (Lovley and Phillips, 1988b). Lovley and Phillips (1988a) reported that the reduction of Fe^{3+} does not begin before all Mn^{4+} is depleted. However, Patrick and Jugsujinda (1992), Du Laing et al. (2007), and Reddy and DeLaune (2008) have observed some overlap.

According to the reactions mentioned above, Fe concentrations in the pore water are high at low E_h, and vice versa. This behavior has been frequently described and can be attributed to the formation of Fe (hydr)oxides at high E_h (e.g., Brümmer, 1974; Yu et al., 2007). At low E_h, Fe is reduced to soluble Fe^{2+} (e.g., Alewell et al., 2008). The stability of Fe^{2+} and Fe (hydr)oxides depends on a combination of the E_h and pH of the sediment. The nearly amorphous $Fe(OH)_3$ minerals (ferrihydrite) are reduced at a higher E_h for a given pH than are the crystalline minerals of FeOOH (goethite) or Fe_2O_3 (hematite) (Du Laing et al., 2009). Generally, Fe occurs in soluble forms under acid conditions. At neutral pH, Fe is soluble at low redox potentials or as a soluble organic complex in oxic soils.

Plenty of studies (e.g., Yu et al., 2007) have shown that concentrations of Mn in the pore water decrease with increasing Eh. Reduced Mn^{2+} is more soluble than its oxidized counterpart (Chadwick and Chorover, 2001). When the oxygen level increases, Mn ions precipitate and form secondary Mn compounds such as Mn (hydr)oxides at neutral to alkaline pH (e.g., Koretsky et al., 2007).

Elements that were fixed to these (hydr)oxides, such as arsenic (As) and heavy metals, are often transformed to a more mobile and plant-available form upon inundation of the soil. On the other hand, these elements can also be immobilized due to co-precipitation with or adsorption to Fe and Mn (hydr)oxides under oxic conditions.

Figure 9.3 shows a concept of a model with selected abiotic and biotic redox processes involving Fe that have an impact on the fate of As (Borch et al., 2010). The mobility, dynamics, bioavailability, toxicity, and environmental fate of As are controlled by biogeochemical transformations that can form or destroy As-bearing carrier phases, or modify the redox state and chemical speciation of As (Figure 9.3; Borch et al., 2010).

As illustrated in Figure 9.3, abiotic processes include (a) oxidation of As(III) and reduction of As(V) (dissolved or surface-bound) by semiquinones and hydroquinones, respectively (as present, for example, in humic substances) and (b) formation of reactive Fe(II/III) mineral phases (e.g., by microbially produced Fe^{2+}). The latter can lead to As(V) reduction or As(III) oxidation (Figure 9.3, abiotic processes, c). Biotic processes include (a) microbial oxidation of Fe(II) leading to Fe(III) mineral formation and sorption or co-precipitation of As(III) or As(V) to/with these Fe(III) minerals (oxidation of Fe(II) can also occur abiotically), (b) microbial reduction of As-loaded Fe(III) minerals leading to release of As(III) or As(V) into solution, (c) formation of secondary Fe minerals by reaction of microbially released Fe^{2+} with remaining Fe(III) minerals leading to sorption/co-precipitation of As, and finally (d) direct microbial oxidation of As(III) or reduction of As(V) for either dissimilation or detoxification.

The elucidation of the release mechanisms of As remains the subject of current research, but a general behavior is often observed. The absence of As in the dissolved phase is related to the presence of poorly soluble iron (hydr) oxides that are able to sorb arsenite (As(III)) and arsenate (As(V)) (Dixit and Hering, 2003). Conversely, Islam et al. (2004) concluded that the simultaneous

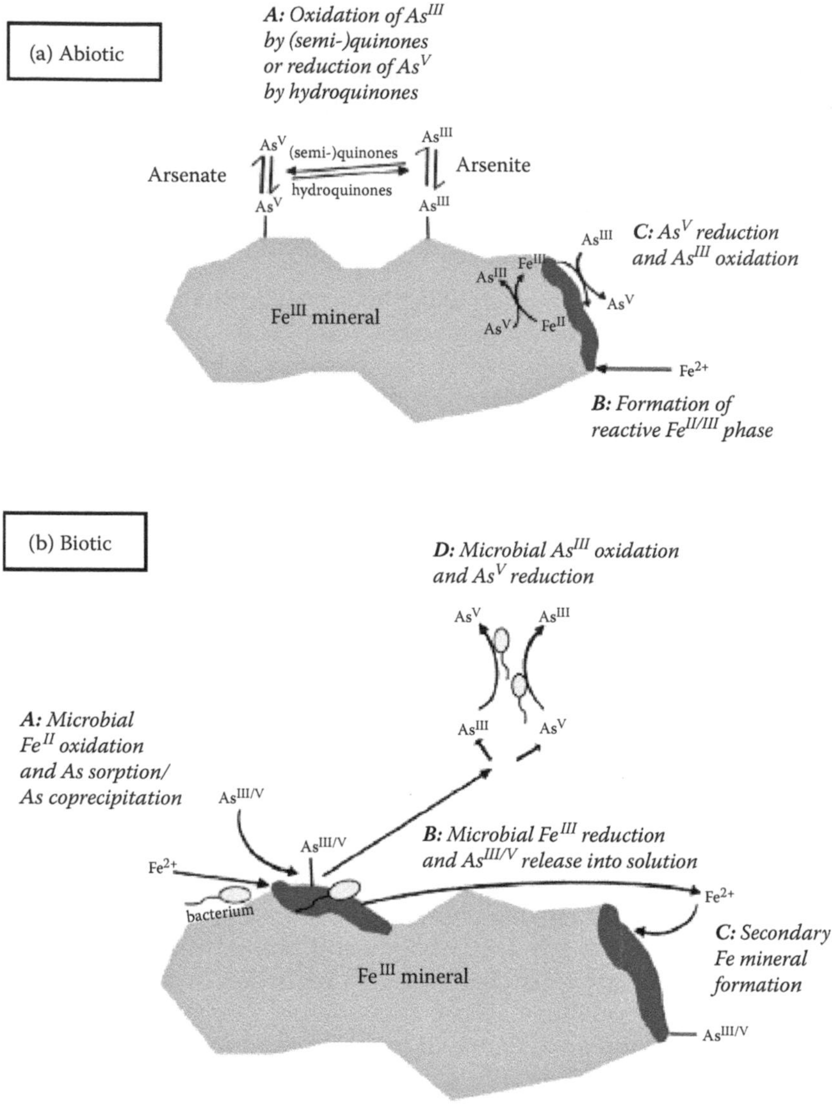

FIGURE 9.3
Concept of a model with selected (a) abiotic and (b) biotic redox processes that influence the fate of arsenic in the environment. (*Source:* Reproduced from Borch, T. et al. 2010. *Environ. Sci. Technol.* 44: 15–23. With permission from Elsevier.)

presence of highly dissolved As and Fe(II) concentrations in anoxic zones caused by a reductive dissolution of As-rich Fe(III) (hydr)oxides mobilizes geogenic As.

Tufano and Fendorf (2008) have found that a direct reduction of Fe(III) by microorganisms can cause As sequestration by sorption of As onto secondary

Fe minerals. A complete depletion of sorbent phases with subsequent As mobilization under reducing conditions can occur in Fe(III) hydroxide-rich soils or sediments (Borch et al., 2010). In general, the mobility of As seems to be controlled by a fine balance between the biogeochemical redox transformations of Fe(III) and Fe(II) (Borch et al., 2010).

Next to Fe (hydr)oxides, Mn (hydr)oxides are also important binding agents for heavy metals (Du Laing et al., 2009). For instance, Ni was reported to primarily co-precipitate with Mn oxides in salt marsh sediments from the Western Scheldt upon oxidation (Zwolsman, Berger, and Van Eck, 1993). Guo, DeLaune, and Patrick (1997) reported an increasing affinity between both Fe and Mn oxides and As, Cd, Cr, and Zn with increasing redox potential.

Sulfur

Sulfur (S) plays an important role in the biogeochemistry of wetlands (Reddy and DeLaune, 2008). The most common form of soluble S in soils is SO_4^{2-} (Neue, 1991). Sulfur occurs in both inorganic and organic forms. According to Reddy and DeLaune (2008), three major categories can be distinguished: (1) oxidized inorganic sulfur (sulfate, sulfite, and thiosulfate), (2) reduced inorganic sulfur (elemental sulfur and sulfide), and (3) gaseous sulfur compounds (SO_2, H_2S, dimethyl sulfoxidase, and dimethylsulfoxide (DMSO)). Reactions of S are closely associated with organic matter transformations and the activity of microorganisms. Upon flooding, sulfates are reduced to sulfides, and proteins are dissimilated after hydrolysis to hydrogen sulfide (H_2S), mercaptans, disulfide, NH_3, and fatty acids (Neue, 1991). Under variable redox conditions, S species can be used as an electron acceptors or donators, inducing speciation changes in surface waters, sediments, floodplains, and wetland soils. These changes are often referred to as "sulfur cycling." Sulfur cycling has been extensively studied in coastal and marine ecosystems.

An important transformation within this sulfur cycle is the formation of sulfides by reduction of sulfates under anaerobic conditions, and the reverse—the oxidation of sulfides to sulfate in presence of oxygen. Sulfate reduction is catalyzed by microbial communities (Burkhardt et al., 2010) at low redox potentials, below –150 mV (Gambrell, DeLaune, and Patrick, 1991a) or –220 mV according to Ross (1989) and result in the formation of sulfides:

$$SO_4^{2-} + 8e^- + 8H^+ \leftrightarrow S^{2-} + 4H_2O$$

Elemental sulfur can be considered an intermediary metabolite in the formation of sulfides.

Iron is often involved into the generation of sulfide precipitates. Iron oxides and oxihydroxides in the layers in which sulfate reduction occurs lead to the

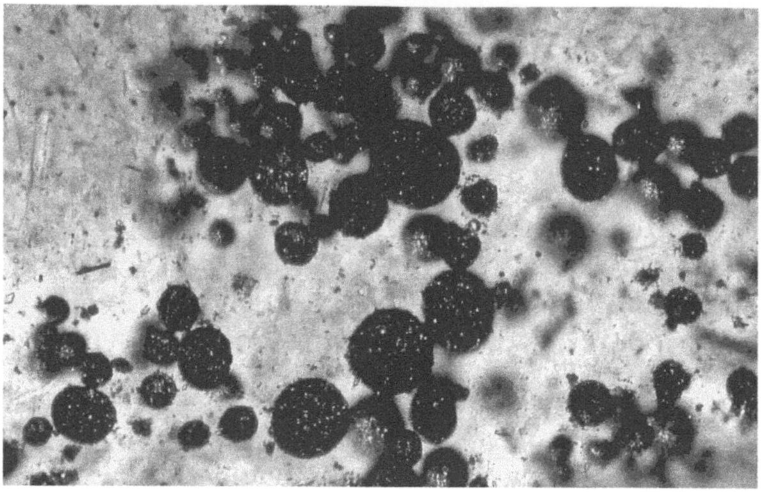

FIGURE 9.4
Sulfides in pyrite form in an anaerobic zone of a floodplain soil (photo taken after using the thin-ground technique; length of the picture: 300 µm). (*Source:* From Rinklebe, J., Ehrmann, O., and Neue, H.U. 2001. Bodenmikromorphologische Studien von fluviatilen Schichtungen, von Pyriten sowie der Verkittung von Quarzen mit Eisenoxiden in einem Gley aus Auensand über tiefem Auenschluffton. UFZ-Bericht. Nr. 8/ 2001. Hrsg.: Scholz, M., Stab, S., Henle, K., pp. 154–155. With permission.)

release of ferrous Fe. Subsequently, there are reactions with the dissolved H_2S to produce amorphous FeS and/or crystallized FeS (such as mackinawite and greigite) in the solid phase; the latter are considered precursors for pyrite formation (FeS_2) (Billon et al., 2001), which is the most ubiquitous sulfide mineral in sediments. Accordingly, sulfides, such as pyrites, can be found in the anaerobic zones of floodplain soils, which can be seen in microscopic pictures (Figure 9.4; Rinklebe, Ehrmann, and Neue, 2001).

Du Laing et al. (2007) found that the sulfates needed for sulfide formation in estuarine tidal marsh soils mainly originate from the flooding water.

The sulfur cycle can have considerable effects on the mobility of metals. During the sulfidization process, some trace metals that are liberated from sedimentary components such as organic matter and metal oxides can adsorb onto or co-precipitate with FeS minerals or can precipitate directly as discrete solid phases if the sedimentary medium becomes (super)saturated with respect to sulfides, according to: $M^{2+} + FeS(s) \leftrightarrow MS(s) + Fe^{2+}$, where, for example, M = Cd, Cu, Ni, Pb, Zn, Co, or Hg. Iron monosulfide (FeS) is often considered the most reactive of the sulfide phases although pyretic sulfide phases (FeS_2) can also be reactive toward trace metals (Simpson, Rosner, and Ellis, 2000; Billon et al., 2001). However, especially pyritization is considered a key process as pyrites are considered to be stable over a broad range of redox conditions. Therefore, pyritization may be decisive when determining whether an element is retained in sediments or floodplain soils or remobilized into the water column and subsequently transported into open waters (Scholz and Neumann, 2007).

Guo, DeLaune, and Patrick (1997) have observed increasing association of Cd and Zn with sulfides, next to carbonates and insoluble large molecular humic substances with decreasing redox potentials, whereas Cr association was restricted to insoluble large molecular humic substances. The influence of the sulfide concentration on the speciation and solubility of Zn, Cu, and Cd in a solution with 5,000 mg Cl L^{-1} and an alkalinity of 50 mmol.L^{-1} was calculated by Salomons et al. (1987). In this experiment the effects of chlorides on the solubility of Cd, Cu, and Zn and of hydroxides and (bi)carbonates on the solubility of Zn at low sulfide concentrations ($<10^{-3}$ μg.L^{-1} for Cd, $<10^{-2}$ μg.L^{-1} for Cu, and $<10^{-1}$ μg.L^{-1} for Zn) were studied. These results demonstrate the stability of metal sulfide precipitates formed at higher sulfide concentrations. Nevertheless, at higher sulfide concentrations ($>10^{5}$ μg.L^{-1} for Cd, $>10^{3}$ μg.L^{-1} for Cu, and >10 μg.L^{-1} for Zn), the formation of soluble bi- and polysulfide complexes increased the total metal concentrations in solution. In pore waters of anoxic environments, (bi)sulfide ions might compete with metals complexed by dissolved organic matter. In pore water concentrations of sorption experiments, no differences could be detected for Cr and As, whether sulfide was present or not. In both cases, dissolved concentrations could be described by adsorption processes only. These results give a strong indication that Cr and As do not form solid sulfide compounds (Salomons et al., 1987). Du Laing et al. (2008a) have reported that the depth at which sulfide precipitation significantly contributed to metal accumulation in the solid phase of sediments of intertidal reed beds along the Scheldt estuary largely depends on the sampling location and sediment characteristics, and varied from less than 5 cm below the surface in clayey, organic sediments to more than 1 m in sandy sediments.

It should be emphasized that very low sulfide concentrations in surface water or pore water of floodplain soils and sediments do not prove the absence of sulfide precipitates in the solid sediment fraction as sulfides of Cd, Pb, and Zn are highly insoluble. The formation of small sulfide amounts that are below the detection limit of common titration-based analysis procedures can still explain the very low levels of Cd, Cu, and Zn amounts in solution in the presence of nonprecipitated Fe and Mn, as calculated by Du Laing et al. (2007). This was confirmed by Salomons et al. (1987), who reported the effects of some ligands (e.g., chlorides, hydroxides, and bicarbonates) on metal solubility to disappear already from very low sulfide concentrations in solution (e.g., <0.1 μg.L^{-1}).

When sulfide-containing flooded soils are oxidized upon contact with water containing dissolved oxygen or with air, sulfides will tend to be gradually oxidized to sulfate. The oxidation of metal sulfides might be caused by chemical reactions at high pH values, whereas it can be microbiologically mediated at lower pH values (Salomons et al., 1987). Once the reactive sulfide phase has been exhausted in the absence of other binding phases, the metals are expected to appear in the pore waters in order of decreasing metal sulfide solubility, that is, Ni, Zn, Pb, Cd, and then Cu (Di Toro et al., 1992). This results in an increased mobility and bioavailability of metals retained within the system (Du Laing et al., 2009). Caetano, Madureira, and Vale (2002) have reported that metals that are co-precipitated with or adsorbed to FeS and MnS can be rapidly oxidized, due to their relative solubility under oxic conditions. Nevertheless, CuS and pyrite, as more stable sulfide-bound metals, are unlikely to be oxidized in the short term due to their slower oxidation kinetics (Caetano, Madureira, and Vale, 2002).

Next to the direct release of metals precipitated as sulfides, the pH can decrease during sulfide oxidation due to proton release during the following reactions (presented for FeS):

$$FeS + 2\,O_2 \rightarrow Fe^{2+} + SO_4^{2-}$$

$$4\,Fe^{2+} + O_2 + 10\,H_2O \rightarrow 4\,Fe(OH)_3 + 8\,H^+$$

For each molecule of FeS oxidized, several protons are produced. In unbuffered sediments, these reactions will cause a drop in pH, which will limit the transfer of trace metals from the water phase to the solid phase and/or cause desorption from the soils or sediments, as discussed in the next section. Following sulfide oxidation, released Fe and Mn can be rapidly reprecipitated and deposited as insoluble oxides or hydroxides, to which released metals can become adsorbed at varying rates and extents (Caetano, Madureira, and Vale, 2002; Eggleton and Thomas, 2004).

Carbonates and pH

Changes in pH can be induced by varying redox conditions that can affect the mobility of metals indirectly (e.g., Kumpiene et al., 2009). During reduction, protons are consumed, whereas during oxidation, acidification tends to occur (Yu et al., 2007; Du Laing et al., 2009). As a consequence, these pH changes might affect the mobility and reaction kinetics of metals. Generally, a higher proton activity reduces the negative surface charge of organic matter, clay particles, and Fe and Al oxides, and increases the solubility of precipitates, such as sulfides (Du Laing et al., 2009). Several authors have proved that pH-induced mobilization is high for Cd, Zn, and Ni (e.g., Miller et al., 2010). Chuan, Shu, and Liu (1996) detected that acidic conditions led to high solubilities of Cd and Zn. Their investigations indicate that metal solubility is significantly higher at pH 5 in comparison with pH 8 and increases drastically at pH 3.3. Lair et al. (2008) found that the amounts of desorbable and weakly bound Cd were higher in floodplain soils with low pH compared to floodplain soils with higher pH. Copper is reported to be mobile at low pH (Kumpiene et al., 2009) with a mobility threshold of pH 5.5 (Graf et al., 2007). It is also proven that Ni becomes mobile under acidic conditions (Carbonell et al., 1999). A pH drop prevents the transfer of many metals to the sediment and/or causes desorption (e.g., Salomons et al., 1987; Gambrell et al., 1991b; Calmano, Hong, and Förstner, 1993).

The presence of carbonates in calcareous floodplain soils or sediments constitutes an effective buffer against a decrease in pH. Furthermore, carbonates may also directly precipitate metals (e.g., Gambrell, 1994; Guo, DeLaune, and Patrick, 1997; Charlatchka and Cambier, 2000). These carbonates can be (bio) geochemically formed and deposited within frequently flooded soils and sediments. On the other hand, decalcification can also occur and a complete decalcification may result in acidification of the pore water. At a certain pH in the pore water, metals can become rapidly mobilized. Because the processes of decalcification and subsequent mobilization of contaminants might be fairly slow in calcareous soils, decalcification of hydric soils is regarded as a chemical time bomb (Van den Berg and Loch, 2000). Du Laing et al. (2007) demonstrate that, in addition to a pH drop, enhanced Ca concentrations in the pore water during partial decalcification can cause an increased release of metals in a calcareous soil layer of a contaminated overbank sedimentation zone.

Oxidation of Fe sulfides under aerobic conditions and increased CO_2 pressure in soils during waterlogging are reasons for decalcification of the top layer of frequently flooded soils. The production of CO_2 can be caused by (1) organic matter breakdown during aerobic conditions, (2) methanogenesis, and (3) root respiration. The diffusion is hindered by the escape of CO_2 to the water column and the atmosphere. Thus, the decalcification processes should be more intensive in anaerobic soils (or layers) than in aerated soils. Increased CO_2 pressure in soils results in calcium carbonate dissolution according to

$$CaCO_3 + CO_2 + H_2O \rightarrow Ca^{2+} + 2\,HCO_3^-$$

The vegetation is able to stimulate decalcification in soils, particularly in the rhizosphere. Living roots and microorganisms are sources of CO_2, and dead roots are a source of easily decomposable organic matter. Oxidation of sulfides may also be an important decalcification process in soils subject to periodic oxidation and reduction due to the release of protons, as discussed in the previous section. These protons react with calcium carbonate as follows:

$$CaCO_3 + H^+ \leftrightarrow Ca^{2+} + HCO_3^-$$

The intensity of decalcification due to sulfide oxidation varies with the amount of sulfide formed in the floodplain soil or sediment, and thus on the duration of waterlogging. Van den Berg and Loch (2000) have demonstrated that, depending on the site-specific hydrological conditions, approximately 0.1% to 0.3% of calcium carbonate may be dissolved per year by a combination of the two processes mentioned above.

Adsorption and Desorption

Ion exchange is the interchange between an ion in solution and another ion in the boundary layer between the solution and a charged soil surface (Sparks, 2003). In soil chemistry, it has been well established for a long time that the major sources of cation exchange in soils are clay minerals, organic matter, and amorphous minerals. According to Evans (1989), electropositive charged elements can be bound to negatively charged surfaces of organic matter, clay particles, as well as Fe and Al oxides, which determine the cation exchange capacity (CEC). The CEC is the maximum adsorption of readily exchangeable ions on soil particle surfaces (Sparks, 2003). In general, a high CEC reduces the mobility of metals and increases metal cation retention. Thus, significant correlations between CEC, clay or organic matter contents, and metal contents have been frequently reported (e.g., Rinklebe et al., 2000).

The sorption mechanism, as a simple assumption, postulates a single stage in which dissolved metals in the solution attain rapid equilibrium with weak binding sites on the surfaces of particles. Additionally, slower stages have also been reported in which the metals migrate into the pores and/or undergo solid-state reactions to higher energy binding sites (Figure 9.5). The time constants of the latter reactions are longer. This indicates that if the supplementary stages are accounted for in estuarine floodplain soils and sediments, it is unlikely that a dissolved constituent will be in equilibrium with the particulate phase during typical estuarine flushing times (Millward and Liu, 2003).

Limousin et al. (2007) have reviewed physical bases, modeling, and measurement of sorption isotherms. They have provided a scheme (Figure 9.6) in which they illustrate different sorption mechanisms of metal ions (Me) on clay minerals by (a) adsorption by outer-sphere surface complexes on exchange sites located on basal planes (hydrated metal) and as inner-sphere surface complexes on the edges (dehydrated metal); (b) inclusion of the metal into crystal structure by co-precipitation; or (c) precipitation of a new solid phase.

An adsorbed compound can react with the solid by (i) slow diffusion inside the solid, or (ii) inclusion of the metal into crystal structure by coprecipitation (Figure 9.6b), or (iii) crystalization of a new solid phase (Figure 9.6c). The duration of desorption is increased by the time needed to reverse these reactions (Limousin et al., 2007). This highlights the importance of

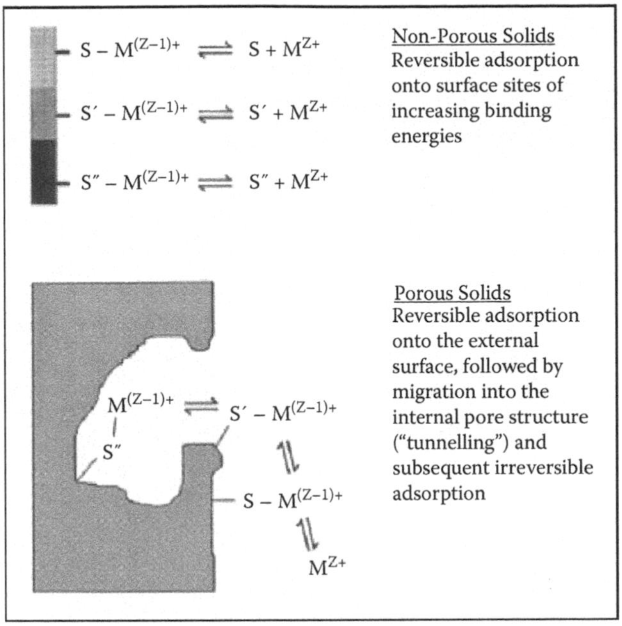

FIGURE 9.5
Sorption mechanisms for uptake of dissolved metals by nonporous and porous solids. S, S′, and S″ are binding sites of increasing energy and M^{z+} is a dissolved metal. (*Source:* From Millward, G.E., and Liu, Y.P. 2003. *Sci. Total Environ.* 314/316: 613–623. With permission from Elsevier.)

knowing the history of a pollution and taking into account the "aging" effect (Strawn and Sparks, 1999), which might be particularly important in frequently flooded soils because their contaminants can originate from different periods of time.

Salinity

Tidal variations in estuaries result in varying salinities of the river water and pore water of the floodplain soils. Seawater of high ionic strength (0.6 mol. L^{-1}) is diluted by river water (ionic strength 1–4 × 10^{-3} mol.L^{-1}), leading to axial salinity gradients and the formation of the fresh water–seawater interface (Millward, 1995). When negatively charged clay particles move from freshwater to salt water, free cations neutralize the negatively charged surfaces, allowing molecular attractive forces to dominate when the particles are brought close enough. They flocculate, that is, they attach to one another, and their settling velocity increases, leading to increased deposition of sedi-

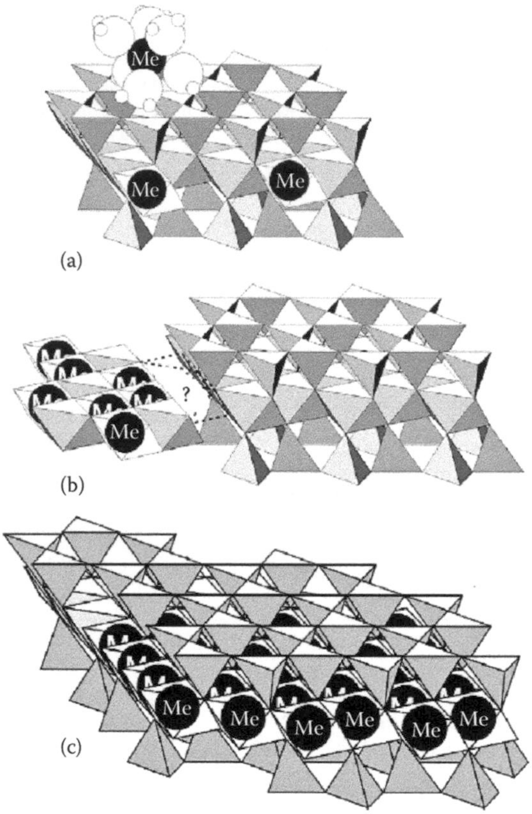

FIGURE 9.6
Illustration of different sorption mechanisms of metal ions (Me) on clay minerals by (a) adsorption by outer-sphere surface complexes on exchange sites located on basal planes (hydrated metal) and as inner-sphere surface complexes on the edges (dehydrated metal); (b) inclusion of the metal into crystal structure by co-precipitation; or (c) precipitation of a new solid phase. (*Source:* Reproduced from Limousin, G. et al. 2007. *Appl. Geochem.* 22: 249–275. With permission from Elsevier.)

ments in the floodplains. Because heavy metals are often strongly sorbed to particles, they will also tend to accumulate in floodplains.

In addition to salinity-induced flocculation of metal-containing particles, salinity can also affect metal fate in different other ways. Due to increases in chloride concentrations when inland fresh river water mixes with seawater, metals may mobilize from sediments as soluble chloride complexes (Hahne and Kroontje, 1973). This is especially the case for Cd as the stability and solubility of cadmium chloride complexes are relatively high and the affinity for sorption to the solid phase is low (Comans and Van Dijk, 1988). Upon formation of these complexes, the activity of free Cd^{2+} in the solution will decrease and desorption will increase. Only Cd that is rather weakly bound to the solid phase will be desorbed as a result of chloro-complexation (Paalman,

Van der Weijden, and Loch, 1994). An increase in the salinity is, however, also associated with an increase in the concentrations of major elements (Na, K, Ca, Mg), which compete with heavy metals for the sorption sites (Tam and Wong, 1999). The addition of Ca salts results in a higher release of exchangeable metals in the solution compared to the addition of Na salts, which are less competitive for sorption (Khattak, Jarrell, and Page, 1989).

Balls, Laslett, and Price (1994) observed increasing Cd desorption from the particulate phase with increasing salinities in the Forth estuary (Scotland), whereas Gerringa et al. (2001) found that high salinities resulted in a faster release of Cd from CdS during the oxidation of reduced sediments of the Scheldt estuary (Belgium and the Netherlands). This release was found to be stimulated by both the formation of chloride complexes and ion exchange. Du Laing et al. (2008b) reported that such effects of salinity on the Cd mobility and availability only occur in oxidized (that is, sulfide-poor) sediments. This is illustrated in Figure 9.7, which presents the evolution of redox potential, Fe, Mn, and Cd contents in the pore water of a soil flooded with water

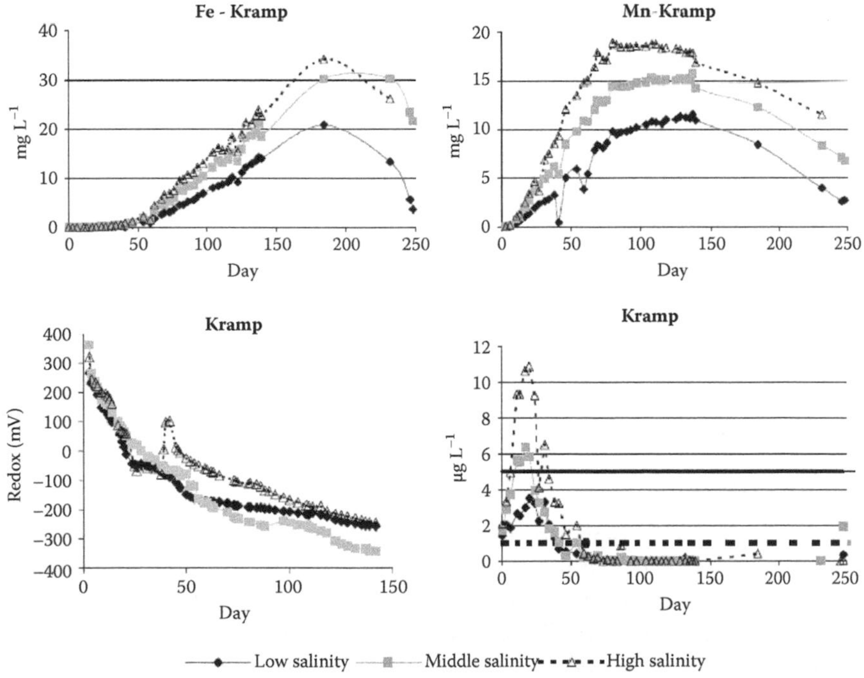

FIGURE 9.7
Evolution of redox potential, Fe, Mn, and Cd contents in the pore water (right) of a flooded soil as a function of time after the beginning of the flood; flooding was conducted with water of three different salinities (low salinity = 0.5 g NaCl L^{-1}, middle salinity = 2.5 g NaCl L^{-1}, and high salinity = 5 g NaCl L^{-1}). (*Source:* From Du Laing, G., De Vos, R., Vandecasteele, B., Lesage, E., Tack, F.M.G., and Verloo, M.G. 2008. *Estuar. Coastal Shelf Sci.* 77: 589–602. With permission from Elsevier.)

of different salinities as a function of time after the beginning of the flood. It can be noted that the soil remains quite oxic during the first 50 days of the experiment: the redox potentials are still above –200 mV, Fe oxide reduction is not yet significant because Fe concentrations in the pore water are still low, but Mn oxide reduction has already started. Under these conditions, Cd concentrations in the pore water are high and significantly affected by salinity. The release of Cd and Mn into the pore water is higher when the salinity increases. However, after day 50, Fe oxide reduction is also significant and Fe release is also promoted by elevated salinities. However, at that time the redox potential also drops below –200 mV, allowing the formation of sulfides. These sulfides seem to precipitate available Cd from the pore water, thus masking the effect of salinity. After day 100, Mn concentrations also decrease, and later on Fe concentrations also start to decrease, which may be related to the fact that sufficient amounts of sulfides are formed at that time to precipitate Fe and Mn and counteract their release into the pore water.

Gambrell et al. (1991b) also studied salinity effects during the oxidation of reduced metal-polluted brackish marsh sediments. Soluble Cd, Cr, and Cu were found to enhance with increasing salinity, whereas the salinity did not significantly affect Ni and Pb mobility. Greger, Kautsky, and Sandberg (1995) have studied the effect of salinity on Cd uptake by the submerged macrophyte *Potamogeton pectinatus* from both sediments and water. Cd uptake from the water was found to decrease when salinity increased up to 10%. However, Cd uptake increased with increasing salinity in the presence of sediments. This clearly illustrates that increasing salinity promotes Cd desorption from the sediments and floodplain soils, and hence increases Cd concentrations in the water column and Cd uptake by organisms. It, however, also stimulates formation of less bioavailable Cd chloride complexes in the water column.

Organic Matter

Soils in recent flood areas are subject to the dynamics of erosion and sedimentation as well as transformation and translocation of matter, which result in permanent influence and alteration, such as accumulation of organic material (Rinklebe et al., 2007). A long duration of flooding with decelerated flow rates (including stagnant water) results in sedimentation of mainly fine-grain sediments as well as considerable amounts of organic matter (Rinklebe, Franke, and Neue, 2007). Additionally, decaying plant material provides litter that reaches floodplain soils and accumulates. The organic material will contribute to the binding of metals by adsorption, complexation, and chelation (Alvim Ferraz and Lourenço, 2000; Du Laing et al., 2009). At the same time, dissolved organic ligands, such as low to medium molecular weight carboxylic acids, amino acids, and fulvic acids, can form soluble metal complexes.

Thus, the net effect of the presence of organic matter can be either a decrease or an increase in metal mobility (Du Laing et al., 2009).

In general, the role of organic material can be considered an important factor in determining the mobility of metals mainly due to the effects of adsorption, complexation, and chelation (e.g., Grybos et al., 2007). Organic matter has a high capacity to complex and adsorb cations due to the presence of many negatively charged groups (Laveuf and Cornu, 2009).

Dissolved organic carbon (DOC) is the most mobile soil organic fraction. DOC is involved in the co-transport of metals through physical and chemical binding and can have a significant impact on the bioavailability of metals (Zsolnay, 1996). Therefore, DOC in soils plays an important role in the transport of pollutants as well as in the biogeochemistry of metals in soils (Kalbitz et al., 2000). Complexation with dissolved organic matter is important for Cu because it reveals a high affinity for organic ligands (Du Laing et al., 2009). Charlatchka and Cambier (2000) found that Pb was complexed by organic acids in a flooded soil. Conversely, high molecular weight organic matter compounds in the solid soil phase can reduce the metal availability (Gambrell, 1994).

Enhanced mobilization of metals as dissolved organic complexes was observed for Pb, Cu, and Zn (Alvim Ferraz and Lourenço, 2000) as well as for As, Cr, Cu, and Hg (Kalbitz and Wennrich, 1998). Nickel can also be mobilized by dissolved organic material (Koretsky et al., 2007; Antic-Mladenovic, et al., in revision). In contrast, Zn is rarely associated with DOC (Kalbitz and Wennrich, 1998; Beesley et al., 2010). Regarding Cd, it is interesting that some authors have observed a mobilization of Cd by dissolved organic material (Koretsky et al., 2007) while others have not (Kalbitz and Wennrich, 1998). Frohne et al. (in press) could also not detect a relationship between Cd and DOC under changing redox conditions; however, in this experiment, DOC was at maximum due to the addition of glucose as an additional source of organic carbon to decrease the redox potential.

DOC was found to be mobilized under reducing conditions (e.g., Yu et al. 2007; Grybos et al., 2009; Antic-Mladenovic et al., in revision). Figure 9.8 shows an anaerobic experiment performed by Grybos et al. (2009) without pH buffer. Herein, E_h decreased rapidly from 500 mV to 100 mV during the first 100 hours; while the pH increased from 5.5 to ca. 7.4 during the incubation (Figure 9.8a). DOC was strongly mobilized into solution during the incubation. DOC increased after about 40 hours of anaerobic incubation. This was more rapid from 40 to about 140 hours, slower between 140 and 250 hours, and very slow after 250 hours (Grybos et al., 2009). An increase in NO_3^- as well as the total concentrations of Fe and Mn during the anaerobic incubation was simultaneously observed (Figure 9.8b). Enhanced DOC concentrations were also found at slightly low pH conditions around pH 5.5 (Grybos et al., 2009; Frohne et al., in press).

Moreover, microorganisms in the rhizosphere of wetland plants can accumulate metals (Scholes et al., 1999). Decho (2000) reported a considerable role

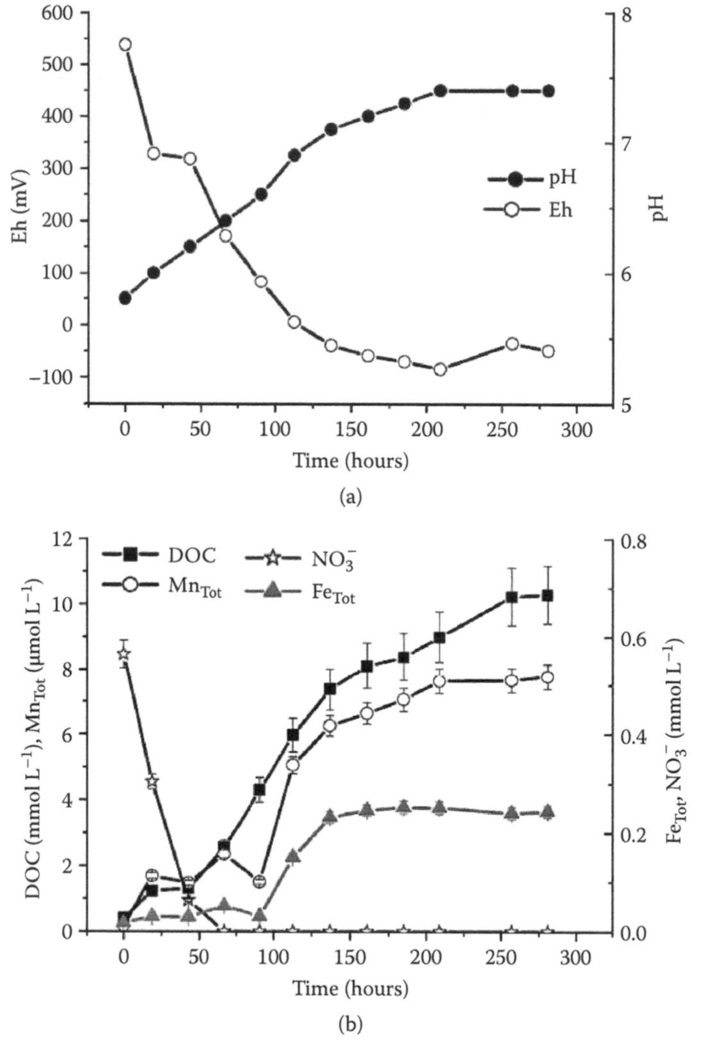

FIGURE 9.8
Time-variation of E_h and pH (a), and dissolved organic carbon (DOC), Fe_{Tot}, Mn_{Tot} and NO_3-concentrations (b) for anaerobic experiment without pH buffer. (Reproduced from Grybos, et al., 2009, with permission from Elsevier.)

of extracellular microbial polymeric secretions (microbial biofilms) in intertidal systems, both in the binding and concentration of metal contaminants, and also in the trophic transfer of metals to the food web. These exopolymers were particularly produced in large amounts by Cu-resistant bacteria in the rhizosphere of wetland plants, which indicates that they might be involved in the detoxification of Cu (Kunito et al., 2001). Organic matter has also an indirect effect on the behavior of metals because it serves as a food source for microorganisms. Microorganisms, in turn, catalyze a series of redox reactions in the presence of electron acceptors. These factors have an impact on the mobility and availability of metals in frequently flooded soils (Du Laing et al., 2009).

References

Alewell, C., Paul, S., Lischeid, G., and Storck, F. 2008. Co-regulation of redox processes in freshwater wetlands as a function of organic matter availability. *Sci. Total Environ.* 404, 335–342.

Alvim Ferraz, M.C.M., and Lourenço, J.C.N. 2000. The influence of organic matter content of contaminated soils on the leaching rate of heavy metals. *Environ. Prog.* 19, 53–58.

Antic-Mladenovic, S., Rinklebe, J., Frohne, T., Stärk, H.J., Wennrich, R., Tomić, Z., and Licina, V. In revision. Impact of controlled redox conditions on nickel in a serpentine soil. *J. Soil Sediments.*

Balls, P.W., Laslett, R.E., Price, N.B. 1994. Nutrient and trace metal distributions over a complete semi-diurnal tidal cycle in the Forth estuary, Scotland. *Neth. J. Sea Res.* 33, 1–17.

Beesley, L., Moreno-Jimenez, E., Clemente, R., Lepp, N., and Dickonsin, N. 2010. Mobility of arsenic, cadmium and zinc in a multi-element contaminated soil profile assessed by in-situ soil pore water sampling, column leaching and sequential extraction. *Environ. Pollut.* 158, 155–160.

Billon, G., Ouddane, B., Laureyns, J., and Boughriet, A. 2001. Chemistry of metal sulfides in anoxic sediments. *Phys. Chem. Chem. Phys.* 3, 3586–3592.

Borch, T., Kretzschmar, R., Kapper, A., Van Cappellen, P., Ginder-Vogel, M., Voegelin, A., and Campbell, K. 2010. Biogeochemical redox processes and their impact on contaminant dynamics. *Environ. Sci. Technol.* 44, 15–23.

Brümmer, G. 1974. Redoxpotentiale und Redoxprozesse von Mangan-, Eisen- und Schwefelverbindungen in hydromorphen Böden und Sedimenten. *Geoderma* 12, 207–222.

Burkhardt, E.-M., Akob, D.M., Bischoff, S., Sitte, J., Kostka, J.E., Banerjee, D., Scheinost, A.C., andKüsel, K. 2010. Impact of biostimulated redox processes on metal dynamics in an iron-rich creek soil of a former uranium mining area. *Environ. Sci. Technol.* 44, 177–183.

Caetano, M., Madureira, M.J., and Vale, C. 2002. Metal remobilisation during resuspension of anoxic contaminated sediment: short-term laboratory study. *Water Air Soil Pollut.* 143, 23–40.

Calmano, W., Hong, J., and Förstner, U. 1993. Binding and mobilisation of heavy metals in contaminated sediments affected by pH and redox potential. *Water Sci. Technol.* 28, 223–235.

Cao, X., Chen, Y., Wang, X., and Deng, X. 2001. Effects of redox potential and pH value on the release of rare earth elements from soil. *Chemosphere* 44, 655–661.

Carbonell, A.A., Porthouse, J.D., Mulbah, C.K., DeLaune, R.D., and Patrick, W.H. 1999. Metal solubility in phosphogypsum-amended sediment under controlled pH and redox conditions. *J. Environ. Qual.* 28, 232–242.

Chadwick, O.A., and Chorover, J. 2001. The chemistry of pedogenic thresholds. *Geoderma* 100, 321–353.

Chuan, M.C., Shu, G.Y., and Liu, J.C. 1996. Solubility of heavy metals in a contaminated soil: Effects of redox potential and pH. *Water Air Soil Pollut.* 90, 543–556.

Charlatchka, R., and Cambier, P. 2000. Influence of reducing conditions on solubility of trace metals in contaminated soils. *Water Air Soil Pollut.* 118, 143–167.

Comans, R.N.J., and Van Dijk, C.P.J. 1988. Role of complexation processes in cadmium mobilization during estuarine mixing. *Nature* 336, 151–154.

Decho, A.W. 2000. Microbial biofilms in intertidal systems: An overview. *Cont. Shelf Res.* 20, 1257–1273.

Di Toro, D.M., Mahony, J.D., Hansen, D.J., Scott, K.J., Carlson, A.R., and Ankley, G.T. 1992. Acid volatile sulfide predicts the acute toxicity of cadmium and nickel in sediments. *Environ. Sci. Technol.* 26, 96–101.

Dixit, S., and Hering, J.G. 2003. Comparison of arsenic(V) and arsenic(III) sorption onto iron oxide minerals: Implications for arsenic mobility. *Environ. Sci. Technol.* 37, 4182–4189.

Du Laing, G., Rinklebe, J., Vandecasteele, B., and Tack, F.M.G. 2009. Trace metal behaviour in estuarine and riverine floodplain soils and sediments: a review. *Sci. Total Environ.* 407, 3972–3985.

Du Laing, G., De Meyer, B., Meers, E., Lesage, E., Van de Moortel, A., Tack, F.M.G., and Verloo, M.G. 2008a. Metal accumulation in intertidal marshes: Role of sulphide precipitation. *Wetlands* 28, 735–746.

Du Laing, G., De Vos, R., Vandecasteele, B., Lesage, E., Tack, F.M.G., and Verloo, M.G. 2008b. Effect of salinity on heavy metal mobility and availability in intertidal sediments of the Scheldt estuary. *Estuar. Coast. Shelf Sci.* 77, 589–602.

Du Laing, G., Vanthuyne, D.R.J., Vandecasteele, B., Tack, F.M.G., and Verloo, M.G. 2007. Influence of hydrological regime on pore water metal concentrations in a contaminated sediment-derived soil. *Environ. Pollut.* 147, 615–625.

Eggleton, J., and Thomas, K.V. 2004. A review of factors affecting the release and bioavailability of contaminants during sediment disturbance events. *Environ. Int.* 30, 973–980.

Evans, L.J. 1989. Chemistry of metal retention by soils. *Environ. Sci. Technol.* 23, 1046–1056.

Fenchel, T., King, G.M., and Blackburn, T.H. 1998. *Bacterial biogeochemistry: The ecophysiology of mineral cycling. 2nd ed.* Academic Press, San Diego.

Frohne, T., Rinklebe, J., Diaz-Bone, R., and Du Laing, G. Controlled variation of redox conditions in a floodplain soil: impact on metal mobilisation and biomethylation of arsenic and antimony. *Geoderma,* in press.

Gambrell, R.P. 1994. Trace and toxic metals in wetlands – A review. *J. Environ. Qual.* 23, 883–891.

Gambrell, R.P., DeLaune, R.D., and Patrick, W.H., Jr. 1991a. Redox processes in soils following oxygen depletion. In Jackson, M.B., Davies, D.D., and Lambers, H., Eds. *Plant life under oxygen deprivation,* SPB Academic Publishing, The Hague, The Netherlands, pp. 101–117.

Gambrell, R.P., Wiesepape, J.B., Patrick, W.H., Jr., and Duff, M.C. 1991b. The effects of pH, redox, and salinity on metal release from a contaminated sediment. *Water Air Soil Pollut.* 57/58, 359–367.

Gerringa, L.J.A., de Baar, H.J.W., Nolting, R.F., and Paucot, H. 2001. The influence of salinity on the solubility of Zn and Cd sulphides in the Scheldt estuary. *J. Sea Res.* 46, 201–211.

Graf, M., Lair, G.J., Zehetner, F., and Gerzabek, M.H. 2007. Geochemical fractions of copper in soil chronosequences of selected European floodplains. *Environ. Pollut.* 148, 788–796.

Greger, M., Kautsky, L., and Sandberg, T. 1995. A tentative model of Cd uptake in *Potamogeton pectinatus* in relation to salinity. *Environ. Exp. Bot.* 35, 215–225.

Grybos, M., Davranche, M., Gruau, G., and Petitjean, P. 2007. Is trace metal release in wetland soils controlled by organic matter mobility or Fe-oxyhydroxides reduction?. *J. Colloid. Interf. Sci.* 314, 490–501.

Grybos, M., Davranche, M., Gruau, G., Petitjean, P., and Pedrot, M. 2009. Increasing pH drives organic matter solubilization from wetland soils under reducing conditions. *Geoderma* 154, 13–19.

Guo, T., DeLaune, R.D., and Patrick, W.H., Jr. 1997. The influence of sediment redox chemistry on chemically active forms of arsenic, cadmium, chromium, and zinc in estuarine sediment. *Environ. Int.* 23, 305–316.

Hahne, H.C.H., and Kroontje, W. 1973. Significance of the pH and chloride concentration on the behavior of heavy metal pollutants Hg(II), Cd(II), Zn(II), Pb(II). *J. Environ. Qual.* 2, 444–450.

Harter, R.D. 1983. Effect of soil pH on adsorption of lead, copper, zinc, and nickel. *Soil Sci. Soc. Am. J.* 47, 47–51.

Islam, F., Gault, A.G., Boothman, C., Polya, D.A., Charnock, J.M., Chatterjee, D., and Lloyd, J.R. 2004. Role of metal-reducing bacteria in arsenic release from Bengal delta sediments. *Nature* 430, 68–71.

Kalbitz, K., Solinger, S., Park, J.H., Michalzik, B., and Matzner, E. 2000. Controls on the dynamics of dissolved organic matter in soils: A review. *Soil Sci.* 165, 277–304.

Kalbitz, K., and Wennrich, R. 1998. Mobilization of heavy metals and arsenic in polluted wetland soils and its dependence on dissolved organic matter. *Sci. Total Environ.* 209, 27–39.

Khattak, R.A., Jarrell, W.M., and Page, A.L. 1989. Mechanism of native manganese release in salt-treated soils. *Soil Sci. Soc. Am. J.* 53, 701–705.

Koretsky, C.M., Haveman, M., Beuving, L., Cuellar, A., Shattuck, T., and Wagner, M. 2007. Spatial variation of redox and trace metal geochemistry in a minerotrophic fen. *Biogeochemistry* 86, 33–62.

Kumpiene, J., Ragnvaldsson, D., Lövgren, L., Tesfalidet, S., Gustavsson, B., Lättström, A., Leffler, P., and Maurice, C. 2009. Impact of water saturation level on arsenic and metal mobility in the Fe-amended soil. *Chemosphere* 74, 206–215.

Kunito, T., Saeki, K., Nagaoka, K., Oyaizu, H., and Matsumoto, S. 2001. Characterization of copper-resistant bacterial community in rhizosphere of highly copper-contaminated soil. *Eur. J. Soil Biol.* 37, 95–102.

Lair, G.J., Graf, M., Zehtner, F., and Gerzabek, M.H. 2008. Distribution of cadmium among geochemical fractions in floodplain soils of progressing development. *Environ. Pollut.* 156, 207–214.

Langer, U., and Rinklebe, J. 2009. Lipid biomarkers for assessment of microbial communities in floodplain soils of the Elbe River (Germany). *Wetlands* 29, 353–362.

Laveuf, C., and Cornu, S. 2009. A review on potentiality of rare earth elements to trace pedogenetic processes. *Geoderma* 154, 1–12.

Limousin, G., Gaudet, J.P., Charlet, L., Szenknect, S., Barthès, V., and Krimissa, M. 2007. Sorption isotherms: A review on physical bases, modeling and measurement. *Appl. Geochem.* 22, 249–275.

Lovley, D.R., and Phillips, E.J.P. 1988a. Manganese inhibition of microbial iron reduction in anaerobic sediments. *Geomicrobiol. J.* 6, 145–155.

Lovley, D.R., and Phillips, E.J.P. 1988b. Novel mode of microbial energy-metabolism — Organic-carbon oxidation coupled to dissimilatory reduction of iron or manganese. *Appl. Environ. Microbiol.* 54, 1472–1480.

Miller, F.S., Kilminster, K.L., Degens, B., and Firns, G.W. 2010. Relationship between metals leached and soil type from potential acid sulphate soils under acidic and neutral conditions in Western Australia. *Water Air Soil Pollut.* 205, 133–147.

Millward, G.E. 1995. Processes affecting trace element speciation in estuaries. *Analyst* 120, 609–614.

Millward, G.E., and Liu, Y.P. 2003. Modelling metal desorption kinetics in estuaries. *Sci. Total Environ.* 314/316, 613–623.

Neue, H.U. 1991. Holistic view of chemistry of flooded soil. In *Soil management for sustainable rice production in the tropics*. International Board for Soil Research and Management. IBSRAM, Monogragh No. 2, pp. 5–32.

Paalman, M.A.A., Van der Weijden, C.H., and Loch, J.P.G. 1994. Sorption of cadmium on suspended matter under estuarine conditions: Competition and complexation with major seawater ions. *Water Air Soil Pollut.* 73, 49–60.

Pardue, J.H., and Patrick, W.H., Jr. 1995. Changes in metal speciation following alternation of sediment redox status. In Allen, H., Ed., *Metal-contaminated aquatic sediments*. Science Publishers, Ann Arbor, MI, pp. 169–185.

Patrick, W.H., Jr., and Jugsujinda, A. 1992. Sequential reduction and oxidation of inorganic nitrogen, manganese, and iron in flooded soil. *Soil Sci. Soc. Am. J.* 56, 1071–1073.

Patrick, W.H., Jr., and Henderson, R.E. 1981. Reduction and reoxidation cycles of manganese and iron in flooded soil and water solution. *Soil Sci. Soc. Am. J.* 45, 855–859.

Paul, E.A., and Clark, F.E. 1996. *Soil microbiology and biochemistry, 2nd ed.* Academic Press, San Diego, CA.

Reddy, K.R., and DeLaune, R.D. 2008. *Biogeochemistry of wetlands: Science and applications*. Taylor & Francis Group, LLC, Boca Raton, FL.

Rinklebe, J., Heinrich, K., Morgenstern, P., Franke, C., and Neue, H.U. 2000. Heavy metal concentrations, distributions and mobilities in wetland soils. In: *Mitteilung Nr. 6 der Bundesanstalt für Gewässerkunde/PG Elbe Ökologie*, Koblenz-Berlin, pp. 227–228.

Rinklebe, J., Ehrmann, O., and Neue, H.U. 2001. Bodenmikromorphologische Studien von fluviatilen Schichtungen, von Pyriten sowie der Verkittung von Quarzen mit Eisenoxiden in einem Gley aus Auensand über tiefem Auenschluffton. UFZ-Bericht. Nr. 8/ 2001. Hrsg.: Scholz, M., Stab, S., Henle, K., pp. 154–155.

Rinklebe, J., Stubbe, A., Staerk, H.J., Wennrich, R., and Neue, H.U. 2005. Factors controlling the dynamics of As, Cd, Zn, Pb in alluvial soils of the Elbe river (Germany). In Lyon, W.G., Hong, J., and Reddy, R.K., Eds. *Proceedings of the First International Conference on Environ. Sci. & Technol.* New Orleans. American Science Press, Vol. 2, pp. 265–270.

Rinklebe, J., Franke, C., and Neue, H.U. 2007. Aggregation of floodplain soils as an instrument for predicting concentrations of nutrients and pollutants. *Geoderma* 141, 210–223.

Roberts, L.C., Hug, S.J., Dittmar, J., Voegelin, A., Saha, G.C., Ali, M.A., Badruzzaman, A.B.M., and Kretzschmar, R. 2007. Spatial distribution and temporal variability of arsenic in irrigated rice fields in Bangladesh. 1. Irrigation water. *Environ. Sci. Technol.* 41, 5960–5966.

Ross, S. 1989. *Soil processes.* Routledge, New York.

Salomons, W., De Rooij, N.M., Kerdijk, H., and Bril, J. 1987. Sediments as a source for contaminants? *Hydrobiologia* 149, 13–30.

Scholes, L.N.L., Shutes, R.B.E., Revitt, D.M., Purchase, D., and Forshaw, M. 1999. The removal of urban pollutants by constructed wetlands during wet weather. *Water Sci. Technol.* 40, 333–340.

Scholz, F., and Neumann, T. 2007. Trace element diagenesis in pyrite-rich sediments of the Achterwasser lagoon, SW Baltic Sea. *Marine Chem.* 107, 516–532.

Simpson, S.L., Rosner, J., and Ellis, J. 2000. Competitive displacement reactions of cadmium, copper, and zinc added to a polluted, sulfidic estuarine sediment. *Environ. Toxicol. Chem.* 19, 1992–1999.

Smith, R.M., and Martell, A.E. 1976. *Critical stability constants, Vol. 4,* Plenum Press, New York.

Sparks, S.L. 2003. *Environmental soil chemistry. Second edition.* Academic Press, San Diego, CA.

Strawn, D.G., and Sparks, D.L. 1999. Sorption kinetics of trace elements in soils and soil materials. In Selim, H.M., and Iskandar, K.I. Eds., *Fate and transport of heavy metals in the vadose zone.* Lewis Publishers, Boca Raton, FL, pp. 1–28.

Tam, N.F.Y., and Wong, Y.S. 1999. Mangrove soils in removing pollutants from municipal wastewater of different salinities. *J. Environ. Qual.* 28, 556–564.

Tufano, K.J., and Fendorf, S. 2008. Confounding impacts of iron reduction on arsenic retention *Environ. Sci. Technol.* 42, 4777–4783.

Vandecasteele, B., Du Laing, G., and Tack, F.M.G. 2007. Effect of submergence-emergence sequence and organic matter or aluminosilicate amendment on metal uptake by woody wetland plant species from contaminated sediments. *Environ. Pollut.* 145, 329–338.

Van den Berg, G.A., and Loch, G.J.P. 2000. Decalcification of soils subject to periodic waterlogging. *Eur. J. Soil Sci.* 51, 27–33.

Van den Berg, G.A., Gustav Loch, J.P., Van der Heijdt, L.M., Zwolsman, J.J.G., and Friese, B.M. 2000. Chapter 21: Geochemical behaviour of trace metals in freshwater sediments. *Trace metals in the environment, Vol. 4,* Elsevier, Amsterdam, p. 517.

Van Griethuysen, C., Luitwieler, M., Joziasse, J., and Koelmans, A.A. 2005. Temporal variation of trace metal geochemistry in floodplain lake sediment subject to dynamic hydrological conditions. *Environ. Pollut.* 137, 281–294.

Yu, K., Böhme, F., Rinklebe, J., Neue, H.-U., and DeLaune, R.D. 2007. Major biogeochemical processes in soils —A microcosm incubation from reducing to oxidizing conditions. *Soil Sci. Soc. Am. J.* 71, 1406–1417.

Zsolnay, A. 1996. Dissolved humus in soil waters. In Piccolo, A. Ed., *Humic substances in terrestrial ecosystems*, Elsevier, Amsterdam, pp. 171–223.

Zwolsman, J.J.G., Berger, G.W., and Van Eck, G.T.M. 1993. Sediment accumulation rates, historical input, post-depositional mobility and retention of major elements and trace metals in salt marsh sediments of the Scheldt estuary SW Netherlands. *Marine Chem.* 44, 73–94.

10

Heavy Metals Forms in Biosolids, Soils, and Biosolid-Amended Soils

Christos D. Tsadilas

CONTENTS

Introduction

Explanations of terms used in the relevant literature will precede the beginning of this chapter. This is considered necessary because for many of these terms there are no clear, scientifically accepted definitions and their use is often inappropriate even when applied in regulations. Starting therefore from the widely used term *heavy metals,* we stress that there is no scientifically precise definition of these elements. This term usually refers to a group of elements with an atomic density greater than 6 $g.cm^{-3}$ (Alloway, 1995). The same term is used to describe metals or metalloids associated with pollution, potential toxicity, or ecological toxicity, although there is no evidence for that (Duffus, 2002). Another widely used term for the same elements is *toxic metals;* this is inappropriate because all metals can become toxic to living organisms when they are present in excess. Another term for the same elements that is gaining wide acceptance is *potentially toxic elements* (PTEs) (Smith, 1996). Even for the metals that are essential for the growth of higher plants, the terms *micronutrients* and *trace elements* are used (Alloway, 1995). In this chapter, basic information regarding the elements contained in treated municipal waste included in European Union (EU) legislation 86/278/EEC will be presented. These elements are the metals cadmium (Cd), lead (Pb), mercury (Hg), zinc (Zn), copper (Cu), and nickel (Ni). This does not mean that other metals cannot become toxic to living organisms; but within this chapter, only those elements included so far in the EU legislation are discussed.

The term *biosolids* is used by the U.S. Environmental Protection Agency (EPA) for sludge whose treatment fulfills the terms of the regulations in 40CFR503 (USEPA, 1994). This term is considered to have evolved from the term *municipal sewage sludge* and is a more gentle and friendly expression for material previously considered useless and harmful waste that should be discarded. Subsequently, for the management of biosolids, different opinions were adopted based on the rationale of reuse. The term *sludge* is a general term used for many different organic and inorganic materials generated by industrial activity or the treatment of domestic sewage. This chapter uses the term *biosolids*, representing treated municipal waste or similar materials contained in the EU Directive 86/278/EEC.

The metals mentioned above are present in soils, biosolids, and their mixtures in various physiochemical forms, allowing for a better assessment of health hazards, toxicity, and plant availability. The physiochemical forms are called *fractions*, *species*, or *phases*, and are usually determined with procedures involving a sequential extraction or "speciation" using a variety of reagents under a wide range of experimental conditions, including storage, preparation, temperature, reagent concentration, sequence of extractants, solid-to-solution ratio, treatment time, and centrifugation and filtration (Sutherland et al., 2000).

The use of biosolids in agriculture is widely practiced all over the world. That is because it is the most economical solution enabling the recycling of nutrients in plants and organic matter. At the same time, biosolids contain substances and elements that are potentially harmful to the environment, including heavy metals (HMs). For this reason, certain rules and restrictions were set up by the EPA in the United States and the EU or by the member states in Europe. Application of biosolids increases the HM concentration of soils. Relative to the effects of long-term biosolid application, an argument has developed including opposite claims. One side claims that soils on which biosolids are added have the ability to immobilize HMs into forms unavailable to plants while the other considers that the mineralization of organic matter of biosolids may release HMs in soluble forms, thus creating an environmental threat. The main forms of HMs in biosolids are soluble, precipitated, coprecipitated with metal oxides, adsorbates, and associated with residues of microorganisms. The distribution of HMs among these forms varies widely, depending on the metal chemical properties and the processing conditions of the biosolids. Methods for the determination of HM forms in biosolids are similar to those applied for soils and involve both single and sequential chemical extractions. Single extraction methods extract soluble, exchangeable, organic, adsorbed, and precipitated forms while sequential extraction methods usually determine exchangeable, adsorbed, organically bound, carbonate bound, and residual bound in clay lattice structure forms. The application of biosolids to soils causes alterations in some soil properties, leading to the redistribution of HMs between various fractions, thus changing their *availability* to plants. The soil properties affected by biosolids application are the pH and redox potential (E_h), organic matter content, ionic strength of the soil solution, concentration of essential elements, and concentration of potentially toxic HMs. Biosolids application to soils generally increases the concentrations of *available* forms of the heavy metals Cd, Cu, Zn, Ni, and Pb extracted. Regarding the HM fractions, it was often found that application of biosolids leads to an increase in the organically bound, carbonate, and residual fractions. Because there is a wide range of procedures for sequential extraction without comparable results, the Community Bureau of Reference (BCR) developed a sequential extraction method to be used as the standard method for determining HM fractions and designated certified reference materials; this is gaining acceptance. The process consists of four steps to determine exchangeable and associated with carbonate phases, the reducible fraction or fraction associated with Fe and Mn oxides, the oxidizable fraction or bound to organic matter, and the residual fraction containing primary and secondary solids occluding metals in their crystalline structures.

Biosolids Management

Municipal waste management has been a concern ever since the organization of human society. Related references have even been made to ancient Greek history (Doxiadis, 1973) and Roman and Byzantine times. In modern times, information discussed the disposal of urban waste in agriculture in Europe, North and South America, and Australia (Epstein, 2003). Later, during the twentieth century, the use of biosolids as fertilizer in agriculture was more systematically utilized (Anderson, 1959). Currently, the use of biosolids in agriculture is widely practiced. In the United States in 1997, 54% of produced sludge, representing approximately 6.9 million tons of dry sludge per year, was applied to soil in liquid or solid form with or without further processing, to agricultural soils and forests, for the improvement of degraded lands, recreational areas, etc. The respective percentage for 1989 was 42%, showing a continuous increase in the amount of biosolids applied in agriculture (Epstein, 2003). In Europe, the percentage of biosolids applied in agriculture is 37%, while in the United

Kingdom this figure is 44% (Smith, 1996). Other management methods applied during the same period in Europe include deposition in landfills (40%), burning (11%), dumping in the sea (6%), depositing in other places (4%), and a very small percentage (2%) for other beneficial uses. The latter refer to the utilization of thermal energy and its chemical composition and are carried out using thermal processes such as pyrolysis, wet oxidation, and gasification (Fytili and Zambaniotou, 2008).

Benefits and Problems of Biosolids Use in Agriculture

The reason the agricultural use of biosolids is the primary alternative method of management is that it is the most economical solution enabling the recycling of nutrients in plants and organic matter, as reported repeatedly in the literature (King and Morris, 1972; Sabrey, Agbim, and Markstrom, 1977; Magdoff and Amadon, 1980; Coker and Carlton-Smith, 1986; Coker et al., 1987; Samaras, Tsadilas, and Stamatiadis, 2009). In Table 10.1 the concentrations of the basic essential for plant growth macro- and micronutrients of sewage sludge are presented.

At the same time, however, biosolids contain substances and elements that are potentially harmful in high concentrations and can cause problems in the environment and in the health of plants, animals, soil microorganisms, and humans (Smith, 1996; Jeyakumar et al., 2010). Among these elements are the HMs, which are discussed in this chapter. In Table 10.3, the usual concentrations of potentially toxic elements in biosolids are given.

TABLE 10.1

Total Concentrations of the Basic Macro- and Micronutrients of Biosolids (% dwt, $mg.kg^{-1}$ dwt)

Nutrient	Range	Mean
N	0.5–17.6	5.0
P	0.5–14.3	3.3
K	0.02–2.64	0.52
Zn	279–27.600	1.144
Cu	69–6.140	589
Mn	55–13.902	376
Fe	2.480–106.812	16.299
B	15–1000	—
Mo	<2–154	5

Source: From Somers, L.E. 1977. *J. Environ. Qual.* 6(2): 225–232.

Note: Values for macronutrients are for anaerobic sludges.

There are a huge number of publications regarding the deposition into the soil of HMs as a result of sewage sludge application and the effects on plants (some examples include Williams et al., 1987; Tsadilas et al., 1995; Walter and Cuevas, 1999; Udom et al., 2004; Shober, Stachouwer, and MacNeal 2006; Su et al., 2008; and Jordan et al., 2008). Vigerust and Selmer-Olsen (1986) published data for a large number of agricultural crops that absorb HMs when grown in soils to which biosolids have been added. From this study it appears that biosolid application significantly increases uptake of HMs from agricultural crops in the order Zn > Cd > Ni > Cu > Pb ≈ Hg ≈ Cr, and in many cases can exceed 250%. In the literature there is also a lot of data on the critical concentrations of potentially toxic elements in plants (MacNicol and Beckett, 1985). By combining similar data with those of Vigerust and Selmer-Olsen (1986) and MacNitol and Beckett (1985), it is possible to check whether the application of biosolids in soil may cause toxicity of HMs in agricultural crops. Take note that the main transport mechanisms of HMs from soil to plants are diffusion and convection (Chaney, 1975, as reported by Epstein, 2003).

Additionally within biosolids, a very large number of organic pollutants and pathogenic organisms are present. Harisson et al. (2006) report that in biosolids, 516 organic compounds have been measured, classified into 15 fractions for which the EPA has risk-based soil screening limits. Similarly, a number of pathogens, including bacteria, viruses, protozoa, nematodes, etc., have been found in biosolids that can cause health problems in animals and humans (Smith, 1996). A very good description of the issues related to the challenging problems of organic pollutants is given by Langenkamp et al. (2001).

Legislation on Biosolids in Agricultural Use

For the aforementioned reasons, the use of biosolids in agriculture must follow certain rules and restrictions set out by the EPA and the EU or by the individual member countries. The terms and conditions set out by the EPA for the use of biosolids can be found in 40CFR503 (USEPA, 1994). This text includes all relevant information and obligations of producers and users of sludge. As far as the EU is concerned, conditions and limitations have been set out in Directive 86/278. This directive has been adopted and serves as the force of law in member states but each member retains the right to establish stricter rules. For several years now there has been rising speculation about revising the directive to make it more austere, and thus a new draft directive has been created that is now being discussed and is expected to be implemented by the EU (Working Document on Sludge, third draft; EU, 2000). In Switzerland, however, the use of biosolids in agriculture is totally forbidden (Smith, 1996). In Greece, the agricultural use of biosolids in actual fact is very limited. The main reasons for this are the reluctance of producers to test its application on their land, as well as the relative administrative weakness of the involved agencies (Kouloubis and Tsadilas, 2007). In the United Kingdom, a code of practice for agricultural use of sewage sludge has been drafted and implemented, and provides users with relevant guidelines (DoE, 1989).

Heavy Metals in Soils and Sludge

HM levels in agricultural soils have been found to be increasing; this is attributed to phosphate fertilizer, agrochemicals application, and atmospheric deposition (Jones, 1991; Billet, Fitzpatrick, and Cresser, 1991; Hovmand, Tjell, and Mosbaek, 1983). This, in turn, has led to the increased uptake of HMs by plants, as reported for the pea, radish, and lettuce crops, which were found with increased levels of Cd due to the phosphate fertilizers used (Reuss et al., 1978). In agricultural soils within Europe, it is estimated that there has been a 10% to 15% increase in the concentration of the heavy metals Cd, Pb, and Hg during the twentieth century, due to factors such as those mentioned previously (McBride, 1995). This increase, however, is very small as compared with the increase in HMs observed after the application of biosolids in agricultural lands and resulting in increased uptake of heavy metals by plants, as indicated by the large number of publications found in the literature (Bingham et al., 1975; Chang et al., 1984; Logan and Feltz, 1985; Heckman et al., 1987; King and Hajjar, 1990; Tsadilas et al., 1995 and references therein).

Relative to the effects of long-term biosolid application, an intense debate has developed, based on the following two points of view. One side claims that soils to which biosolids are added have the ability to immobilize HMs into forms unavailable to plants (Chaney and Ryan, 1993). Defenders of this view argue that the solubility of HMs released by the decomposition of the organic matter of biosolids can be kept at low levels due to the organic matter and inorganic components of biosolids that bind heavy metals, such as phosphates, silicates, and oxides of Fe, Al, and Mn. Likewise, they argue that the uptake of HMs by plants reaches a maximum with increasing amounts of biosolids applied (Ryan and Chaney, 1993). The other, more pessimistic view considers that the mineralization of organic matter of biosolids leads to the release of HMS in relatively soluble forms that can pose an environmental threat (McBride, 1995). According to this view, given that the ability of a soil to immobilize metals by adsorption and precipitation is not unlimited without the protective effect of the biosolids themselves, there is the possibility that the relationship between metal content and their solubility in soil with added biosolids is not quadratic or linear, but of the Langmuir type, that is, continuous increasing solubility with increasing metal concentration in the soil. This latter view was designated the "sludge time bomb hypothesis," contrary to the previous view, which was termed the "sludge protection hypothesis" (McBride, 1995, 2003). A related deliberation of these aforementioned opinions was attempted by Antoniadis et al. (2006). The safe solution for which of these two opinions is correct gives us the answer of how and to what extent biosolids can be used safely in agriculture in relation to the risk caused by HMs.

To assess the effect of HMs in biosolids on crops, it is necessary to determine the types of metals that are present in the biosolids and in the soil and their relationships with various plant species. Knowledge of these forms of HMs also allows for the risk assessment of their transfer to the subsoil and the possible contamination of groundwater. Moreover, the possible redistribution of the metals when biosolids are applied to soils should be identified. In this chapter, we summarize the types of HMs found in biosolids, in soils, and in biosolids-amended soils.

Forms of Heavy Metals in Biosolids

Biosolids contain organic and inorganic compounds. The content of organic compounds and their chemical composition depends on various factors and mainly on the type of digestion applied. Anaerobically digested sludge contains 25% to 30% organic carbon on a dry mass basis, although the content can vary widely. During the process of anaerobic digestion, organic solids are stabilized by the almost complete microbial fermentation of carbohydrates, apart from cellulose, resulting in a reduction of volatile solids at a rate of about 60% to 75%. The remaining organic material consists of a mixture of microbial tissue, lignin, cellulose, lipids, organic nitrogen compounds, and humic compounds (Miller, 1974). When biosolids are applied to soil there is decomposition of organic matter by the action of various factors, especially microorganisms using carbon (C) and nitrogen (N). The half-life of the decomposition of organic matter has been estimated at about ten years (Bell et al., 1991). During organic matter mineralization of biosolids, compounds such as proteins, cellulose, and hemicellulose that easily decompose are used primarily as a source of C and N by microorganisms. More stable and not so easily decomposed are humic substances. A better indicator, however, of the changes taking place in the organic matter of biosolids when applied to soils is the alteration in the composition of dissolved organic matter (DOM), the most active part of biosolids both biologically and chemically. Chefetz et al. (1998), studying the alterations undergone by organic matter in biosolids during composting, indicated that DOM decreased sharply by 84.4% in 33 days and by 8% more until 105 days and then remained constant until the end of the experiment. The components that decomposed during composting were mainly from the hydrophilic fraction, resulting in an increased proportion of hydrophobic aromatic and aliphatic structures in the residual DOM, displaying the humification progress. This study also showed that the hydrophobic fraction is more mobile and therefore more active. The fact that hydrophilic compounds decreased and hydrophobic compounds increased by the end of composting may be correlated with the fact that compost enhances plant growth through the increase in hydrophobic compounds. Antoniadis et al. (2007), working with two typical Greek soils amended with sewage sludge, found that dissolved organic carbon (DOC) and water-extractable Zn increased in the first 23 days of incubation and decreased later to values similar to those of unamended soils and attributed that to a flush of microbial activity.

Anaerobically digested sludge is a complex of certain minerals and organic matter, with remnants of bacteria and colloidal inorganic and organic materials. After digestion, the sludge is subjected to rapid oxidation and alterations in microbial activity. Thus, the chemical forms of HMs are the result of a balance between the solids precipitated, the complexes and hydrated ions in the solution, and the same ions that are held in the organic materials, in the bacterial residues, and on the surfaces and interstices of the minerals (Lake, Kirk, and Lester, 1984). The interactions of all these are so complicated that the distribution of the HMs among the different phases mentioned above can only be described through quantitative models that require accurate values of equilibrium constants for all the reactions of precipitation, complex formation, and redox processes that occur (Fletcher and Beckett, 1987a). These later researchers, studying the formation of Cu complexes with DOM concluded that ion exchange with protons is a dominant component of Cu binding and that the DOC has a cation exchange capacity of approximately 9 $meq.g^{-1}$. The same researchers found that Cu and Pb can replace more protons from the DOC in comparison with the heavy metals Ni(II), Zn(II), Mn(II), and Fe(II) (Fletcher and Beckett, 1987b). They put forth the view that there are two distinct groups of exchange sites. In the one group, the heavy metals Ni(II), Zn(II), Mn(II), and Fe(II) are bound, and in the other metals Cu(II) and Pb(II) are bound.

Holtzclaw et al. (1978), studying the distribution of heavy metals between humic acids, fulvic acids, and fractions extracted with 0.5 N NaOH that precipitate by adjusting the pH, found that almost all the Cu is associated with the humic acid fraction, while Cd and Ni are associated with the precipitate and the fulvic acid fraction, and thus concluded that Cd, Ni, and Zn are the most mobile HMs in soil with added sludge. Lagerwerff, Biersdorf, and Brower, (1976) demonstrated that Cd and Zn are leached easily with water or dilute $CaCl_2$ in comparison with Cu and Pb.

The main forms of HMs in biosolids are soluble, precipitated, co-precipitated with metal oxides, adsorbates, and associated with residues of microorganisms (Lester, Sterritt, and Kick, 1983). The distribution of HMs among these main forms varies widely, and depends on the chemical properties of each metal and on the characteristics of the biosolids, which are related to the processing conditions of the biosolids. Gould and Genetelti (1978) performed adsorption experiments and found that the affinity with the solid phase of biosolids (expressed in moles bound per unit weight) followed the order Cu > Zn > Cd > Ni and that the equilibrium pH significantly affected complexation. At lower pH levels, the concentration of soluble metals was higher. Other parameters that affect complexation were temperature, redox potential, the presence of complexing agents, and the concentrations of precipitant ligands. In a similar study, Alibahai, Mehrota, and Forster (1985) found that the mechanism of attachment involved metal and surface ligand interactions.

Methods of Determination of Heavy Metal Forms

Single Chemical Extraction

The forms of HMs in biosolids are usually studied by selective chemical extraction techniques. The HM fractions defined by these methods are those representing soluble, exchangeable, organic, adsorbed, and precipitated forms. The soluble forms of heavy metals in biosolids can be determined using a simple extraction with water. The procedure has been performed with variations by Jenkins and Cooper (1964), and Bloomfield and Pruden (1975), who applied repeated percolation through columns that contained biosolids with water; by Lagerwerff, Biersdorf, and Brower (1976), who used tap water; and by Emmerich et al. (1982), who used river water. The concentration of heavy metals (percent of total) found in the above cases ranged from <0.1% to 6.2% for Cu, from <0.03% to 0.1% for Pb, from 1.5% to 14.3% for Ni, from <0.1% to 11% for Cd and from <0.01% to 36% for Zn. Anaerobic incubation reduced the levels of soluble forms of Cu and Ni, increased the solubility of Cd and Pb, and left the solubility of Zn unaffected (Bloomfield and Pruden, 1975). Aeration that followed increased the solubility of Cu and Zn but did not affect the solubility of Cd, Ni, and Pb. It follows that Cd and Ni, and in some cases Zn, are the most soluble HMs within biosolids.

Another group of procedures determining the so-called "available" forms that have been added to biosolids include the use of dilute acids such as acetic or hydrochloric acid and/or in combination with chelates, for example, ethylenediaminetetraacetic acid (EDTA). Berrow and Webber (1972), using 0.42 M acetic acid, extracted greater than 95% of the total Zn. The heavy metals Cu and Ni also exhibited a similar high solubility in 0.095 M acetic acid, because 50% to 75% of the total was extracted (Jenkins and Cooper, 1964). Stover et al. (1976), using 0.5 M HCl, extracted 60% to 73% of the metals Cd, Ni, and Zn and 18% to 24% of Cu and Pb. In general, anaerobic incubation reduced the percentage of extractable heavy metals.

Sequential Chemical Extraction

Another technique widely used for soils has also been applied for determining the types of HMs in biosolids. This is a sequential extraction using substances that are believed to extract metals from specific fractions of biosolids. Emmerich et al. (1982b), working with anaerobically digested sludge, applied the same sequential extraction that they applied to soils, using solutions of 0.5 M KNO_3 (exchangeable forms), water (adsorbed forms), 0.5 M NaOH (organically bound forms), 0.05 M Na_2-EDTA (carbonate-bound forms), and 4.0 M HNO_3 at 70 to 80°C (residual forms, bound in clay lattice structure). They made it clear that the quantities extracted by the procedure they followed did not consist of specific forms of metals, but of chemically similar forms that are extracted by particular extractants. This study showed that the "exchangeable" fraction was found to be <0.1% to 0.1% for the metals Cd, Cu, and Zn, but 10.9% of the total for Ni. The "adsorbed" fraction was found to have very low levels (<0.1%) for Cd and Zn, and between 0.5% and 1.5% for Ni and Cu. Most of the "organically bound" fraction was found to be for Cu (60.4% of total) while the other three HMs ranged from 22% to 27% of the total. The "carbonate-bound" fraction was higher in Zn (57.5% of total), followed by Cd (51.6% of total), Ni (31.9%), and Cu (23.2%). The "residual" fraction (the fraction bound in the clay lattice structure) was highest in Cd (35.6%), followed by Ni (26.4%), Cu (16%), and Zn (9.8%). From the above it seems that the largest amount within the most stable forms was found to be Cd (36%), whereas Cu appears to have a greater affinity for organic matter (>65%). The more available forms (exchangeable and adsorbed) were found to be considerable only in Zn (>11%), whereas the remaining metals comprised an insignificant percentage (<0.2% to 1.5%).

Forms of Heavy Metals in Soils

The sources of HMs in soils are both natural and anthropogenic. The natural sources are the parent materials of soils derived from the weathering of rocks. In igneous rocks, HMs are found as components of primary minerals, in the structure of which are entering with isomorphous substitution, occurring during the period of crystallization of the rocks between ions that differs in electric charge by up to one unit and in size up to 15% (Krauskopf, 1967). HMs described in Directive 278/86 of the EU that are of interest in this chapter are mainly found in the primary minerals olivine (Ni, Zn, Cu); hornblende (Ni, Zn, Cu); augite (Ni, Zn, Pb, Cu); biotite (Ni, Zn, Cu); apatite (Pb); anorthite (Cu); andesine (Cu); albite (Cu); orthoclaste (Cu); ilmenite (Ni); and magnetite (Zn, Ni) (Krauskopf, 1967). In sedimentary rocks, which constitute about 75% of the earth's surface, the concentration of HMs depends on the mineralogy and the adsorptive properties of their components, which are resistant primary minerals, secondary minerals, or precipitates such as $CaCO_3$. Generally, clays and shales contain higher concentrations of HMs due to their greater absorption capacity. In Table 10.2, the mean concentrations of

TABLE 10.2

Mean Concentrations of Heavy Metals of Major Rock Types

Heavy Metal	Igneous Rocks[a]	Sedimentary Rocks[b]	Earth's Crust
Cd	0.11	0.10	0.1
Cu	48.3	24.8	50
Ni	718	28	80
Pb	19.3	12.9	14
Zn	70	56.7	75
Hg	0.03	0.21	0.05

Source: Adapted from Alloway, B.J. 1995. The origins of heavy metals in soils. In B.J. Alloway (Ed.) *Heavy metals in soils, second edition.* Blackie Academic & Professional, London.

[a] Mean values for ultramafic, mafic, and granitic rocks.

[b] Mean values for limestone, sandstone, and shales.

heavy metals of interest in this chapter according to the main rock types are shown. Anthropogenic sources contributing significantly to the input of heavy metals in soils are fertilizers, pesticides and lime, sewage sludge, animal waste, coal residues, mining and milling wastes, etc.

Table 10.3 shows the concentrations of HMs in uncontaminated soils, in various anthropogenic sources of HMs, and in soils contaminated with HMs due to anthropogenic activities.

Heavy metals in soils exist in various chemical forms that show different availability to plants. These various forms are determined using sequential extraction techniques, as previously described for biosolids. Many such techniques have been used thus far, using different chemicals to extract fractions of HMs that are usually distinguished into exchangeable, carbonate bound, organic matter bound, iron and manganese oxide bound, and residual fractions (Shuman, 1979; Iyengar, Martins, and Miller, 1981; Hickey and Hittrick, 1984; Tsadilas, et al., 1995). From these, the exchangeable fraction, usually extracted with KNO_3, is considered the most plant-available (Pierzinski and Schwab, 1993; LeClaire et al., 1984; Sims, 1986). The distribution of these fractions in soils is unstable and strongly affected by many factors, such as the addition of organic matter, the addition of metals, the pH, and redox potential, which may all cause redistribution. A simple routine practice that causes the redistribution of plant-available forms of HMs to unavailable forms is the increase in soil pH by the addition of lime, which is effectively used as a method of remediation of soils contaminated with HMs (Tsadilas, 2001; Sims, 1986). In assessing the availability of the essential for plant growth heavy metals in soil, various methods are used, including that proposed by Lindsay and

TABLE 10.3

Heavy Metal Content (mg.kg^{-1} dry wt) of Unpolluted Soils, Sewage Sludges, Fertilizers, Lime, Fly Ash, Animal Manures and Soils Amended with Sewage Sludge

	Heavy Metal						
Material	**Cd**	**Cu**	**Ni**	**Pb**	**Zn**	**Hg**	**Source**
Typical soil	0.06	20	40	10	50	0.03	Sterritt and Lester, 1980
Sewage sludge	3–3.000	200–8000	20–530	120–300	700–490	0.1–50	Adriano, 1986
Phosphate fertilizers	0.9–50	16.6–49	1.1–64	0.5–962	1–673	—	Adriano, 1986
Limestone	—	<0.3–89	—	—	<1–425	—	Adriano, 1986
Fly ash	0.3–1.3	<0.3–89	1.8–15	3.1–16	14–39	0.04–0.18	Adriano, 1986
Animal wastes	0.05–0.8	3–62	29	0.36–168	20–198	<0.01–0.2	Adriano, 1986
Biosolid-amended soils	1.3	37	29	78	109	0.38	Alloway, 1995

Norvell (1978) and which seems to be the most acceptable. This method uses as the extractant solution 0.005 M diethylenetriaminepentaaceticacid (DTPA) in 0.01 M $CaCl_2$ and 0.1 M triethanolamine adjusted to pH 7.30 and was developed for the determination of available forms of the heavy metals Fe, Mn, Zn, and Cu in soils containing calcium carbonate. From its publication onward, this method has been widely used for this purpose. The same determination procedure was extended to include the potentially toxic metals Ni and Cd by Baker and Amacher (1982) because stability constants for Ni and Cd are between those for Cu and Mn, thus making DTPA equally suitable for these two metals and now widely used for the determination of plant-available forms of these elements (Pierzinski and Schwab, 1993).

Heavy Metal Fractions in Biosolid-Amended Soils

The application of biosolids to soils causes alterations in some soil properties, which in turn may lead to the redistribution of HMs between various fractions and thus change their availability to plants. The soil properties that appear to be affected by the application of biosolids are pH and redox potential, the organic matter content, the concentration of minerals in the soil solution and ionic strength, the concentration of essential elements to plant growth, and the concentration of potentially toxic heavy metals. Tsadilas et al. (1995) reported that the application of biosolids on acidic soil increased the pH to about 7.0 and then tended to retain it at this value. They also found that the application of biosolids increased the soil electrical conductivity and organic matter content. A significant reduction in soil pH due to the application of biosolids was reported by several other researchers (Hinesly, Jones, and Ziegler, 1972; Emmerich et al., 1982, and references therein). Epstein, Taylor, and Chaney, (1976) reported that biosolid application increased the water retention capacity of soil, the salinity, and cation exchange capacity, and reduced the redox potential (E_h). Gerritse et al. (1982) found a significant effect of biosolids in the E_h of biosolid solutions. Williams et al. (1987) reported that biosolid application on soil changed its pH, while reducing its bulk density. Regarding the influence of biosolids on the increase in the total concentration of HMs, the number of publications is vast (Chang et al., 1984; Emmerich et al., 1982b; Tsadilas et al., 1995; Walter and Cuevas, 1998; Udom et al., 2004; Behel, Melson, and Somers, 1983; Wang, Li, and Shuman, 1997; Williams et al., 1987).

The decrease in soil pH with the application of biosolids may result in an increase in solubility of HMs and their mobility in soil (Emmerich et al., 1982b). Despite this, Williams et al. (1987) observed in long-term experiments that the increase in soil acidification, and consequently in the availability of heavy metals Cd, Zn, and Ni, as estimated by the DTPA method, did not result in any significant movement of these metals within the soil profile. The reduction in E_h caused by the application of biosolids had a significant effect on the solubility of HMs (Gerritse et al., 1982). The increase in organic matter as a result of biosolid application seriously affects the forms and behavior of HMs in soils. Organic matter increases the adsorptive capacity of soils. As a consequence, the addition of biosolids, which contain a large proportion of stabilized organic matter that is resistant to decomposition, is likely to cause an increase in the adsorption capacity of soils (Gerritse et al., 1982). In contrast, Lieu et al. (2007) believe that the addition of biosolids to soil, due to the induced increase in dissolved DOM that has the ability to create stable and soluble complexes with metals, can contribute to the leaching of HMs. The same researchers showed that if the estimates for determining the allowable doses of biosolids on soil took into account the parameters of adsorption resulting from dissolved organometallic complexes, the calculated quantities were much smaller than when the estimates did not take into account the effect of DOM. For this reason, they suggested that the effect of sewage sludge-derived DOM on heavy metal sorption and soil properties should be considered in the course of regulating the safe application rates of sewage sludge to soil. Antoniadis et al. (2007b), studying the adsorption of Cd, Ni, and Zn in a biosolids-amended Greek Entisol, found that after a one-year incubation, the sorption capacity of amended soils decreased in monometal systems in the order Zn > Cd > Ni. In competitive systems, Cd exhibited decreased sorption as the soil organic matter content was reduced, suggesting that in such environments Cd is likely to be of the most environmental significance. The increase in cation exchange capacity resulting from the application of biosolids to soil, as was shown by Epstein, Taylor, and Chaney (1976), may affect the adsorption of heavy metals in a manner similar to that reported for organic matter.

Effect of Biosolids on Forms of Heavy Metals

All attempts to determine the forms of HMs in soils or in biosolid-amended soils are, in fact, ultimately aimed at the identification of plant-*available* forms. Despite the fact that until now several methods have been developed and used in the determination of metal forms, both simple and sequential, none has been unconditionally accepted by the scientific community. This is because, thus far, a clear definition of the term *available,* based on using a standard assay procedure that uses one or more extractants, is unavailable. The *available* forms of HMs are therefore determined by several processes, which may include single chemical extractants or processes using sequential extractions with various extractants that allegedly extract specific forms of metals (sequential chemical extraction). Furthermore, there are techniques that may determine with relative success the forms of HMs in soil solutions, for example, with the use of chromatography or by using resins. In such cases, appropriate computer programs such as GEOCHEM® are useful tools in determining the forms of metals in soil solutions (Lake, Kirk, and Lester, 1984).

There are many single chemical extractants that were used for the determination of *available* forms of HMs in soil. The simplest extractant used for soil was water and the amounts of metals extracted were very low (Gupta and MacKay, 1965). In biosolids or in mixtures of biosolids was used by Jenkins and Cooper (1964) and Bloomfield and Pruden (1975). Berrow and Webber (1972) used 0.42 M acetic acid to determine the available forms of HMs and compared the quantities extracted in unamended soils and sludges. In all cases, the concentrations of all metals except Al were significantly lower in soils as compared to the sludges. Beckett, Warr, and Brindley (1983) used a solution of 0.05 M EDTA and 0.5 M acetic acid as the extractant of available forms of HMs and found that the extracted quantities of Cu, Ni, and Zn were higher in sludge-treated soils than in the native soils. An extractant containing DTPA has been widely used in the estimation of available forms of HMs in biosolid-amended soils (Kelling et al., 1977; Schauer et al., 1980; Tsadilas et al., 1995). In nearly all cases, it was found that the addition of biosolids increased the concentrations of available forms of the heavy metals Cd, Cu, Zn, Ni, and Pb.

Sequential extraction has been widely used by the scientific community to study the forms of HMs and assess their mobility, plant availability, and thus ecotoxicity. In the literature there are a large number of sequential extraction procedures whose results are often not comparable, depending on the extraction method used (Lake, Kirk, and Lester, 1984). Emmerich et al. (1982a) used a sequential extraction protocol in soils amended with biosolids to determine the exchangeable, adsorbed, organically bound, carbonate bound, and residual fractions of the metals Cd, Cu, Ni, and Zn. The largest amounts of metals were found in the organically bound, carbonate, and residual fractions, while in biosolids less than 36% were found in the residual fraction, and in soils the corresponding percentage was greater than 65%. Cd, Ni, and Zn shifted to the residual form during the research period (about two years), while Cu was not affected. Emmerich et al. (1982b), Sposito et al. (1982), and Sposito, Lund, and Chang (1982), using the same extractants, found that the percentage of the total concentration of heavy metals extracted in $KHNO_3$ was on the order of less than 3.7%. Tsadilas et al. (1995), using the same procedure for the same heavy metals in acidic soil amended with sewage sludge, found that only the total concentration of Cu and Zn increased significantly but remained below the limits of the relevant EU Directive 86/278. Of the determined fractions, the organically bound, carbonate, and residual fractions for the metals Cd, Ni, Cu, and Pb increased significantly while Zn was not affected. Jeyakumar et al. (2008) reported for biosolids amended with Cu and Zn that after 117 days of incubation, Cu in the unamended biosolid solid phase was mainly found in the organic and residual fractions (85% to 95%); but in the amended biosolids, Cu addition decreased the percentage of Cu in these fractions and increased the percentage of Cu in the oxide and specifically adsorbed fractions. In the solution phase, relatively all the Cu was complexed with organic matter. Zinc in the solid phase was mainly associated with the oxide (35% to 65%), specifically adsorbed (25% to 30%), and exchangeable fractions (10% to 40%).

Given the wide range of procedures used in sequential extraction without comparable results, as noted above, the BCR undertook a harmonization of sequential extraction methods and the designation of certified reference materials (Fuentes et al., 2004). The process used by the BCR consists of four steps that determine the exchangeable fraction and that associated with carbonate phases (Step 1), the reducible fraction or fraction associated with Fe and Mn oxides (Step 2), the oxidizable fraction or bound to organic matter (Step 3), and the residual fraction containing primary and secondary solids occluding metals in their crystalline structures (Step 4). Using the method of the BCR, Scancar et al. (1999) found that applying biosolids on Slovenian soil resulted in increased mobility of the metals Ni and Zn, while Cd, Cu, and Pb were rarely found in soluble fractions, thus showing reduced mobility.

Concluding Remarks

Biosolid use in agriculture is widely practiced all over the world because it is the most economical solution enabling the recycling of nutrients and increasing soil organic matter. However, the application of biosolids increases the HM concentration of soils, thus creating a risk to the environment. Relative to the effects of long-term biosolid application, an argument has developed: one side is optimistic, claiming that added biosolids have the ability to immobilize HMs; the other side considers that the mineralization of organic matter of biosolids may release HMs in soluble form, thus creating an environmental threat. A huge number of publications can be found in the literature concerning HM forms in biosolids, soils, and biosolid-amended soils. The main forms of HMs in biosolids are soluble, precipitated, co-precipitated with metal oxides, adsorbates, and associated with residues of microorganisms. Their distribution varies widely, depending on the metal chemical properties and the processing conditions of the biosolids. Methods for determining HM forms in biosolids involve both single and sequential chemical extractions, which determine soluble, exchangeable, organic, adsorbed, and precipitated forms or exchangeable, adsorbed, organically bound, carbonate bound, and residual bound in clay lattice structure forms, respectively. The application of biosolids to soils causes alterations in some soil properties, leading to the redistribution of HMs among various fractions, thus changing their availability to plants. Among the soil properties affected by biosolids application are the pH and redox potential, organic matter content, ionic strength of soil solution, concentration of essential elements, and concentration of potentially toxic HM. Biosolids application to soils in general increases the concentrations of available forms of the heavy metals Cd, Cu, Zn, Ni, and Pb extracted. Regarding the HM fractions, application of biosolids leads to an increase in organically bound, carbonate, and residual fractions. A procedure for sequential extraction was recently developed by the BCR to be used as a standard method for determining HM fractions and is gaining widespread acceptance. The process used consists of four steps to determine exchangeable and associated with carbonate phases, the reducible fraction or the fraction associated with Fe and Mn oxides, the oxidizable fraction or bound to organic matter, and the residual fraction containing primary and secondary solids occluding metals in their crystalline structures.

Acknowledgments

The author expresses his warm thanks to Dr. C. Christofides for her valuable help in editing the manuscript. Also the support of Dr. L. Evangelou and Dr. S. Shaheen in literature selection is acknowledged.

References

Adriano, D.C. 1986. *Trace elements in the terrestrial environment.* Springer-Verlag New York Inc

Alibahai, K.R., I. Mehrota, and C.F. Forster. 1985. Heavy metal binding to digested sludge. *Water Res.* 19(12): 1483–1488.

Alloway, B.J. 1995. The origins of heavy metals in soils. In B.J. Alloway (Ed.) *Heavy metals in soils, second edition.* Blackie Academic & Professional, London. pp. 38–57.

Anderson, M.S. 1959. Fertilizing characteristics of sewage sludge. *Sewage Ind. Wastes* 31: 678–682.

Antoniadis, V., C.D. Tsadilas, V. Samaras, and J. Sgouras. 2006. Availability of heavy metals applied to the soil through sewage sludge. In M.N.V. Prasad, K.S. Sajwan, and R. Naidu (Eds.) *Trace elements in the environment.* CRC Press, Boca Raton, FL, pp. 39–61.

Antoniadis, V., C.D. Tsadilas, and S. Stamatiadis. 2007a. Effect of dissolved organic carbon on zinc solubility in incubated biosolid-amended soils. *J. Environ. Qual.* 36: 379–385.

Antoniadis, V., C.D. Tsadilas, and D.J. Ashworth. 2007b. Monometal and competitive adsorption of heavy metals by sewage sludge-amended soil. *Chemosphere* 68: 489–494.

Baker, D.E., and M.C. Amacher. 1982. Nickel, copper, zinc, and cadmium. In A.L. Page, R.H. Miller, and D.R. Keeney (Eds.) *Methods of soil analysis. 2. Chemical and microbiological properties, second edition.* ASA, SSSA, Madison, WI. pp. 323–336.

Beckett, P.H.T., E. Warr, and P. Brindley. 1983. Changes in the extractabilities of the heavy metals in water-logged extracted from sewage sludges. *Water Pollut. Control* 82: 107–113.

Behel, D., Jr., D.W. Nelson, and L.E. Somers. 1983. Assessment of heavy metals equilibria in sewage sludge-treated soils. *J. Environ. Qual.* 12(2): 181–186.

Bell, P.F., B.R. James, and R.L. Chaney. 1991. Heavy metal extractability in long-term sewage sludge and metal salt-amended soils. *J. Environ. Qual.* 20: 481-486.

Berrow, M.L., and J. Webber. 1972. Trace elements in sewage sludges. *J. Sci. Food Agric.* 23: 93–100.

Billet, M.F., E.A. Fitzpatrick, and M.S. Cresser. 1991. Long-term changes in copper, lead, and zinc content of forest and soil organic horizons form north-east Scotland. *Water Air Soil Pollut.* 59: 179–192.

Bingham, F.T., A.L. Page, R.J. Mahler. And T.J. Ganje. 1975. Growth and cadmium accumulation of plants grown on a soil treated with a cadmium-enriched sewage sludge. *J. Environ. Qual.* 4: 207-211

Bloomfield, C., and G. Pruden. 1975. The effects of aerobic and anaerobic incubation on the extractability of heavy metals in digested sewage sludge. *Environ. Pollut.* 8: 217–232.

Chaney, R.L., and J.A. Ryan. 19993. Heavy metals and toxic organic pollutants in MSW-composts: Research results on phytoavailabilty, bioavailability, fate, etc. pp. 451–506. In H.A.J. Hoitink, and H.M. Keener (Eds.) *Science and engineering of composting: design, environmental, microbiological and utilization aspects.* Renaissance Publishing, Worthington, OH.

Chang, A.C., J.E. Warneke, A.L. Page, and L.J. Lund. 1984. Accumulation of heavy metals in sewage-treated soils. *J. Environ. Qual.* 13(1): 87–91.

Chefetz, B. Y. Chen, and Y, Hadar. 1998. Purification and characterization of the laccase from *Chaetomium thermophilium* and its role in humification. *Appl. Environ. Microbiol.* 64: 3175–3179.

Commission of European Communities. Council Directive 91/271/EEC of March 21, 1991 concerning urban waste-water treatment (amended by the 98/15/EC of February 27, 1998).

Coker, E.G., and C.H. Carlton-Smith. 1986. Phosphorus in sewage sludge as a fertilizer. *Waste Manage. Res.* 4: 303–319.

Coker, E.G., J.E. Hall, C.H. Carlton-Smith, and R.D. Davis. 1987. Field investigations into the manorial value of lagoon-matured digested sewage sludge. *J. Agri. Sci. Cambridge* 109: 767–478.

DoE; Department of the Environment. 1989. Code of Practice for Agricultural Use of Sewage Sludge, HMSO, London.

Doxiadis, C.A. 1973. Ancient Greek settlements. Third report. *Ekistics* 35(11).

Duffus, J.H. 2002. "Heavy metals" a meaningless term? *Pure Appl. Chem.* 2002, 74(5) 793–807.

Emmerich, W.E., L.J. Lund, A.L. Page, and A.C. Chang. 1982a. Movement of heavy metals in sewage sludge-treated soils. *J. Environ. Qual.* 11(2): 174–178.

Emmerich, W.E., L.J. Lund, A.L. Page, and A.C. Chang. 1982b. Solid phase forms of heavy metals in sewage sludge-treated soils. *J. Environ. Qual.* 11(2): 178–181.

Epstein, E., J.M. Taylor, and R.L. Chaney. 1976. Effects of sewage sludge and sludge compost applied to soil on some soil physical and chemical properties. *J. Environ. Qual.* 5: 422–426.

Epstein, E. 2003. *Land application of sewage sludge and biosolids.* Lewis Publishers, Boca Raton, FL. p. 201.

EU; European Union. (2000). Working Document on Sludge, Third Draft. Unpublished, p. 19.

Fletcher, P., and P.H. Beckett. 1987a. The chemistry of heavy metals in digested sewage sludge. I. Copper(II) complexation with soluble organic matter. *Water Res.* 21(10): 1153–1161.

Fletcher, P., and P.H. Beckett. 1987b. The chemistry of heavy metals in digested sewage sludge. II. Heavy metal complexation with soluble organic matter. *Water Res.* 21(10): 1163–1172.

Fuentes, A., M. Liorens, J. Saez, A. Soler, M.I. Aguliar, J.F. Ortuno, and V.F. Meseguer. 2004. Simple and sequential extractions of heavy metals from different sewage sludges. *Chemosphere* 54: 1039–1047.

Fytili, D., and A. Zabaniotou. 2008. Utilization of sewage sludge in EU application of old and new methods — A review. *Renewable Sustainable Energy Rev.* 12: 116–140.

Gerritse, R.G., R. Vriesema, J.W. Dalenberg, and H.P. De Roos. 1982. Effect of sewage sludge on trace element mobility in soils. *J. Environ. Qual.* 11(3): 359–364.

Gould, M.S., and E.J. Genetelti. 1978. Heavy metal complexation behavior in anaerobically digested sludges. *Water Res.* 12: 505–512.

Gupta, U.C. and D.C. MacKay. 1965. Extraction of water soluble copper and molybdenum from mineral soils. *Soil Sci. Soc. Am. Proc.* 29: 323-326.

Harrison, J.R., S.R. Oakes, M. Hysell, and A. Hay. 2006. Organic chemicals in sewage sludges. *Science Total Environ.* 367: 481–497.

Heckman, J.R., J.S. Angle, and R.L. Chaney. 1987. Residual effects of sewage sludge on soybean: I. Accumulation of heavy metals. *J. Environ. Qual.* 16(2): 113-117.

Hickey, M.G., and J.A. Kittrick. 1984. Chemical partitioning of cadmium, copper, nickel and zinc in soils and sediments containing high levels of heavy metals. *J. Environ. Qual.* 13, 372–376.

Hinesly, T.D., R.L. Jones, and E.L. Ziegler. 1972. Effects on corn by application of heated anaerobically digested sludge. *Compost Sci.* 13: 26–30.

Holtzclaw, M.H., D.A. Keech, A.L. Page, G. Sposito, T.J. Ganje, and N.B. Ball. 1978. Trace metal distribution among the humic acid, fulvic acid, and precipitatable fractions extracted with NaOH from sewage sludges. *J. Environ. Qual.* 7(1): 124–127.

Hovmand, M.F., J.C. Tjell, and M. Mosbaek. 1983. Plant uptake of airborne cadmium. *Environ. Pollut.* 30: 27–38.

Iyengar, S.S., D.C. Martins, and V.P. Miller. 1981. Distribution and plant availability of soil zinc fractions. *Soil Sci. Soc. Am. J.* 45: 735–739.

Jenkins, S.H., and J.S. Cooper. 1964. The solubility of heavy metal hydroxides in water, sewage, and sewage sludge. III. The solubility of heavy metals present in digested sewage sludge. *Int. J. Air Water Pollut.* 8: 695–703.

Jeyakumar, P., P. Loganathan, S. Sivakumaran, C.W.N. Anderson, and R.G. McLaren. 2008. Copper and zinc spiking of biosolids: Effect of incubation period on metal fractionation and speciation and microbial activity. *Environ. Chem.* 5: 347–354.

Jeyakumar, P., P. Loganathan, S. Sivakumaran, C.W.N. Anderson, and R.G. McLaren. 2010. Bioavailability of copper and zinc to poplar and microorganisms in a biosolids-amended soil. *Aus. J. Soil Res.* 48: 459–469.

Jones, K.C. 1991. Contamination trends in soils and crops. *Environ. Pollut.* 69: 311-326.

Jordan, M.M., M.A. Montero, S. Pina, and E. Garcia-Sanchez. 2009. Mineralogy and distribution of Cd, Ni, Cr, and Pb in biosolids-amended soils from Castelln Province (NE, Spain). *Soil Sci.* 174(1): 14-20.

Kelling, K.A., D.R. Keeney, L.M. Walsh, and J.A. Ryan. 1977. A filed study of the agricultural use of sewage sludge: III. Effect on uptake and extractability of sewage sludge-borne metals. *J. Environ. Qual.* 6: 352-358.

King, L.D., and L.M. Hajjar. 1990. The residual effects of sewage sludge on heavy metal content of tobacco and peanut. *J. Environ. Qual.* 19: 738-748.

King L.D., and H.D. Morris. 1972. Land disposal of liquid sewage sludge. III. The effect on soil nitrate. *J. Environ. Qual.* 1(4): 442–446.

Kouloubis, P., and C. Tsadilas. 2007. *Manual of sound agricultural practice for the proper utilization of municipal sewage sludge, 1st ed.* Ministry of Rural Development and Foods. p. 172. ISBN: 978-960-00-2015-I (in Greek with summary in English).

Krauskopf, K.B. 1967. *Introduction to Geochemistry.* McGraw-Hill, New York.

Lake, D.L., P.W.W. Kirk, and J.N. Lester. 1984. Fractionation, characterization, and speciation of heavy metals in sewage sludge and sludge-amended soils: A review. *J. Environ. Qual.* 13(2): 175–183.

Lagerwerff, J.V., G.T. Biersdorf, and D.L. Brower. 1976. Retention of metals in sewage sludge. I. Constituent heavy metals. *J. Environ. Qual.* 5(1): 19–23.

Langenkamp H., P. Part, W. Erhardt, and A. Prüeß. 2001. Organic Contaminants in Sewage Sludge for Agriculture Use. Report, European Commission, Joint Research Centre, Ispra.

LeClaire, J.P., A.C. Chang, C.S. Levesque, and G. Sposito. 1984. Trace metal chemistry in arid-zone field soils amended with sewage sludge. IV. Correlation between zinc uptake and extracted soil zinc fractions. *Soil Sci. Soc. Am. J.* 48: 509–513.

Lester, J.N., R.M. Sterritt, and P.W.W. Kick. 1983. Significance and behavior of heavy metals in waste water treatment processes. II. Sludge treatment and disposal. *Sci. Total Environ.* 30: 45–83.

Lindsay, W.L., and W.A. Norvell. 1978. Development of a DTPA test for zinc, iron, manganese, and copper. *Soil Sci. Soc. Am. J.* 42: 421–428.

Lieu, X., S. Zhang, W. Wu, and H. Liu. 2007. Metal sorption on soils as affected by dissolved organic matter in sewage sludges and the relative calculation of sewage sludge application. *J. Hazard. Mater.* 149: 399–407.

Logan, T.J., and R.E. Feltz. 1985. Plant uptake of cadmium from acid-extracted anaerobiaclly digested sewage sludge. *J. Environ. Qual.* 14(4): 495-500.

MacNicol, R.D., and Beckett, P.H.T. 1985. Critical tissue concentration of potentially toxic elements. *Plant Soil* 85: 107–129.

Magdoff, F.R., and J.F. Amadon. 1980. Nitrogen availability from sewage sludge. *J. Environ. Qual.* 9(3): 451–455.

McBride, M.B. 1995. Toxic metal accumulation from agricultural use of sludge: Are USEPA regulations protective? *J. Environ. Qual.* 24: 5–18.

McBride, M.B. 2003. Toxic metals in sewage sludge-amended soils: Has promotion of beneficial use discounted the risks? *Adv. Environ. Res.* 8: 5–19.

Miller, R.H. 1974 : Factors affecting the decomposition of anaerobically digested sewage sludge in soil. *J. Environ. Qual.* 3(4) : 376-380.

Pierzinski, G.M., and A.P. Schwab. 1993. Bioavailability of zinc, cadmium, and lead in a metal contaminated alluvial soil. *J. Environ. Qual.* 22: 247–254.

Reuss, J.O., H.L. Dooley, and W. Griffits. 1978. Uptake of cadmium from phosphate fertilizers by peas, radishes, and lettuce. *J. Environ. Qual.* 7: 128-133.

Ryan, J.A., and R.L. Chaney. 1993. Regulation of municipal sewage sludge under the Clean Water Act section 503: A model for exposure and risk assessment for MSW-compost. In H.A.J. Hoitink and H.M. Keener (Eds.) *Science and engineering of composting: Design, environmental, microbiological and utilization aspects.* Renaissance Publishing, Worthington, OH. pp. 422–450.

Sabrey, B.R., N. Agbim, and D.C. Markstrom. 1977. Land application of sewage sludge. IV. Wheat growth, N content, N fertilizer value, and N use efficiency as influenced by sewage sludge and wood waste mixtures. *J. Environ. Qual.* 6(1): 52–58.

Samaras, V., C.D. Tsadilas, and S. Stamatiadis. 2008. Effects of repeated application of municipal sewage sludge in a cotton field: Effects on soil fertility, crop yield, and nitrate leaching. *Agron. J.* 100: 477–483.

Scancar, J., R. Milacic, M. Strazar, and O. Burica. 1999. Total metal concentrations and partitioning of Cd, Cr, Cu, Fe, Ni and Zn in sewage sludge. *Sci. Tot. Environ.* 250: 9–19.

Shober, A.L., R.C. Stachouwer, and K.E. MacNeal. 2006. Chemical fractionation of trace elements in biosolid-amended soils and correlation with trace elements in crop tissue. *Commun. Soil Sci. Plant Anal.* 38: 1029–1046.

Shuman, L.M. 1979. Zinc, manganese, and copper in soil fraction. *Soil Sci.* 127(1): 10–17.

Sims, J.T. 1986. Soil pH effects on the distribution and plant availability of manganese, copper, and zinc. *Soil Sci. Soc. Am. J.* 50: 367–373.

Smith, S.R. 1996. *Agricultural recycling of sewage sludge and the environment*. CAB International, Wallingford, UK. p. 382.

Somers, L.E. 1977. Chemical composition of sewage sludges and analysis of their potential use as fertilizers. *J. Environ. Qual.* 6(2): 225–232.

Sposito, G.L., L.J. Lund, and A.C. Chang. 1982. Trace metal chemistry in arid-zone field soils amended with sewage sludge. I. Fractionation of Ni, Cu, Zn, Cd, and Pb in solid phases. *Soil Sci. Soc. Am. J.* 46: 260–264.

Sterritt, R.M., and J.N. Lester. 1980. The value of sewage sludge to agriculture and effects of the agricultural use of sludge contaminated with toxic elements: a review. *Sci. Total Environ.* 16: 55–90.

Stover, R.C., L.E. Sommers, and D.J. Silvera. 1976. Evaluation of metals in wastewater sludge. *J. Water Pollut. Control Fed.* 48: 2165-2175.

Su, J., H. Wang, M.O. Kimberley, K. Beecroft, G.N. Magesan, and C. Hu. 2008. Distribution of heavy metals in a sandy forest soil repeatedly amended with biosolids. *Austr. J. Soil Res.* 46: 502–508.

Sutherland, R.A., F.M.G. Tack. C.A. Tolosa, and M.G. Verloo. 2000. Operationally defined metal fractions in road deposited sediment, Honolulu, Hawaii. *J. Environ. Qual.* 29: 1431–1439.

Tsadilas, C.D., T. Matsi, N. Barbayiannis, and D. Dimoyiannis. 1995. The influence of sewage sludge application on soil properties and on the distribution and availability of heavy metal fractions. *Commun. Soil Sci. Plant Anal.* 26(15-16): 2603–2619.

Tsadilas, C.D. 2001. Soil pH effect on the distribution of heavy metals among soil fractions. In I.K. Iskandar (Ed.) *Environmental restoration of metals-contaminated soils*. Lewis Publishers, Boca Raton, FL. pp. 107–119.

Udom, B.E., J.S.C. Mbagwu, J.K. Adesodum, and N.N. Agbim. 2004. Distribution of zinc, copper, cadmium and lead in a tropical ultisol after long-term disposal of sewage sludge. *Environ. Int.* 30: 467–470.

USEPA; United States Environmental Protection Agency. 1994. A Plain English Guide to the EPA Part 503 Biosolids Rule. EPA/832/R-93/003.

Vigerust, E., and A.R. Semler-Olsen. 1986. Basis for metal limits relevant to sludge utilisation. In R.D. Davis, H. Haeni, and P. L 'Hermite (Eds.) *Factors influencing sludge utilization practices in Europe*. Elsevier Applied Science Publishers Ltd., Barking. pp. 26–42.

Walter I., and G. Cuevas. 1998. Chemical fractionation of heavy metals in a soil amended with repeated sewage sludge application. *Sci. Total Environ.* 226: 113–119.

Wang, P., E. Qu, Z. Li, and L.M. Shuman. 1997. Fractions and availability of nickel in loessian soil amended with sewage sludge. *J. Environ. Qual.* 26: 795–801.

Williams, D.E., J. Vlamis, A.H. Pukite, and J.E. Corey. 1987. Metal movement in sludge-amended soils: A nine year study. *Soil Sci.* 143(2): 124–131.

Index

T

U

W